Python for ArcGIS Pr

Automate cartography and data analysis using ArcPy, ArcGIS API for Python, Notebooks, and pandas

Silas Toms
Bill Parker

BIRMINGHAM—MUMBAI

Python for ArcGIS Pro

Senior Publishing Product Manager: Dr. Shailesh Jain
Acquisition Editor – Peer Reviews: Saby Dsilva
Project Editor: Amisha Vathare
Content Development Editor: Lucy Wan
Copy Editor: Safis Editing
Technical Editor: Aditya Sawant
Proofreader: Safis Editing
Indexer: Subalakshmi Govindhan
Presentation Designer: Pranit Padwal

First published: April 2022

Production reference: 1220422

Published by Packt Publishing Ltd.
Livery Place
35 Livery Street
Birmingham
B3 2PB, UK.

ISBN 978-1-80324-166-1

`www.packt.com`

Forewords

Geography is a gift that allows us to see over the horizon, or around the corner; to explore distant lands, or the complexities of the underground environments beneath our own feet. Geography, as both an art and a science, helps us understand our ever-changing world in ways that let us address our pressing challenges at local, regional, national, and global scales. The ability to rigorously quantify our world has been the quest of Geography since Eratosthenes first surveyed his corner of the planet, resulting in an amazingly accurate approximation of the Earth's circumference, more than 2,000 years ago.

It was this rich quantitative tradition that was supercharged when Geographic Information Systems became broadly accessible – enabling the industrious to characterize our world in infinite detail with points, lines, and polygons; gridded rasters; quantized mesh; and other quantitative frameworks that allow the rigorous interrogation of our planet as it changes geographically, over time. This transformation of the field of Geography, which has been several decades in the making, has seen Geographic Information Systems become more commonplace, applied to an ever-increasing range of business and mission challenges. GIS has been key to the digital transformation of industry, government, academe, and the social sector.

This transformation has been hastened, in no small part, by GIS's embrace of scripting languages that empower a wide community of practitioners to continuously innovate as they explore their world in new and innovative ways. These scripts have extended and augmented our collective GIS capabilities as they have been crafted, posted on the Web, shared with others, modified, improved, reshared, and road-tested by countless geospatial practitioners and scholars across our global community. These scripts have thrown fuel on the fire of geospatial innovation, helping all of us better understand our world. Everything that has ever happened on Planet Earth, after all, by definition, has existed in space and time – creating a mammoth amount of work that only thoughtful and creative automation can help us address.

It is books like this one that will help expand the pool of geospatial innovators, expand geographic understanding, and help us all, as a global community, address the vital issues of the day, through a geographic lens. As you read this book, I encourage you to think about how these new superpowers might help you leave our world just a little better off than it was when you found it.

Dr. Christopher Tucker
Chairman, American Geographical Society

A popular question that comes up in the GIS community is "How do I become a GIS developer?". It's a valid question; lots of people in GIS might be making maps, doing analysis, and looking to further their knowledge. I probably have a lot to say on this topic, but one of the first things I would recommend is to learn Python.

One of my duties in a previous career was to prepare a series of about two dozen maps every year that would get printed in the newspaper. They weren't complicated maps, but preparing the data, updating the layouts, and generating the maps usually took two to three days. It wasn't difficult, but it was tedious. I figured there had to be a better way to do it. This was one of the first tasks I had where I really embraced Python. I was able to use Python to prepare the latest data, update dates and information on the layouts, and generate all the documents. Writing and testing the script took me maybe a day, and saved me numerous hours over the following years. I was sold after that.

As professionals working in the spatial industry, it's important to stay up to date with technology that can help us in our daily tasks. Many people working in GIS will have stories of having to gather data from various sources, doing a little copying and pasting, maybe even editing in their favorite spreadsheet software, and producing something useful for stakeholders. At the end of the day, data is data; it's our job to give it meaning.

Python is that Swiss army knife in a developer's toolbox. It can parse data, analyze imagery, convert file types, upload/download, iterate, massage, and output various results. There's a strong possibility that if you need to perform some analysis, there's a Python library available for it. Combine that with ArcGIS API for Python and you can take your skills to the next level. This book will walk you through the basics, as well as working with data; even publishing it. This is the kind of book I wish I had when I was a wide-eyed technician and I'm excited for what you can do with it today!

René Rubalcava
SoftWhere Development Engineer, Esri

Contributors

About the authors

Silas Toms is a geospatial data expert and data engineer with over 15 years of experience in the field of geographic data systems. A graduate of Cal Poly Humboldt, he has gone from environmental data analysis to building a GIS for the Super Bowl, to his current role as the Director of Data Engineering for an electric vehicle charging company. This is his fourth book, including two other ArcGIS and ArcPy books, and *Mastering Geospatial Analysis with Python 3*.

I would like to thank my partner, Laura, and daughter, Sloane, who brighten my day and whose support and love helped me to write this book. I would like to thank my parents and sister for their years of support. I would also like to thank Gabriel Paun, Dara O'Beirne, Josh Butler, and Beth Stone, who taught me and believed in me.

Bill Parker is a GIS Professional with over 15 years of GIS and Python experience. He previously worked at ICF as a GIS lead on large-scale environmental projects, using ArcPy to automate GIS analyses and map production. His project highlights include being the GIS lead for the Caltrain Modernization EIR/S, the California High Speed Rail Project San Jose to Merced, and the San Francisco to San Jose EIR/S. He is now working at Volta Charging, automating spatial analyses and ArcGIS Online workflows using Python.

I would like to thank my wife, Natalie, and children, Teddy and Jack, for supporting me and being understanding of the time I needed to work on this. I would also like to thank my co-author for inviting me to work on this with him and believing that I could do it.

Josh Bonifield is an experienced geospatial analyst with a demonstrated history of working in the marketing, energy, and agriculture industries. He holds an undergraduate degree from Loyola University Chicago in Environmental Science and recently finished his master's in GIS with a focus on data science and predictive analytics from Johns Hopkins University. He now works at OpenTeams as a Geospatial Data Engineer, organizing and applying mobile data with precision to aid their customers in solving unique problems.

Though he is not credited on the cover, Josh kindly contributed the crop yield prediction case study that constitutes *Chapter 13*.

About the reviewer

Gregory Brunner is an experienced data scientist and award-winning professor. He is an expert in the fields of geographic information systems, geospatial application development, remote sensing, big data analytics, Python, and ArcGIS. Since 2011, he has worked for Esri, where he supervises a team of data scientists and consultants and supports customers by helping them develop and implement geospatial solutions using ArcGIS. He has authored several series of lessons on Python and ArcGIS Notebooks that are hosted on the Learn ArcGIS website. Greg also teaches programming for GIS and remote sensing at Saint Louis University, where he has been an adjunct professor since 2017. In 2019, he was the recipient of the award for *Excellence in Adjunct Teaching*.

Table of Contents

Chapter 2: Basics of ArcPy 53

Chapter 3: ArcGIS API for Python 95

Chapter 9: Raster Analysis with Python 335

Chapter 10: Geospatial Data Processing with NumPy 373

Part IV: Case Studies 411

Chapter 11: Case Study: ArcGIS Online Administration and Data Management 413

Chapter 12: Case Study: Advanced Map Automation 457

Preface

This book will welcome you to the wide world of ArcGIS Pro automation, which will elevate your skillset and career to new heights. We will teach you how to optimize and streamline data management, analysis, and map-making processes in ArcGIS Pro and ArcGIS Online by using Python. The tips and tricks you'll learn make it easy to manage data for entire cities or huge companies, to create or edit entire map series, or to generate analysis results quickly from big data series, making your life as a GIS professional easier – whether you work in a government organization, private industry, or are an aspiring student.

Who this book is for

This book is for ArcGIS professionals, intermediate ArcGIS Pro users, ArcGIS Pro power users, students, and people who want to move from being a GIS Technician to GIS Analyst; GIS Analyst to GIS Programmer; or GIS Developer/Programmer to a GIS Architect.

Basic familiarity with geospatial/GIS syntax, ArcGIS, and data science (Pandas) is helpful, though not necessary.

What this book covers

Part 1: Introduction to Python Modules for ArcGIS Pro

Chapter 1, Introduction to Python for GIS, introduces the core components of Python required for the automation of ArcGIS Pro and other Esri productions. This chapter also includes an overview of Python syntax to introduce the required data structures and scripting concepts.

Chapter 2, Basics of ArcPy, explains the syntax and modules available for ArcPy, a Python package that it is very important to be familiar with for map production and data management for ArcGIS Pro. You will explore the functions and modules available in ArcPy, and do some geoprocessing in the ArcGIS Pro window using ArcPy.

Chapter 3, ArcGIS API for Python, introduces ArcGIS API for Python. This is a Python package designed to work with web GIS and allows you to work directly with data on ArcGIS Online or ArcGIS Enterprise. We will cover how to set up and manage a virtual environment within ArcGIS Pro, and introduce ArcGIS Pro Notebooks, which are similar to Jupyter Notebooks. The ArcGIS Notebooks will be used throughout the book as a way to write and run Python in ArcGIS Pro.

Part 2: Applying Python Modules to Common GIS Tasks

Chapter 4, Data Access Module and Using Cursors, covers how to use the Data Access module to assist in automating import steps in geoprocessing tasks. The `Walk` function will be used to walk through directories to find datasets. The search, insert, and update cursors will be used for finding and updating data within feature classes.

Chapter 5, Publishing to ArcGIS Online, covers how to publish and organize data on ArcGIS Online from within ArcGIS Pro. We will use ArcGIS API for Python in ArcGIS Pro Notebooks to publish, append, and edit data. We will also show you how repetitive tasks involved in managing ArcGIS Online content can be automated using Python.

Chapter 6, ArcToolbox Script Tools, demonstrates the process of turning a Python script into a script tool. A script tool is stored in a custom toolbox and runs like an ArcGIS tool. Creating script tools is a great way to share your scripts, as it allows non-Python users in your organization to run tools you developed for specific tasks.

Chapter 7, Automated Map Production, introduces the `arcpy.mp` module that is used to automate map production tasks. We will see how to use Python to update broken data source links, add/move/remove data layers from a map, adjust the symbology of a layer, work with the different layout elements, and export maps.

Part 3: Geospatial Data Analysis

Chapter 8, Pandas, Data Frames, and Vector Data, introduces you to using Pandas for geospatial data analysis. We cover some Pandas basics, how to get data in and out of Pandas DataFrames, and look at a GeoJSON-to-CSV file-to-shapefile workflow that relies on Pandas.

Chapter 9, Raster Analysis with Python, demonstrates how to use the raster tools in the `arcgis` and `arcpy` modules to work with raster and imagery layers. We look at creating rasters, saving them, and accessing their properties, as well as how the Spatial Analyst toolset enables more advanced spatial modeling and analysis.

Chapter 10, Geospatial Data Processing with NumPy, covers how and when you can use the NumPy module when working with raster data. We look at some basic NumPy array manipulations and how they are used in the context of geospatial analysis.

Part 4: Case Studies

Chapter 11, Case Study: ArcGIS Online Administration and Data Management, contains case studies showing you how to create Notebooks within ArcGIS Pro to administer your ArcGIS Online account. These Notebooks will allow you to manage users, report credits, reassign items, and download and rename photos. All of this can be done from within ArcGIS Pro using the ArcGIS API for Python, and ArcGIS Pro Notebooks.

Chapter 12, Case Study: Advanced Map Automation, is a case study showing you how to create a map automation from start to finish. We walk through the different map settings that cannot be changed with `arcpy.mp` to help create a good template for your map automation. Then, we use `arcpy.mp` to create a map series showing the different minority status of the block groups around bus lines that were suspended in 2020 to identify any potential environmental justice issues.

Chapter 13, Case Study: Interactive Data Science Web Map, is a case study that demonstrates an extract, transform, load (ETL) workflow applied to the problem of predicting crop yields using agricultural data from around the world. We write a Notebook that performs data collection, data cleaning, and fits a random forest model to make our predictions, and then create a simple JavaScript web app on top of our Python code.

To get the most out of this book

To follow along with the exercises in this book, you need to have ArcGIS Pro 2.7 or higher installed, along with the Python version that is installed with ArcGIS Pro. Don't worry, though - in *Chapter 2*, we will guide you through how to check your environment is set up properly before you begin.

Download the example code files

The code bundle for the book is hosted on GitHub at `https://github.com/PacktPublishing/Python-for-ArcGIS-Pro`. We also have other code bundles from our rich catalog of books and videos available at `https://github.com/PacktPublishing/`. Check them out!

Download the color images

We also provide a PDF file that has color images of the screenshots/diagrams used in this book. You can download it here: `https://static.packt-cdn.com/downloads/9781803241661_ColorImages.pdf`

Conventions used

There are a number of text conventions used throughout this book.

`CodeInText`: Indicates code words in text, database table names, folder names, filenames, file extensions, pathnames, dummy URLs, user input, and Twitter handles. For example: "ArcGIS Pro comes with a default environment called `arcgispro-py3`."

A block of input code is set as follows:

```
from arcgis.gis import GIS
from IPython.display import display
gis = GIS('home')
```

When we wish to draw your attention to a particular part of a code block, the relevant lines or items are highlighted:

```
from arcgis.gis import GIS
from IPython.display import display
gis = GIS('home')
```

Any Notebook output is written as follows:

```
<Item title:"Farmers Markets in Alameda County" type:Feature Layer
Collection owner:billparkermapping>
<Item title:"Farmers Markets in Alameda County" type:CSV
owner:billparkermapping>
```

Bold: Indicates a new term, an important word, or words that you see on the screen, for example, in menus or dialog boxes. For example: "However, it is not just a Python package: it is also an **application programming interface** (API)."

Warnings or important notes appear like this.

Tips and tricks appear like this.

Get in touch

Feedback from our readers is always welcome.

General feedback: Email `feedback@packtpub.com`, and mention the book's title in the subject of your message. If you have questions about any aspect of this book, please email us at `questions@packtpub.com`.

Errata: Although we have taken every care to ensure the accuracy of our content, mistakes do happen. If you have found a mistake in this book, we would be grateful if you would report this to us.

Please visit `http://www.packtpub.com/submit-errata`, selecting your book, clicking on the Errata Submission Form link, and entering the details.

Piracy: If you come across any illegal copies of our works in any form on the Internet, we would be grateful if you would provide us with the location address or website name. Please contact us at `copyright@packtpub.com` with a link to the material.

If you are interested in becoming an author: If there is a topic that you have expertise in and you are interested in either writing or contributing to a book, please visit `http://authors.packtpub.com`.

Share your thoughts

Once you've read *Python for ArcGIS Pro*, we'd love to hear your thoughts! Scan the QR code below to go straight to the Amazon review page for this book and share your feedback.

https://packt.link/r/1803241667

Your review is important to us and the tech community and will help us make sure we're delivering excellent quality content.

Part I

Introduction to Python Modules for ArcGIS Pro

1

Introduction to Python for GIS

Programming with computers is one of the most rewarding and frustrating of human endeavors.

Those rewards can be in the form of money, as we can see with today's high-tech salaries. I would argue, however, that the most rewarding part of mastering programming is to make yourself into a computer power user who can execute both simple and complex applications and analyses, written in reusable code, with ease.

The frustrations will come and go, and it is a good thing: you, like me and millions before you, will learn from each mistake. You will grow and learn with each exercise in this book, and by asking the right questions and paying close attention you can avoid some of these issues.

If you are an ArcGIS expert or novice, seeking to expand on your skillsets, congratulations – you are in the right place. In this book, you will learn how to take your existing GIS expertise (or interest) and multiply its potential using a deceptively simple programming language called Python.

Computer programming is a vast field of knowledge, about which whole books have been written. In this chapter, we will explain the basic knowledge necessary to read, write, and run Python scripts. We'll leave the ArcGIS tools for later chapters and focus on Python: its beginnings, its current state, how to use it, and importantly, what Python is and what it is not.

We will cover the following topics:

- The basics of Python
- The basics of computer programming
- Installing and importing modules
- Writing and executing scripts

Python: Built different

Guido Van Rossum, the creator of the Python programming language, was frustrated with the state of computer programming in the late 1980s. Programming languages were too complex and, at the same time, too loose with their formatting requirements. This led to large codebases with complex scripts poorly written and rarely documented.

Merely running a simple program could take a long time, as the code would need to be **type-checked** (variables declared correctly and assigned to the correct data type) and **compiled** (converted from high-level code written in text files into the assembly language or machine code understood by the CPU).

As this Dutch programmer had completed professional work on the ABC programming language, where he had learned much about language design, he decided he wanted to turn his gripes about the limits of ABC and other languages into a hobby.

With a master's degree in mathematics and computer science from the University of Amsterdam, his hobbies tended towards the computer, but he did have a love for Monty Python, the British comedy series. So, he combined his passions and created Python, which is now used for all kinds of programmatic solutions. Today Python is everywhere, used to power the internet, kitchen appliances, cars, and so much more. Because of its ubiquity and its simplicity, it has been adopted by the GIS software ecosystem as a standard programming tool.

Thanks to Van Rossum's extensive experience with the state of computer languages in the 1980s, he was well positioned to create a language that solved many of their deficiencies. He added features that he admired from many other languages and added a few of his own. Here is an incomplete list of Python features built to improve on other languages:

Issue	Improvement	Python feature
Memory overrun	Built-in memory management	Garbage collection and memory management
Slow compiler times	One-line testing, dynamic typing	Python interpreter
Unclear error messages	Messages indicating the offending line and affected code	Error traceback
Spaghetti code, i.e. code with unclear internal logic	Clean importation and modularization	Importation

Unclear code formatting and spacing, making code unreadable	Indentation rules and reduced brackets	Forced whitespace
Too many ways to do something	There should be only one way: the Pythonic way	The Zen of Python, a philosophy of programming that is unique to Python, which expects clean and simple implementations. Type `import this` into a Python interpreter and explore the built-in "easter egg" poem.

Python versions

The original Python version released in 1991 by Van Rossum, Python 1.0 and its successors, was eventually superseded by the widely popular Python 2.x. Care was taken to ensure that version 2.0 and beyond were backward-compatible with Python 1.x. However, for the new Python 3.0 and beyond, backward compatibility with Python 1 and Python 2 was broken.

This break has caused a divergence in the Python ecosystem. Some companies chose to stick with Python 2.x, which meant that the "sunset" date, or retirement date, for the older version was extended from 2015 until April 2020. Now that the sunset date has passed, there is no active work by the **Python Software Foundation** (**PSF**) on Python 2.x. Python 3.x development continues and will continue into the future, overseen by the PSF.

Van Rossum served as the Benevolent Dictator for Life of the PSF until he resigned from the position in 2018.

Check out more about the history of Python here: `https://docs.python.org/3/faq/general.html`

ArcGIS Python versions

Since ArcMap version 9.x, Python has been integrated into the ArcGIS software suite. However, ArcGIS Desktop and ArcGIS Pro now both depend on different versions of Python:

- **ArcGIS Pro**: Python 3.x

  ArcGIS Pro, which was designed after the decision to sunset Python 2.0 was announced, was divorced from the Python 2.x ecosystem and instead ships with Python 3.x.

  Along with the `arcpy` module, ArcGIS Pro uses the `arcgis` module, known as the ArcGIS API for Python.

- **ArcGIS Desktop**: Python 2.x

  ArcGIS Desktop (or ArcMap) version 9.0 and above ships with Python 2.x included. The installer for ArcGIS will automatically install Python 2.x and will add the `arcpy` module (originally `arcgisscripting`) to the Python system path variable, making it available for scripting.

  ArcMap, ArcCatalog, ArcGIS Engine, and ArcGIS Server all depend on `arcpy` and the Python 2.x version included when the ArcGIS Desktop or Enterprise software is installed.

The sunsetting of ArcGIS Desktop has been extended to March 2025, meaning that Python 2.7 will be included by Esri until that time, despite it being officially retired by the Python Software Foundation. With the sunsetting of ArcGIS Desktop approaching, users are now writing scripts in Python 3 to work with ArcGIS Pro.

What is Python?

In short, Python is an application: `python.exe`. This application is an executable file, meaning it can be run to process lines of code, or it can be called from other applications to run custom scripts. When ArcGIS Pro is installed, Python is also installed on your computer, along with a series of supporting files and folders, at this default location:

```
C:\Program Files\ArcGIS\Pro\bin\Python\envs\arcgispro-py3
```

Python includes a large standard library of tools, or **modules**. These include support for internet requests, advanced math, CSV reading and writing, JSON serialization, and many more modules included in the Python core. While these tools are powerful, Python was also built to be extensible, meaning that third-party modules can be easily added to a Python installation.

The ArcGIS Python modules, `arcpy` and `arcgis`, are both good examples of extending the capabilities of Python. There are hundreds of thousands of others, covering almost any type of programming need, of varying quality.

Python is written in the programming language C. There are variants of Python written in other languages for a variety of technical reasons, but most implementations of Python are built on top of C. This means that Python is often expanded through modules built on top of C code, usually for speed improvement reasons.

A Python code **layer** or **wrapper** is put on top of C code to make it work with normal Python packages, gaining the simplicity of Python and the processing speed boosts of precompiled C code. NumPy and SciPy (which are included with the ArcGIS installation of Python) are examples of this type of module.

Python is free and open software, which is another reason it is packaged with so many other software applications for automation purposes. While Python is already installed with ArcGIS Pro, it can also be installed separately, using a free installer from the Python Software Foundation.

Check out the Python Software Foundation on the internet: `https://www.python.org/psf`

Download Python versions directly from the PSF: `https://www.python.org/downloads/`

Where is it installed?

On Windows machines, Python is not included by default; it will be installed along with ArcGIS Pro or separately using an installer from the Python Software Foundation.

Once the ArcGIS Installer is run, a few versions of Python will be installed. For our use in this book, the main version is the Python 3 virtual environment installed at this folder location:

```
C:\Program Files\ArcGIS\Pro\bin\Python\envs\arcgispro-py3
```

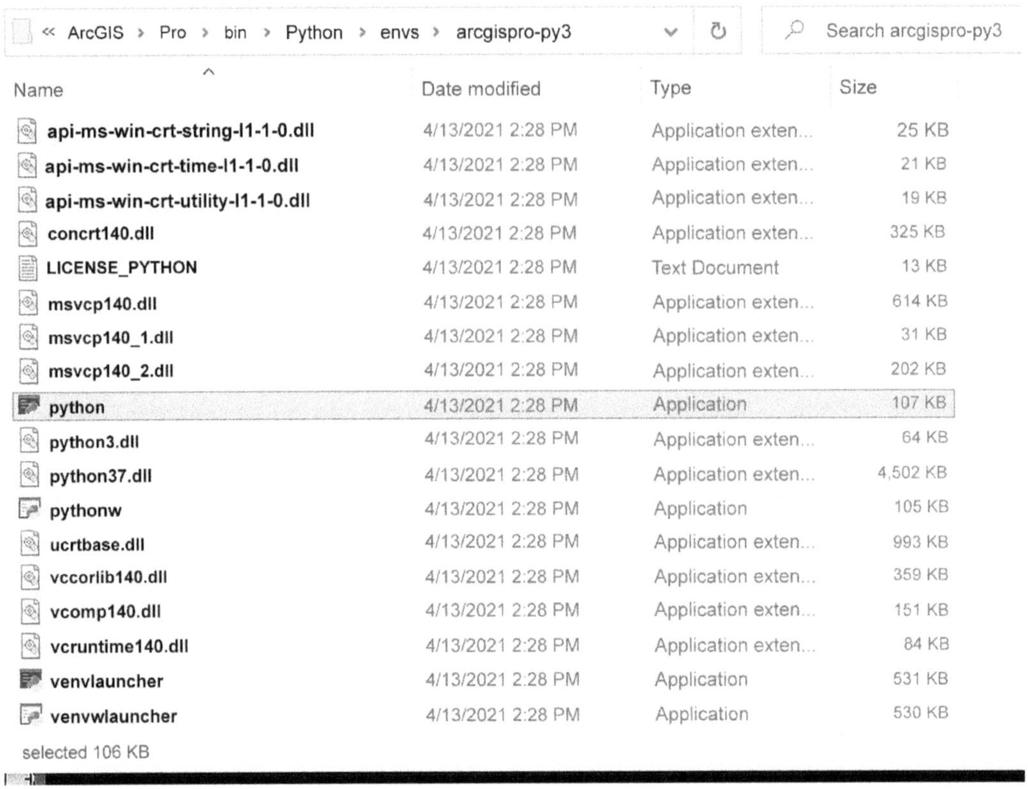

« ArcGIS › Pro › bin › Python › envs › arcgispro-py3

Search arcgispro-py3

Name	Date modified	Type	Size
api-ms-win-crt-string-l1-1-0.dll	4/13/2021 2:28 PM	Application exten...	25 KB
api-ms-win-crt-time-l1-1-0.dll	4/13/2021 2:28 PM	Application exten...	21 KB
api-ms-win-crt-utility-l1-1-0.dll	4/13/2021 2:28 PM	Application exten...	19 KB
concrt140.dll	4/13/2021 2:28 PM	Application exten...	325 KB
LICENSE_PYTHON	4/13/2021 2:28 PM	Text Document	13 KB
msvcp140.dll	4/13/2021 2:28 PM	Application exten...	614 KB
msvcp140_1.dll	4/13/2021 2:28 PM	Application exten...	31 KB
msvcp140_2.dll	4/13/2021 2:28 PM	Application exten...	202 KB
python	4/13/2021 2:28 PM	Application	107 KB
python3.dll	4/13/2021 2:28 PM	Application exten...	64 KB
python37.dll	4/13/2021 2:28 PM	Application exten...	4,502 KB
pythonw	4/13/2021 2:28 PM	Application	105 KB
ucrtbase.dll	4/13/2021 2:28 PM	Application exten...	993 KB
vccorlib140.dll	4/13/2021 2:28 PM	Application exten...	359 KB
vcomp140.dll	4/13/2021 2:28 PM	Application exten...	151 KB
vcruntime140.dll	4/13/2021 2:28 PM	Application exten...	84 KB
venvlauncher	4/13/2021 2:28 PM	Application	531 KB
venvwlauncher	4/13/2021 2:28 PM	Application	530 KB

selected 106 KB

Figure 1.1: Structure of the Python folder, containing the python.exe executable

Python interpreter

When you run `python.exe` (see below for multiple ways to run the executable), it starts what is known as the **Python interpreter**.

This is a useful interface, allowing you to enter, one line at a time, bits of code for testing and confirmation. Once the line is entered, hit *Enter/Return* and the code will be executed. This tool helps you both to learn to code and to test code in the same environment.

Double-clicking on `python.exe` from the folder or starting Python (command line) from the Start menu will start the interpreter, which allows for one-line commands to be executed:

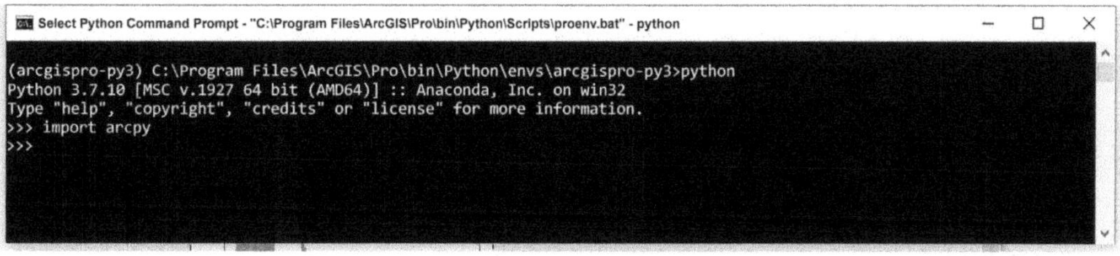

```
Select Python Command Prompt - "C:\Program Files\ArcGIS\Pro\bin\Python\Scripts\proenv.bat" - python

(arcgispro-py3) C:\Program Files\ArcGIS\Pro\bin\Python\envs\arcgispro-py3>python
Python 3.7.10 [MSC v.1927 64 bit (AMD64)] :: Anaconda, Inc. on win32
Type "help", "copyright", "credits" or "license" for more information.
>>> import arcpy
>>>
```

Figure 1.2: Python interpreter for Python 3.7

What is a Python script?

The `python.exe` executable file, along with being a program where code can be run, will also execute Python scripts. These scripts are simple text files that can be edited by any text editing software. Python scripts are saved with the `.py` extension.

When a Python script is **run**, it is passed as the first command-line argument to the Python executable (`python.exe`). This program will read and then execute the code from the top to the bottom, as long as it is valid Python and it contains no errors. If there is an error encountered, the script will stop and return an error message. If there is no error, nothing will be returned unless you have added "print" statements to return messages from the main loop to the Python window as the script is running.

Executables included

Python comes with two versions of the `python.exe` file. These are the same version of Python, to be clear, but each file has a different role. `python.exe` is the main file, and the other version is `pythonw.exe`. This file will not open an interpreter if double-clicked, as the normal `python.exe` will. No interpreter is available from `pythonw.exe`, which is the point: it is used to execute scripts more "silently" than `python.exe` (for example, when called by another application such as ArcGIS to run a Python script).

Use `python.exe` to start the interpreter.

« ArcGIS › Pro › bin › Python › envs › arcgispro-py3 Search arcgispro-py3

Name	Date modified	Type	Size
api-ms-win-crt-string-l1-1-0.dll	4/13/2021 2:28 PM	Application exten...	25 KB
api-ms-win-crt-time-l1-1-0.dll	4/13/2021 2:28 PM	Application exten...	21 KB
api-ms-win-crt-utility-l1-1-0.dll	4/13/2021 2:28 PM	Application exten...	19 KB
concrt140.dll	4/13/2021 2:28 PM	Application exten...	325 KB
LICENSE_PYTHON	4/13/2021 2:28 PM	Text Document	13 KB
msvcp140.dll	4/13/2021 2:28 PM	Application exten...	614 KB
msvcp140_1.dll	4/13/2021 2:28 PM	Application exten...	31 KB
msvcp140_2.dll	4/13/2021 2:28 PM	Application exten...	202 KB
python	4/13/2021 2:28 PM	Application	107 KB
python3.dll	4/13/2021 2:28 PM	Application exten...	64 KB
python37.dll	4/13/2021 2:28 PM	Application exten...	4,502 KB
pythonw	4/13/2021 2:28 PM	Application	105 KB
ucrtbase.dll	4/13/2021 2:28 PM	Application exten...	993 KB
vccorlib140.dll	4/13/2021 2:28 PM	Application exten...	359 KB
vcomp140.dll	4/13/2021 2:28 PM	Application exten...	151 KB
vcruntime140.dll	4/13/2021 2:28 PM	Application exten...	84 KB
venvlauncher	4/13/2021 2:28 PM	Application	531 KB
venvwlauncher	4/13/2021 2:28 PM	Application	530 KB

Figure 1.3: pythonw.exe in the Python folder

How to call the executable

The Python executable (`python.exe`) is accessed to run the Python interpreter or to run a custom Python script. There are many different ways to **call** or start the Python executable:

- **Double-click on** `python.exe` (`"C:\Program Files\ArcGIS\Pro\bin\Python\envs\arcgispro-py3\python.exe"`): This starts the Python interpreter.
- **Run Python inside ArcGIS Pro**: ArcGIS Pro has a built-in Python interpreter that you will use in *Chapter 2* to run custom lines of code. In *Chapter 3*, you will see how to use ArcGIS Pro Notebooks as a way to test, store, and share custom scripts as Notebooks.

- **Open IDLE**, the included integrated development environment (IDE): It can be run directly:

```
C:\Program Files\ArcGIS\Pro\bin\Python\envs\arcgispro-py3\Scripts\
idle.exe
```

In *Chapter 2*, you will see how to create a shortcut on your Desktop to the IDLE associated with your Python 3.x install.

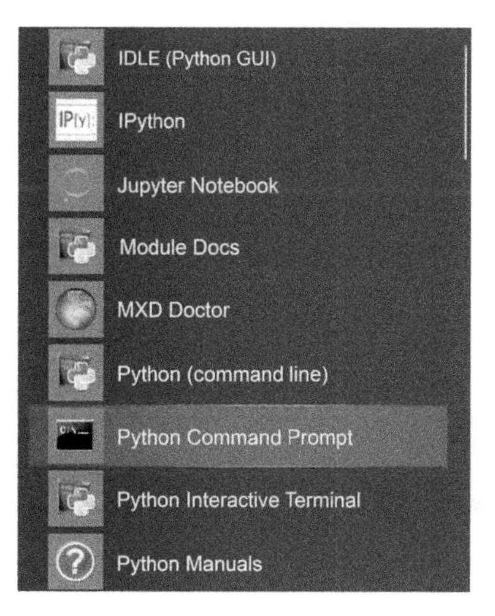

Figure 1.4: Python applications available through the Start/ArcGIS Menu

If you have ArcGIS Desktop and ArcGIS Pro along with other versions of Python installed, always pay attention to which version of Python you are opening from the Start menu. Not all versions may be associated with ArcGIS and therefore may not have the `arcpy` module accessible.

- **Open a CMD terminal and type** `python`: This only works if the Python executable is in the Windows PATH environment variable. If you get an error that says `'python' is not recognized as an internal or external command, operable program or batch file`, the `python.exe` program is not in the Windows PATH environment variable.

Check out this blog for a discussion on how to add your executable to the Path variable: `https://www.educative.io/edpresso/how-to-add-python-to-path-variable-in-windows`

- **Use a third-party IDE such as PyCharm:** Each PyCharm project can have its own virtual environment, and therefore its own executable, or it can use the one installed by Esri when ArcGIS is installed (`C:\Program Files\ArcGIS\Pro\bin\Python\envs\arcgispro-py3\python`). There are a lot of IDEs, but PyCharm is the one we recommend for a variety of reasons: clean interface, easy downloading of modules, built-in virtual environments, and more.
- **Use a Jupyter Notebook:** This requires the installation of Jupyter, which is not included in the standard Python installation.

You will be using ArcGIS Pro Notebooks starting in *Chapter 3*. These are based on Jupyter Notebooks and are very similar, but are stored and run in ArcGIS Pro.

- **Run Python in the command line** by using the whole path to the executable:

```
"C:\Program Files\ArcGIS\Pro\bin\Python\envs\arcgispro-py3\python.
exe"
```

There are multiple ways to directly run the script using the executable, but we find that IDEs make it easier to edit and execute code.

IDLE development environment

The included IDE, called IDLE, is a useful environment that comes standard with every Python instance:

```
File  Edit  Shell  Debug  Options  Window  Help
Python 3.7.10 [MSC v.1927 64 bit (AMD64)] on win32
Type "help", "copyright", "credits" or "license()" for more information.
>>> import arcgis
>>> import arcpy
>>> |
```

Figure 1.5: The Python IDLE interpreter environment is similar to a shell environment. Code can be run one line at a time.

You can create and execute scripts in this environment easily by opening a new script from the **File** menu, and then using the script's **Run** menu to execute the script:

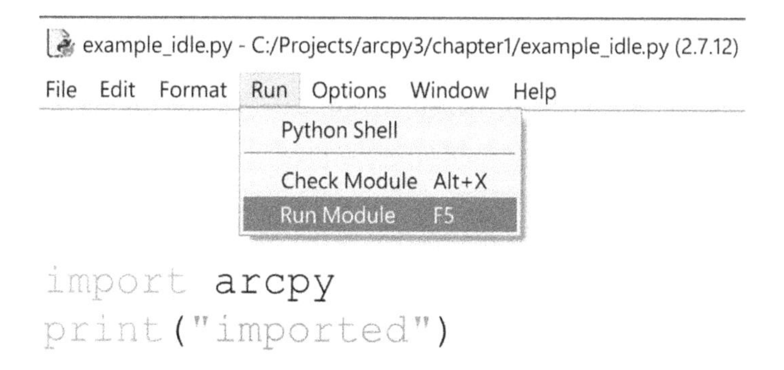

Figure 1.6: Running a script in IDLE

Windows file path issues

Because Python was developed in a Unix/Linux environment, it expects file paths to use forward slashes (/). However, Windows uses backslashes (\) in its file paths.

Windows:

```
'C:\Python\python.exe'
```

Linux:

```
'C:/Python/python.exe'
```

This has consequences in a Python script, because of the presence of a number of special string combinations made with backslashes. For instance, to create a tab character in a string, Python uses a combination of a backslash and a "t" to create this character: `\t`.

The backslashes can be **escaped**; in other words, Python can be told to ignore the special characters in a string, by doubling up the backslash. However, this is inconvenient. The easiest way to address the backslashes inherent in Windows file paths (when passing a shapefile file path to an `arcpy` function, for instance) is to make them into **raw strings** by putting an "r" in front of the string.

The following would cause an error when passed to an `arcpy` function, because of all the `\t` characters:

```
'C:\test\test.shp'
```

To avoid this, you have three options. If you are copying a folder path from Windows Explorer, use an "r" in front of the script to transform it into a **raw string**:

```
r'C:\test\test.shp'
```

You can also use forward slashes:

```
'C:/test/test.shp'
```

Escaping the backslashes by doubling them up also works:

```
'C:\\test\\test.shp'
```

The operating system and Python system modules

Two important modules or code libraries built into Python to know about are the os and sys modules. The first, os, is also called the operating system module. The second, sys, is the Python system module. They are used to control Windows system operations and Python system operations respectively.

The os module

The os module is used for many things, including folder path operations such as creating folders, removing folders, checking if a folder or file exists, or executing a file using the operating system-associated application used to run that file extension. Getting the current directory, copying files, and more, are made possible with this module. The os module will be used throughout this book in examples to do all of the above.

In the following code snippet, we first import the os module since we intend to use it. A string, "C:\Test_folder", is passed to the os.path.exists method, which returns a Boolean value (either True or False). If it returns False, the folder does not exist, and is then created using the os.mkdir method:

```
import os
folderpath = r"C:\Test_folder"
if not os.path.exists(folderpath):
   os.mkdir(folderpath)
```

Read about the os module here: https://www.geeksforgeeks.org/os-module-python-examples/

The sys module

The sys module, among other functions, allows you to accept arguments to a script at **runtime** (meaning when the script is executed). This is done by using the sys.argv method, which is a list containing all arguments made to Python during the executing of the script.

If a name variable is using the sys module to accept parameters, here is what the script looks like:

```
import sys
name = sys.argv[1]
print(name)
```

Note again that the sys.argv method is a list, and the second element in the list (assigned to the variable name above) is the first parameter passed. Python uses zero-based indexing, which we explore in further detail later in the chapter. The first element in the list is the file path of the script being run.

The system path

The sys module contains the Python path or system path (system in this case means Python). The Python system path, available from the sys module at sys.path, is a list that Python uses to search for importable modules, after accessing the Windows Path variable. If you can't edit the Windows Path (due to permissions, usually), you can alter the Python path at runtime using the system path.

The sys.path list is a part of the sys module built into Python:

```
Python Interactive Terminal
Type "help", "copyright", "credits" or "license" for more information.
>>> import sys
>>> sys.path
['', 'C:\\Program Files\\ArcGIS\\Pro\\bin\\Python\\envs\\arcgispro-py3\\python35.zip', 'C:\\Pro
gram Files\\ArcGIS\\Pro\\bin\\Python\\envs\\arcgispro-py3\\DLLs', 'C:\\Program Files\\ArcGIS\\P
ro\\bin\\Python\\envs\\arcgispro-py3\\lib', 'C:\\Program Files\\ArcGIS\\Pro\\bin\\Python\\envs\
\arcgispro-py3', 'C:\\Program Files\\ArcGIS\\Pro\\bin\\Python\\envs\\arcgispro-py3\\lib\\site-p
ackages', 'C:\\Program Files\\ArcGIS\\Pro\\bin', 'C:\\Program Files\\ArcGIS\\Pro\\Resources\\Ar
cPy', 'C:\\Program Files\\ArcGIS\\Pro\\Resources\\ArcToolbox\\Scripts', 'C:\\Program Files\\Arc
GIS\\Pro\\bin\\Python\\envs\\arcgispro-py3\\lib\\site-packages\\setuptools-27.2.0-py3.5.egg']
>>>
```

Figure 1.7: Inspecting the sys.path list

Read more about the sys module here: https://www.geeksforgeeks.org/python-sys-module/

We have given you a lot of information about what Python is, how the Python folder is structured, how the Python executable is run, and how to execute and run scripts. This will help you run Python scripts to automate your analyses. In the next section, we will be zooming out to gain a wider view of computer programming.

This will help you to gain more insight into why Python was chosen to be the language of automation for ArcGIS Pro, and help you to be a better programmer in general.

As well as an introduction to Python programming, the rest of the chapter will be a useful reference for you to come back to as you work through the book. If you'd like to get hands-on with writing code straightaway, start with *Chapter 2, Basics of ArcPy*.

The basics of programming

Computer programming varies from language to language in terms of implementation, but there are remarkable similarities among these languages in how their internal logic works. These programming basics are applicable for all programming languages, with specific code implementations shown in Python:

Concept	Description	Examples in code

Variables	Names assigned to Python objects of any data type. Variables must start with a letter. Underscores are encouraged.	`x=0`
		`y=1`
		`xy = x+y`
		`xy_str = str(xy)`

Data types	Strings are for text. Integers are for whole numbers. Floats are for floating-point numbers. Data containers such as lists, tuples, and dictionaries are used extensively to organize data. Booleans are used for true or false situations.	`str_var = "string"`
		`int_var = 4`
		`float_var = 5.7`
		`list_var = [45,43,24]`
		`tuple_var = (87,'a',34)`
		`dict_var = {'key':'value'}`
		`bool_var = True`

Iteration	For loops are used to iterate through an iterable data object (an **iterator**, such as a data list). While loops are used to loop until a condition has been met.	`for item in datalist:`
		`    print(item)`
		
		`x=0`
		`while x < 1:`
		`    x+=1`

<table>
<tr>
<td>Counters / Enumerators</td>
<td>Using a variable to keep track of the number of loops performed by a for loop or while loop is a good idea. Some languages have built-in enumeration functionality. In Python, this is the <code>enumerate()</code> function. Counters are reassigned to themselves after being increased.

In Python, the shortcut <code>x += y</code> is the same as <code>x = x + y</code>.</td>
<td><pre><code>counter = 0
list_var = [34,54,23,54]
for item in list_var:
    print(item, counter)
    counter += 1

l_var = [34,54,23,54]
for c,i in enumerate(l_var):
    print(i, c)</code></pre></td>
</tr>
<tr>
<td>Conditionals</td>
<td>If/Elif/Else statements that interpret whether an object meets a condition.</td>
<td><pre><code>list_var = [1,'1',1.0]
for item in list_var:
  if type(item) == type(0):
    print('Integer')
  elif type(item) == type('a'):
    print('String')
  else:
    print('Float')</code></pre></td>
</tr>
<tr>
<td>Zero-based indexing</td>
<td>Data containers are accessed using indexes that start with 0. The indexes are passed to the list or tuple using square brackets <code>[]</code>. String characters can be accessed using the same pattern.</td>
<td><pre><code>list_var = ['s','m','t']
m_var = list_var[0]

name_var = "logan"
l_var = name_var[0]</code></pre></td>
</tr>
</table>

Code comments	Comments in code are encouraged. They help explain your thinking to both other readers and yourself. Comments are created by using the # symbol. Comments can be on a line by themselves or can be added to the end of a statement, as anything after the # symbol will be ignored.	`# This is a comment`
		
		`x = 0 # also a comment`

Errors	Error messages of many types are built into Python. The **error traceback** shows the affected lines of code and the type of error. It's not perfect.	`>>> str_var = 'red"`
		`  File "<stdin>", line 1`
		`    str_var = 'red"`
		`                  ^`
		`SyntaxError: EOL while scanning string literal`

In the following sections, we take a look at some of these in more detail, as well as introducing you to functions and classes.

Variables

Variables are used to assign objects to labels or identifiers. They are used to keep track of pieces of data, to organize the flow of the data through the script, and to help programmers read the script.

```
variable = 1 # a variable assignment
```

We recommend you use descriptive variables that are neither too long nor too short. When variables are too short, they can become confusing to read. When they are too long, they can be confusing to write. Using underscores to separate words in variables is a common practice.

Read more about Python variable naming conventions here: https://www.python.org/dev/peps/pep-0008/#function-and-variable-names

Variable formatting rules

Variables must start with a letter. They cannot start with a number or other symbol, otherwise a `SyntaxError` will occur. However, numbers and underscores can be used in them:

```
>>> 2var = 34
  File "<stdin>", line 1
    2var = 34
     ^
SyntaxError: invalid syntax
>>> two_var = 34
>>> two_var
34
```

Read more about variables here: `https://realpython.com/python-variables/`

Assigned to vs is equal to (value comparison)

In Python, variables are assigned to an object using the equals sign (=). To check if a value is equal to another value (in other words, to compare them), use a double equals sign (==):

```
variable = 1 # a variable assignment
variable == 1 # a comparison
```

Data types

The data type of a variable determines its behavior. For instance, the character 5 could be an integer type (5), a float (`5.0`), or a string (`"5"`). Each version of 5 will have different available tools, such as the `replace()` method for strings, which can replace characters in the string with other characters.

The following table presents key data types in Python, along with the corresponding data type object in Python:

Data type	Python data type object
Text data is stored as a String data type.	`str`
Numeric data is stored as an Integer, Float, or Complex type.	`int`, `float`, `complex`
Sequence data (lists or arrays) can be stored as a list or tuple. In Python 3, `range` is a **generator**, a special object that returns a **lazy** iterator which, when called, returns one member of the desired list.	`list`, `tuple`, `range`
Mapping or **key/value pair** data types are also known as dictionaries in Python.	`dict`
A **set** is a data type that contains distinct, immutable objects.	`set`, `frozenset`
Boolean is either `True` or `False`, 1 or 0.	`bool`
Binary data types are used to access data files in binary mode.	`bytes`, `bytearray`, `memoryview`

Checking the data type

To check the data type of a Python variable, use the `type()` function:

```
>>> x = 0
>>> type(x)
<class 'int'>
```

Strings

All text data is represented as the String data type in Python. These are known as **strings**. Common data stored as strings includes names, addresses, or even whole blog posts.

Strings can also be templated in code to allow for "fill-in-the-blank" strings that are not set until the script is run. Strings are technically immutable but can be manipulated using built-in Python string tools and the separate String module.

Here are some of the key concepts relating to strings:

Quotation marks	Single or double quotation marks can be used to designate a string, as long as the same number is used at the beginning and end. Quotes *within* a string can be indicated using the opposite mark as the one opening and closing the string. Triple quotation marks are used for strings with multiple lines.
String addition	Strings can be "added" together to form a larger string. Strings can also be "multiplied" by an integer *N* to repeat the string *N* times.
String formatting	String templates or placeholders can be used in code and filled in at runtime with the data required.
String manipulation	Strings can be manipulated using built-in functionality. Characters can be replaced or located. Strings can be split or joined.

Quotation marks

Strings must be surrounded by quotation marks. In Python, these can be either single or double quotes, but they must be consistent. If a single quote is used to start the string, a single quote must be used to stop it, or you will get an error:

```
>>> string_var = 'the red fox"
  File "<stdin>", line 1
    string_var = 'the red fox"
                             ^
SyntaxError: EOL while scanning string literal
```

A correct way:

```
>>> string_var = 'the red fox'
>>> string_var
'the red fox'
```

Multiple line strings

Multiple line strings are created by a pair of three single quotes or double quotes at the beginning of the string, and three at the end.

In the following example, the variable `string_var` is a multiple-line string (`\n` is a Python character representing a new line):

```
>>> string_var = """the red fox chased the
... dog across the yard"""
```

```
>>> string_var
'the red fox chased the\ndog across the yard'
```

String addition (and more)

Strings can be "added" together to create a new string. This process allows you to build strings from smaller strings, which can be useful for populating new fields composed of other fields in a data file, and other tasks.

In this example, the string `"forest"` is assigned to `string_var`. Another string is then added to `string_var` to create a longer string:

```
>>> string_var = "forest"
>>> string_var += " path" # same as string_var = string_var+ " path"
>>> string_var
'forest path'
```

String formatting

Strings in code often make use of "placeholders" for data that will be filled in later. This is known as **string formatting**, and there are multiple ways to perform string formatting using Python.

Here are the key concepts:

Format function	All strings have a built-in function called `format()` that allows the string to have arguments passed. It will accept all data types and format the string from a template.
String literals	For Python 3.6+, there is a new tool called **string literals**, which allow you to insert variables into strings directly. An `f` is placed in front of the string.
Data type string operators	An older but still useful tool is the string operators, which are used in strings as placeholders for specific data types (either strings, floats, or integers).

String format function

This method of formatting is the preferred form for Python 3. It allows you to pass the variables to the `format()` function, which is built into all strings, and to have them fill up placeholders within the string. Any data type can be passed to the `format()` function.

In the following example, the string template is filled with details contained in other variables using the `format()` string function. The placeholders are filled in the order that the variables are listed, so they must be in the correct order.

The curly brackets are the placeholders, and the `format()` function will accept arguments and fill in the string:

```
>>> year = 1980
>>> day = "Monday"
>>> month = "Feb"
>>> template = "It was a cold {} in {} {}"
>>> template.format(day, month, year)
'It was a cold Monday in Feb 1980'
```

In the next example, the placeholders are named, and are passed to keyword arguments in the `format()` function. The arguments are named and do not need to be in order in the `format()` function:

```
>>> template = 'It was a cold {day} in {month} {year}'
>>> template.format(month=month, year=year, day=day)
'It was a cold Monday in Feb 1980'
```

In this last example, the placeholders are numbered, which makes it much easier to repeat a string:

```
>>> template = "{0},{0} oh no,{1} gotta go"
>>> template.format("Louie", "Me")
'Louie,Louie oh no,Me gotta go'
```

String literals

There is a new (as of Python 3.6) method of formatting strings, known as formatted string literals. By adding an `f` before strings, placeholder variables can become populated by variables without using the `format()` function.

In this example, the variables are formatted directly into the string literal, which has an `f` before the string to indicate that it is a string literal:

```
>>> year = 1980
>>> day = "Monday"
>>> month = "Feb"
>>> str_lit = f"It was a cold {day} in {month} {year}"
>>> str_lit
'It was a cold Monday in Feb 1980'
```

Read more about string formatting here: `https://realpython.com/python-string-formatting/`

String placeholder operators

An older but still useful method for inserting data into strings is the data type string operators. These use placeholders that will format inserted strings in specific ways. However, they are data-specific, meaning that a number inserted into the string must use a number placeholder, and a string being inserted must use a string placeholder, or an error will result.

The placeholders are `%s` for strings and `%d` or `%f` for numbers. They each have optional features specific to the data type. For instance, the `%f` number placeholder can be manipulated to hold only a specific number of decimal points:

```
>>> month = '%0.2f' % 3.1415926535
>>> month
3.14
```

To use them, you place the placeholder in the string template, and then follow the string with a percent sign (%) and the values to be passed into the string template in a tuple after the percent sign:

```
>>> year = 1980
>>> month = "February,"
>>> str_result = "It was a cold  %s %d" % month, year
>>> str_result
'It was a cold February, 1980'
```

Read more about string placeholders here: `https://pyformat.info/`

String manipulation

String manipulation is common and lots of tools are built into the String data type. These allow you to replace characters in a string or find their index location in the string.

The `find()` and `index()` methods are similar, but `find()` can be used in conditional statements. If the character is not found in the string, `find()` will return -1, while `index()` will return an error.

The `join()` method is used to join together a list of string data. The `split()` method is the opposite: it splits a string into a list based on a supplied character or the default empty space.

Here is a non-exhaustive list of methods and examples of their use:

Method	Example
join()	`string_list = ['101 N Main St','Eureka','Illinois 60133']` `address = ', '.join(string_list)`
replace()	`address = '101 N Main St'.replace("St","Street")`

find(), rfind()	`str_var = 'rare'`
	`str_index = str_var.find('a') # index 1`
	`str_index = str_var.find('r') # index 0`
	`str_index = str_var.rfind('r') # index 2`
	`str_index = str_var.rfind('d') # index -1`

upper(), lower(), title()	`name = "Laura"`
	`name_upper = name.upper()`
	`name_lower = name.lower()`
	`name_title = name_lower.title()`

index(), rindex()	`str_var = 'rare'`
	`str_index = str_var.index('a') # index 1`
	`str_index = str_var.index('r') # index 0`
	`str_index = str_var.rindex('r') # index 2`
	`str_var.index('t') # this will cause an error`

split()	`latitude,longitude = "45.123,-95.321".split(",")` `address_split = '101 N Main St'.split()`

String indexing

String indexing is similar to list indexing, as we will see later on. Individual characters, or groups of characters, can be selected from a string by passing the index of the character needed to the string in square brackets, where 0 is the index of the first character.

In the following example, the `d` from `readiness` is accessed by passing the index `[3]` to square brackets next to the string:

```
>>> str_var = "readiness"
>>> d_var = str_var[3]
>>> d_var
'd'
```

Groups of characters are selected by passing a start and end index, where the end index is the index of the first character you do not want to include:

```
>>> str_var = "readiness"
>>> din_var = str_var[3:6] # index 6 is e
>>> din_var
'din'
>>> dine_var = str_var[3:7] # index 7 is s
>>> dine_var
'dine'
```

Integers

The Integer data type represents whole numbers. It can be used to perform addition, subtraction, multiplication, and division (with one caveat as noted below):

```
>>> int_var = 50
>>> int_var * 5
250
>>> int_var / 5
10.0
>>> int_var ** 2
2500
```

Starting in Python 3, you could divide two integers and get a float. In previous versions of Python 2.x, whenever you divided two integers you would only get an integer, with no remainder. Because of the way Python 2.x did integer division, you will come across code where integers are converted to floats for divison. You are encouraged to do the same in your own code.

Read more about integers in Python here: `https://realpython.com/python-numbers/`

Floating numbers

Floating-point numbers in Python are used to represent real numbers as 64-bit double-precision values. Sometimes, using binary systems to represent decimal-based numbers can be a bit odd, but in general, these will work as expected:

```
>>> x = 5.0
>>> x * 5
25.0
>>> x ** 5
3125.0
>>> x/2.3
2.173913043478261
```

One unique result from floating-point division is the case of 1/3. Because it is a binary representation, the assumption that (1/3)* 3= 1 is `True`, even though the value 0.3333333333333333 (produced by the division operation) would never add up to 1 when added together three times in a base 10 system. Here are some examples of binary math in action:

```
>>> 1/3
0.3333333333333333
>>> (1/3) * 3
1.0
>>> (1/3) + (1/3)
0.6666666666666666
>>> (1/3) + (1/3) + (1/3)
1.0
>>> (1/3) + (1/3) + 0.3333333333333333
1.0
>>> (1/3) + (1/3) + 0.3333
0.9999666666666667
>>> (1/3) + (1/3) + 0.3333333333
0.9999999999666667
>>> (1/3) + (1/3) + 0.333333333333333
0.9999999999999996
>>> (1/3) + (1/3) + 0.33333333333333333 1.0
```

Read more about floating-point numbers in Python here: `https://www.geeksforgeeks.org/python-float-type-and-its-methods`

Conversion between strings, integers, and floats

Conversion between data types is possible in Python using built-in functions that are part of the standard library. As we saw earlier, the `type()` function is useful to find the data type of an object. Once identified, the data object can be converted from Integer (`int()` function) to String (`str()` function) to Float (`float()` function), as long as the character would be valid in that data type.

In these examples, a character is converted from String to Integer to Float to String using the `int()`, `str()`, and `float()` functions:

```
>>> str_var = "5"
>>> int_var = int(str_var)
>>> int_var
5
>>> float_var = float(int_var)
>>> float_var
5.0
>>> str_var = str(float_var)
>>> type(str_var)
'<class 'str'>'
```

Data structures or containers

Data structures, also called data containers and data collections, are special data types that can hold, in a retrievable order, any data item of any data type (including other data containers). Data containers are used to organize data items by index in tuples or lists, or by key:value pairs in dictionaries.

To get data out of data containers, square brackets are used to pass either indexes (lists and tuples) or keys (dictionaries). If there is more than one level of data container (in other words, one container contains another), first the data container inside is referenced using an index or key inside a first square bracket, and then the data inside the container is accessed using a second.

The following table summarizes the different types of data containers and how data is retrieved from each of them:

Data container	Example

Tuple	`tuple_var = ("blue", 32,[5,7,2],'plod',{'name':'magnus'})`
	`plod_var = tuple_var[-2]`
	`magnus_var = tuple_var[-1]['name']`

List	`list_var = ['fast','times',89,4.5,(3,8),{'we':'believe'}]`
	`times_var = list_var[1]`
	`dict_var = list_var[-1]`
	`believe_var = list_var[-1]['we']`

Set	`list_var = [1,1,4,6,7,6]`
	`set_var = set(list_var) # removes duplicates`
	`{1, 4, 6, 7}   # result`

Dictionary	`dict_var = {"key": "value"}`
	`dict_info = {"address": "123 Main Street", "name": "John"}`
	`name = dict_info["name"] # gets the name value from the key`
	`address = dict_info["address"] # gets the address value`

Tuples

Tuples are ordered lists that can hold any data type, even in the same tuple. They are **immutable**, meaning they cannot be altered, and data cannot be added to or removed from the tuple once it has been created. They have length and the built-in `len()` function can be used to get the length of the tuple.

In Python, they are declared by using round brackets, `()`, or the `tuple()` function. Data is accessed using zero-based indexing by passing the index to square brackets next to the tuple.

In the following example, a tuple is assigned to the variable name `tuple_var()`, and data is accessed using indexing:

```
>>> tuple_var = ("red", 45, "left")
>>> type(tuple_var)
<class 'tuple'>
>>> ("red",45,"left")[0]
'red'
>>> tuple_var[0]
'red'
```

Read more about tuples in Python here: `https://www.geeksforgeeks.org/python-tuples/`

Lists

Lists (often called arrays in other programming languages) are data containers that can hold any other type of data type, even in the same list, just like tuples. Unlike tuples, though, lists can be altered after they are created. In Python, they are declared by using square brackets, `[]`, or the `list()` function. Data is accessed using zero-based indexing by passing the index to square brackets next to the list.

In this example, a list is assigned to the variable name `list_var`, and data is accessed using indexing:

```
>>> list_var = ["blue",42,"right"]
>>> type(list_var)
<class 'list'>
>>> ["blue",42,"right"][0]
'blue'
>>> list_var[0]
'blue'
```

Read more about lists in Python here: `https://www.geeksforgeeks.org/python-list/`

Conversion between lists and tuples

Lists can be copied into a new tuple object using the `tuple()` function. Conversely, tuples can be copied into a list data type using the `list()` function. This does not convert the original data item but instead creates a copy of the data item in the new data type.

In the following example, the list is copied into a tuple data type, and then the tuple is copied into a list data type. Note that the brackets change with each new data type created:

```
>>> tuple_copy = tuple(list_var)
>>> tuple_copy
('blue', 42, 'right')
>>> list_copy = list(tuple_copy)
>>> list_copy
['blue', 42, 'right']
```

List operations for lists only

Using the append() method, a list can be appended to, which means one data item is added to the list. Using the extend() method, a list can also be extended, which is where all data items in a second list are all added to the first list:

```
>>> list_orig = [34, 'blanket', 'dog']
>>> list_orig.append(56)
>>> list_orig
[34,'blanket','dog',56]

>>> list_first = [34, 'blanket', 'dog']
>>> list_second = ['diamond', '321', 657]
>>> list_orig.extend(list_second)
>>> list_orig
[34,'blanket','dog','diamond','321'.657]
```

The items in a list can be reversed or sorted, using the reverse() method or the sort() method respectively:

```
>>> list_var = [34,'blanket','dog']
>>> list_var.reverse()
>>> list_var
['dog','blanket',34]
```

In Python 3, sorting is only allowed on lists that do not have mixed data types:

```
>>> list_var = [34,5,123]
>>> list_var.sort()
>>> list_var
[5, 34, 123]
```

In Python 2, sorting is allowed on mixed lists, with numbers placed first.

List operations for both tuples and lists

Lists and tuples can be iterated over using `for` loops, which we will look at shortly. They can both be **sliced** as well, creating a subset of the list or tuple that will be operated on for the `for` loop or another operation. Built-in functions allow for the calculation of the maximum (using the `max()` function) or minimum (the `min()` function) value of a list/tuple, or even the sum of a list or tuple, given the data type of the items in the list is correct.

Slicing

Slicing a list or tuple will create a new list or tuple. The slice is created by passing indexes to the list or tuple in square brackets, separated by a colon. The first index is the start index, and it can be ignored if it is index 0 (the beginning of the original list). The second index is the index of the first value that you do *not* want to include (it can be blank if you want the rest of the original list).

In this first example, we see a tuple with three data items sliced to only include the first two items. The string `"left"` is at index 2 in the tuple, meaning that the last index in the slice will be 2. The slice is assigned to variable name `tuple_slice`:

```
>>> tuple_var = ("red", 45, "left")
>>> tuple_slice = tuple_var[:2]
>>> tuple_slice
('red', 45)
```

In this next example, we see a list with four data items sliced to only include the last two items. The first index is the index of the first data item we want (the string `"right"`). The last index is blank:

```
>>> list_var = ["blue", 42, "right", "ankle"]
>>> list_slice = list_var[2:]
>>> list_slice
['right', 'ankle']
```

Sets

Sets represent a collection of distinct objects. In Python, sets are unordered, no duplicates are allowed, and all data items inside a set must be immutable.

Set operations

Sets are especially useful for getting all distinct members of a list:

```
>>> orig_list = ["blue", "pink", "yellow", "red", "blue", "yellow"]
>>> set_var = set(orig_list)
```

```
>>> set_var
{'pink', 'yellow', 'blue', 'red'}
```

Sets cannot be accessed using indexing, because they are unordered and therefore are not **subscriptable**:

```
>>> set_var[0]
Traceback (most recent call last):
  File "<stdin>", line 1, in <module>
TypeError: 'set' object is not subscriptable
```

However, they can be iterated over using looping:

```
>>> for item in set_var:
...     print(item)
...
pink
yellow
blue
red
```

Dictionaries

Dictionaries are key:value stores, meaning they are data containers that use unordered key and value pairs to organize data. Keys are used as reference points for organization and retrieval. When a key is supplied to a dictionary in square brackets, the value is returned:

```
>>> dict_var = {"key":"value"}
>>> dict_var['key']
'value'
>>> dict_var = {"address":"123 Main St", "color":"blue"}
>>> dict_var["address"]
'123 Main St'
>>> dict_var["color"]
'blue'
```

Read more about dictionaries in Python here: `https://www.geeksforgeeks.org/python-dictionary/`

Keys and values

Keys can be any immutable data type, meaning lists cannot be used as keys, but strings, integers, floats, and tuples can. Values can be any type of data, including other dictionaries.

All keys in a dictionary can be accessed as a list using the dictionary `keys()` function. In Python 3.x, the function is a generator, which means it must be called over and over to get each key. This generator can also be passed to the `list()` function to convert it into a list.

All values in a dictionary can be accessed as a list using the dictionary `values()` function. In Python 3.x, the function is a generator.

In Python 2.x, the `keys()` and `values()` functions return a list. In older code written for ArcGIS Desktop, you may see this.

Iteration

The core of computer programming is **iteration**: recursively performing the same action, analysis, function call, or whatever your script is built to process. Computers excel at this type of task: they can quickly iterate through a dataset to perform whatever action you deem necessary, on each data item in the set.

Iteration is run on **iterators**. An iterator is a Python object that contains other objects, each of which can be processed in a **loop**. Iterators can be lists or tuples or even strings, but not integers.

For loops

A **for loop** is an iteration implementation that, when presented with a data list, will perform an operation on each member of the list.

In the following example, a list of integers is assigned to the variable name `data_list`. The list is then used to construct a for loop using the format `for {var} in {iterable}`, where `{var}` is a variable name that is assigned to each object in the list, one at a time as the loop progresses. One convention is to use `item`, but it can be any valid variable name:

```
data_list = [45,56,34,12,2]
for item in data_list:
    print (item * 2)
```

This is the output:

```
90
112
68
24
4
```

While loops

A **while loop** is an iteration implementation that will loop until a specific threshold is met. While loops can be dangerous as they can cause an infinite loop in a script if the threshold is never met.

In the following example, the `while` loop will run, doing nothing but adding `1` to `x` until it reaches `100`, upon which the threshold is met and the `while` loop will end:

```
x = 0
while x < 100:
    x = x + 1   # same as x += 1
```

Read more about loops here: `https://www.geeksforgeeks.org/loops-in-python/`

Counters and enumerators

Iteration in `for` loops or `while` loops often requires the use of counters (also called enumerators) to track loops in an iteration.

`for` loops have the option to use the `enumerate()` function by passing the iterator to the function and using a count variable (which can be any valid variable name, but `count` is logical) in front of the item variable. The count variable will keep track of the loops, starting at index zero:

```
>>> data_list = ['a','b','c','d','e']
>>> for count,item in enumerate(data_list):
...     print(count, item)
...
0 a
1 b
2 c
```

```
3 d
4 e
```

In Python, the shortcut `x += y` is used to increase the value of x while keeping the same variable name, and is the same as `x = x + y`:

```
>>> x = 0
>>> while x <100:
...     x = x + 1
>>> x
100

>>> x = 0
>>> while x <100:
...     x += 1
>>> x
100
```

Conditionals

`if` statements, `elif` statements (short for `else if`), and `else` statements are used to create conditions that will be used to evaluate data objects. If statements can be used by themselves (elif and else are optional) and are used by declaring the keyword `if` and then the condition the data must meet.

In the following example, the data type of objects in a list is compared (notice the two equals signs, meaning it is a comparison) to the data type for `integers`, shown here as `type(0)`, or for `strings`, shown as `type('a')`. If an object in the list meets one of the conditions, a specific `print()` statement is triggered:

```
list_var = [1,'1',1.0]
for item in list_var:
    if type(item) == type(0):
        print('Integer')
    elif type(item) == type('a'):
        print('String')
    else:
        print('Float')
```

Read more about conditionals here: `https://realpython.com/python-conditional-statements/`

If versus else

`if` statements are usually specific to one condition, whereas `else` statements are used as catch-alls to ensure that any data that goes through the `if` statement will have some way of being dealt with, even if it doesn't meet the condition of the `if` statement. `elif` statements, which are dependent on the `if` statement existing and are also condition-specific, are not catch-all statements.

Zero-based indexing

As we have seen, iteration occurs over lists or tuples that contain data. Within the list, these data are differentiated by list order or position. Items in a list are retrieved by item index, the (current) position of the data in the list.

In Python, like most computer programming languages, the first item in a list is at index 0, not index 1.

This is a bit confusing to beginners but is a programming standard. It is slightly more computationally efficient to retrieve an item in a list that starts with 0 than a list that starts with 1, and this became the standard in C and its precursors, which meant that Python (written in C) uses zero-based indexing.

Data extraction using index position

This is the basic format of data retrieval from a list. This list of strings has an order, and the string `"Bill"` is the second item, meaning it is at index 1. To assign this string to a variable, we pass the index into square brackets:

```
names = ["Silas", "Bill", "Dara"]
name_bill = names[1]
```

Data extraction using reverse index position

This is the second format of data retrieval from a list. List order can be used in reverse, meaning that the indexing starts from the last member of the list and counts backward. Negative numbers are used, starting at -1, which is the index of the last member of the list, -2 is the second-to-last member of the list, and so on.

This means that, in the following example, the `"Bill"` and `"Silas"` strings are at indexes -2 and -3 respectively when using reverse index position, and so -2 (or -3) must be passed to the list in square brackets:

```
names = ["Silas", "Bill", "Dara"]
name_bill = names[-2]
name_silas = names[-3]
```

Read more about indexing here: `https://realpython.com/lessons/indexing-and-slicing/`

Functions

Functions are subroutines defined by code. When **called**, or run, functions will do something (or nothing, if written that way). Functions often accept parameters, and these can be required or optional.

Functions make it easy to perform the same action over and over without writing the same code over and over. This makes code cleaner, shorter, and smarter. They are a good idea and should be used often.

Components of a function

Here are the main parts that make up a function in Python:

- `def` **keyword**: Functions are defined using the `def` keyword, which is short for "define function." The keyword is written, followed by the name of the function and round brackets, `()`, into which expected parameters can be defined.
- **Parameters:** Parameters or arguments are values expected by functions and supplied by the code at runtime. Some parameters are optional.
- **Return statement:** Functions allow for data to be returned from the subroutine to the main loop using `return` statements. These allow the user to calculate a value or perform some action in the function and then return back a value to the main loop.

- **Docstrings**: Functions allow for a string after the definition line that is used to declare the purpose of the function:

  ```
  def accept_param(value=12):
      'this function accepts a parameter' # docstring
      return value
  ```

Note that optional parameters with default values must always be defined *after* the required parameters within functions.

Namespaces

In Python, there is a concept called **namespaces**. These are refined into two types of namespaces: global and local.

All variables defined in the main part of a script (outside of any functions) are considered to be in the global namespace. Within the function, variables have a different namespace, meaning that variables inside a function are in a **local namespace** and are not the same as variables in the main script, which are in the **global namespace**. If a variable name inside a function is the same as one outside of the function, changing values inside the function (in the local namespace) will not affect the variable outside the function (in the global namespace).

Read more about namespaces here: `https://realpython.com/python-namespaces-scope/`

Function examples

In this first example, a function is defined and written to return `"hello world"` every time it is called. There are no parameters, but the `return` keyword is used:

```
def new_function():
    return "hello world"

>>> new_function()
'hello world'
```

In this next example, an expected parameter is defined in the brackets. When called, this value is supplied, and the function then returns the value from the local namespace back to the global namespace in the main loop:

```
def accept_param(value):
    return value

>>> accept_param('parameter')
'parameter'
```

In this final example, an expected parameter has a default value assigned, meaning it only has to be supplied if the function uses a non-default parameter:

```
def accept_param(value=12):
    return value

>>> accept_param()
12
>>> accept_param(13)
13
```

Read more about functions here: `https://realpython.com/defining-your-own-python-function/`

Classes

Classes are special blocks of code that organize multiple variables and functions into an object with its own methods and functions. Classes make it easy to create code tools that can reference the same internal data lists and functions. The internal functions and variables are able to communicate across the class so that variables defined in one part of the class are available in another.

Classes use the idea of **self** to allow for the different parts of the class to communicate. By introducing `self` as a parameter into each function inside a class, the data can be called.

Here is an example of a class:

```
class ExampleClass():

    def __init__(self, name):
```

```
        'accepts a string'
        self.name = name

    def get_name(self):
        'return the name'
        return self.name
```

Classes are called or **instantiated** to create a class object. This means the class definition is kind of like a factory for that class, and when you want one of those class objects, you call the class type and pass the correct parameters if required:

```
>>> example_object = ExampleClass('fred')
>>> example_object.get_name()
'fred'
```

Read more about classes here: https://www.geeksforgeeks.org/python-classes-and-objects/

Installing and importing modules

Python was built to be shipped with a basic set of functionalities known as the **standard library**. Knowing that all programming needs would never be covered by the standard library, Python was built to be open and extensible. This allows programmers to create their own modules to solve their specific programming needs. These modules are often shared under an open-source license on the **Python Package Index**, also known as **PyPI**. To add to the capabilities of the standard Python library of modules, third-party modules are downloaded from PyPI using either the built-in `pip` program or another method. For us, modules such as `arcpy` and the ArcGIS API for Python are perfect examples: they extend the capabilities of Python to be able to control the tools that are available within ArcGIS Pro.

ArcGIS Pro comes with a **Python Package Manager** which will allow you to install additional packages to any virtual environments you have set up. You will learn in *Chapter 3* how to use this, creating your own virtual environments in ArcGIS Pro and installing additional packages that you may need. The following sections offer more detail about installing packages and creating virtual environments through the command line in the terminal. Don't worry if you aren't comfortable with the command line, as the Python Package Manager in ArcGIS Pro can manage much of the same and you will work through that in more detail in *Chapter 3*.

If you don't plan on working in the command line, you can skip the next section. But as you get more comfortable as a Python programmer, come back to this, as you will find it very useful in helping you learn how to work from the command line and install more packages. The Python Package Manager does not have access to all the packages available in PyPI. If you need a package that is not listed in the Python Package Manager, you will need the information below to install it.

Using pip

To make Python module installation easier, Python is now installed with a program called **pip**. This name is a recursive acronym that stands for **Pip Installs Programs**. It simplifies installation by allowing for one-line command line calls, which both locate the requested module on an online repository and run the installation commands.

Here is an example, using the open-source PySHP module:

```
pip install pyshp
```

You can also install multiple modules at a time. Here are two separate modules that will be installed by pip:

```
pip install pyshp shapely
```

Pip connects to the Python Package Index. As we mentioned, stored on this repository are hundreds of thousands of free modules written by other developers. It is worth checking the license of the module to confirm that it will allow for your use of its code.

Pip lives in the **Scripts** folder, where lots of executable files are stored:

This PC > Acer (C:) > PythonArcGIS > ArcGIS10.5 > Scripts

Name	Date modified	Type	Size
fio	4/5/2018 1:33 PM	Application	88 KB
geojsonio	4/5/2018 1:33 PM	Application	88 KB
iptest	4/5/2018 1:24 PM	Application	88 KB
iptest2	4/5/2018 1:24 PM	Application	88 KB
ipython	4/5/2018 1:24 PM	Application	88 KB
ipython2	4/5/2018 1:24 PM	Application	88 KB
jsonschema	4/5/2018 1:24 PM	Application	88 KB
jupyter	4/5/2018 1:23 PM	Application	88 KB
jupyter-bundlerextension	4/5/2018 1:24 PM	Application	88 KB
jupyter-console	4/5/2018 1:24 PM	Application	88 KB
jupyter-kernel	4/5/2018 1:24 PM	Application	88 KB
jupyter-kernelspec	4/5/2018 1:24 PM	Application	88 KB
jupyter-migrate	4/5/2018 1:23 PM	Application	88 KB
jupyter-nbconvert	4/5/2018 1:24 PM	Application	88 KB
jupyter-nbextension	4/5/2018 1:24 PM	Application	88 KB
jupyter-notebook	4/5/2018 1:24 PM	Application	88 KB
jupyter-qtconsole	4/5/2018 1:24 PM	Application	84 KB
jupyter-run	4/5/2018 1:24 PM	Application	88 KB
jupyter-serverextension	4/5/2018 1:24 PM	Application	88 KB
jupyter-troubleshoot	4/5/2018 1:23 PM	Application	88 KB
jupyter-trust	4/5/2018 1:24 PM	Application	88 KB
ndg_httpclient	4/5/2018 1:33 PM	Application	88 KB
pip	4/5/2018 1:25 PM	Application	88 KB
pip2.7	4/5/2018 1:25 PM	Application	88 KB
pip2	4/5/2018 1:25 PM	Application	88 KB
pygmentize	4/5/2018 1:24 PM	Application	88 KB

Figure 1.8: Locating pip in the Scripts folder

Installing modules that are not in PyPI

Sometimes modules are not available in PyPI, or they are older modules that don't understand the `pip install` method. This means that available modules have different ways of being installed that you should be aware of (although most now use pip).

The setup.py file

Often in Python 2.x, and sometimes in Python 3.x, a module includes a `setup.py` file. This file is not run by pip; instead, it is run by Python itself.

These `setup.py` files are located in a module, often in a downloadable zipped folder. These zip files should be copied to the `/sites/packages` folder. They should be unzipped, and then the Python executable should be used to run the `setup.py` file using the `install` command:

```
python setup.py install
```

Wheel files

Sometimes modules are packaged as **wheel** files. Wheel files use the extension `.whl`. These are essentially zip files that can be used by pip for easy installation of a module.

Use pip to run the wheel file and install the module, by downloading the file and running the `pip install` command in the same folder as the wheel file (or you can pass the whole file path of the wheel file to `pip install`):

```
pip install module.whl
```

Read more about wheel files here: `https://realpython.com/python-wheels/`

Installing in virtual environments

Virtual environments are a bit of an odd concept at first, but they are extremely useful when programming in Python. Because you will probably have two different Python versions installed on your computer if you have ArcGIS Desktop and ArcGIS Pro, it is convenient to have each of these versions located in a separate virtual environment.

The core idea is to use one of the Python virtual environment modules to create a copy of your preferred Python version, which is then isolated from the rest of the Python versions on your machine. This avoids path issues when calling modules, allowing you to have more than one version of these important modules on the same computer. In *Chapter 3*, you will see how to use the Python Package Manager provided in ArcGIS Pro to create a virtual environment and install a package that you want to run only in that environment.

Here are a few of the Python virtual environment modules:

Name	**Description**	**Example virtual environment creation**
`venv`	Built into Python 3.3+.	`python3 -m venv`
`virtualenv`	Must be installed separately. It is very useful and my personal favorite.	`virtualenv namenv --python=python3.6`
`pyenv`	Used to isolate Python versions for testing purposes. Must be installed separately.	`pyenv install 3.7.7`
Conda/Anaconda	Used often in academic and scientific environments. Must be installed separately.	`conda create --name snakes python=3.9`

Read more about virtual environments here: `https://towardsdatascience.com/python-environment-101-1d68bda3094d`

Importing modules

To access the wide number of modules in the Python standard library, as well as third-party modules such as `arcpy`, we need to be able to import these modules in our script (or in the interpreter).

To do this, you will use `import` statements, as we have seen already. These declare the module or sub-modules (smaller components of the module) that you will use in the script.

As long as the modules are in the `/sites/packages` folder in your Python installation, or in the Windows PATH environment variable (as `arcpy` is after it's been installed), the import statements will work as expected:

```
import csv
from datetime import timedelta
from arcpy import da
```

You will see in *Chapter 2* what happens when you attempt to import `arcpy` from a Python install that does not have the module in the `site/packages` folder. That is why it is important to know which version of Python has the `arcpy` module and use that one when working with IDLE or in the command line. When working in ArcGIS Pro using the Python window or ArcGIS Notebooks, this is not an issue, as they will automatically be directed to the correct version of Python.

Three ways to import

There are three different and related ways to import modules. These import methods don't care if the module is from either the standard library or from third parties:

- **Import the whole module**: This is the simplest way to import a module, by importing its top-level object. Its sub-methods are accessed using dot notation (for example, `csv.reader`, a method used to read CSV files):

  ```
  import csv
  reader = csv.reader
  ```

- **Import a sub-module**: Instead of importing a top-level object, you can import only the module or method you need, using the `from X import Y` format:

  ```
  from datetime import timedelta
  from arcpy import da
  ```

- **Import all sub-modules**: Instead of importing one sub-object, you can import all the modules or methods, using the `from X import *` format:

  ```
  from datetime import *
  from arcpy import *
  ```

Read more about importing modules here: `https://realpython.com/python-import/`

Importing custom code

Modules don't have to just come from "third parties": they can come from you as well. With the use of the special `__init__.py` file, you can convert a normal folder into an importable module. This file, which can contain code but is most of the time just an empty file, indicates to Python that a folder is a module that can be imported into a script. The file itself is just a text file with a `.py` extension and the name `__init__.py` (that's two underscores on each side), which is placed inside a folder. As long as the folder with the `__init__.py` is either next to the script or in the Python Path (e.g. in the site-packages folder), the code inside the folder can be imported.

In the following example, we see some code in a script called `example_module.py`:

```
import csv
from datetime import timedelta

def test_function():
    return "success"

if __name__ == "__main__":
    print('script imported')
```

Create a folder called `mod_test`. Copy this script into the folder. Then, create an empty text file called `__init__.py`:

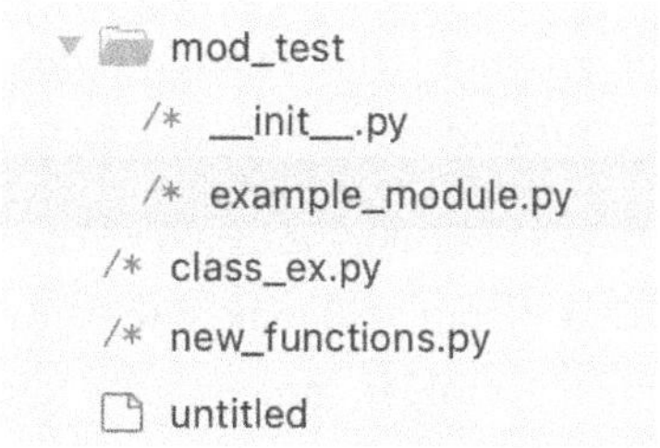

Figure 1.9: Creating an __init__.py file

Now let's import our module. Create a new script next to the mod_test folder. Call it module_import.py:

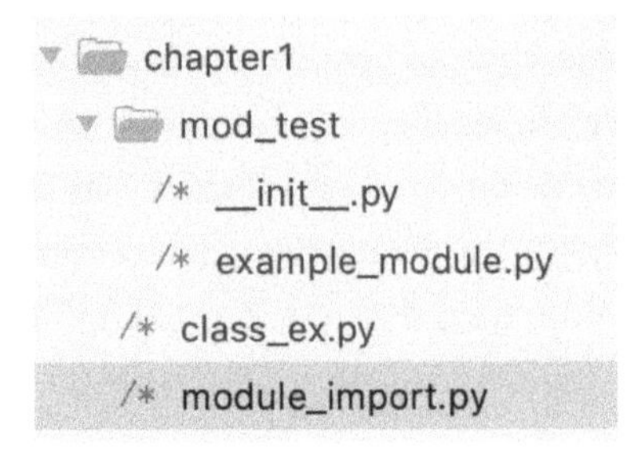

Figure 1.10: Creating a new script

Inside the script, import the function test_function from the example_module script in the mod_test folder using the format below:

```
from mod_test.example_module import test_function

print(test_function())
```

Scripts inside the module are accessed using dot notation (for instance, mod_test.example_module). The functions and classes inside the script called example_module.py are able to be imported by name.

Because the module is sitting next to the script that is importing the function, this import statement will work. However, if you move your script and don't copy the module to somewhere that is on the Python system path (aka sys.path), it won't be a successful import.

That is because the way import statements work is based on the Python system path. This is the sys.path list of folder locations that Python will look in for the module that you are requesting. By default, the first location is the local folder, meaning the folder containing your script. The next location is the site-packages folder.

The site-packages folder

Most modules are installed in a special folder. This folder is inside the folder that contains the Python executable. It is called the site-packages folder and it sits at *\Lib\sites-packages.

To make your module available for import without needing it to be next to your script, put your module folder in the site-packages folder. When you run from mod_test.example_module import test_function, it will locate the module called mod_test in the site-packages folder.

This PC > Acer (C:) > PythonArcGIS > ArcGIS10.5 > Lib > site-packages

Name	Date modified
_dummy_thread	7/18/2017 7:09 AM
_markupbase	7/18/2017 7:09 AM
_thread	4/5/2018 1:25 PM
arcgis	4/2/2018 2:41 PM
arcgis_rest_query-0.14.dist-info	4/2/2018 12:55 PM
arcrest	8/2/2017 7:33 PM
arcresthelper	8/2/2017 7:33 PM
ArcREST-master	8/2/2017 7:32 PM
argparse-1.4.0.dist-info	4/2/2018 12:55 PM

Figure 1.11: The site-packages folder

These tips will make it easier to add your custom code to the Python installation and to import reusable code in other scripts. In the last section, we will explore tips about writing good code.

Basic style tips for writing scripts

To make clean, readable code, it is encouraged to follow these basic tips about how the code should be written and organized. The main rule enforced by Python is the indentation required, which is intended to make the code easier to read and write. The major Python style suggestions and implementations are collectively contained in the Python Enhancement Proposal 8, also known as **PEP8**. We have included our own recommendations as well, based on lots of experience.

Read more about Python code style here: `https://realpython.com/python-pep8/`

Find the PEP8 style guide here: `https://www.python.org/dev/peps/pep-0008/`

Indentation

Python code has strict indentation rules that are enforced by all IDEs. These rules relate to functions and loops especially.

As a standard, four spaces are used after a function is declared, a loop is created, or a conditional is used. This is just a standard, as it could be only one space or however many spaces you want, but that indentation level becomes important when scripts get big. It helps to have four spaces for all indented lines so that they can be more easily read.

Do not mix tabs and spaces when indenting, as this will make it impossible to execute scripts in some IDEs.

Read more about indentation here: `https://www.python.org/dev/peps/pep-0008/#indentation`

Using print statements

The built-in function called `print()` is used to send messages from the script to the command window while the script is running. Pass any valid data to the `print()` statement and use it to track progress or to debug if there are issues:

```
>>> print("blueberry")
blueberry
>>> x = 0
>>> print(x)
0
```

Debugging using print statements is very common, and I encourage it as you learn to code. Well-placed print statements will help you understand how the code execution is progressing, and will help you to find the source of bugs by telling you which part of the script has executed and which part has not. It is not a requirement to use print statements, but they really are a programmer's friend.

Read more about print statements here: `https://realpython.com/python-print/`

Structuring a script

We suggest the following guidelines for good script structure:

- **Add a comment at the top with script details**: This is an optional but recommended way to start your scripts: write a comment at the top with your name, the date, and a quick explanation about what the script is supposed to do. This is especially nice when other people have to read your code.

  Add lots of other comments throughout the script as well, to make sure you know what is happening throughout the script.

- **Follow with import statements**: It is encouraged, but not required, to put the import statements at or near the top of the script. Imports must happen before the module objects are called in the script, but the import statements can be placed anywhere. It is best to put them at the top so that people reading the script can understand what is being imported.
- **Define global variables**: After the import statements, define the necessary variables that will be used in this script. Sometimes it is necessary to define variables later in the script, but it is best to put major variables near the top.
- **Define functions**: By placing function definitions below the global variables, it is easy to read and understand what the functions do when reading them. It is sometimes hard to find a function that is called in another part of the script if the function is not in a known location in the script.
- **Write the executable parts of the script**: After importing modules and defining functions, the next part of the script is where the action takes place. The `for` loops are run, the functions are called, and the script is then done.

  Make sure to add lots of comments to help yourself understand what is happening throughout the script, and `print` statements as well to help while the script is running.

- `if __name__ == '__main__'`: Often at the end of scripts you will see this line. What it means is that the indented code below this line will be run if the script is executed directly, but if the code in the script is imported by another script, the code blocks will not execute until called in the second script.

Read more about this here: `https://www.geeksforgeeks.org/what-does-the-if-__name__-__main__-do/`

Summary

In this chapter, we had a concise but comprehensive overview of computer programming and the Python programming language. We reviewed the basics of computer programming, including variables, iteration, and conditionals. We explored the data types of Python, including integers, strings, and floats, and the data containers of Python, such as lists, tuples, and dictionaries. We learned about importing and installing modules. We learned some basic code structure for scripts, and how to execute those scripts.

Don't worry if this was too theoretical for you – we will get very hands-on in the rest of the book. In the next chapter, we will discuss the basics of `arcpy`. We will learn how to ensure your Python environment is set up for ArcPy, create a shortcut to the Python IDLE associated with ArcGIS Pro, and begin to write some Python in the Python window in ArcGIS Pro by examining the environment settings and doing some simple geoprocessing.

2

Basics of ArcPy

Now that you have an understanding of Python syntax, you can start working with the **ArcPy** package. ArcPy is the Python package provided by ArcGIS to perform and automate geoprocessing and map production. In addition to the **geoprocessing** tools available in ArcGIS, ArcPy gives you access to additional modules, functions, and classes. When these are combined, you can create workflows and standalone tools that simplify and automate complex analysis and map production.

This chapter will cover:

- Ensuring your Python environment is set up for ArcPy
- Accessing **environment settings** in ArcPy
- ArcPy tools and how to use them in ArcGIS Pro
- **Functions** in ArcPy
- ArcPy **modules**

To complete the exercises in this chapter, please download and unzip the `Chapter2.zip` folder from the GitHub repository for this book: `https://github.com/PacktPublishing/Python-for-ArcGIS-Pro/tree/main/Chapter2`.

Checking your ArcPy install

Python is the official scripting language of ArcGIS and ArcPy is a site package designed to automate analysis and map production workflows. The ArcPy package allows you access to the geoprocessing functionality of ArcGIS Pro.

Python packages contain multiple modules, functions, and classes set up with a hierarchical structure. The hierarchical structure allows properties and tools to nest within modules within the ArcPy package.

ArcPy is installed with **ArcGIS Pro** and **ArcGIS Desktop**. ArcPy has been used to write Python scripts in ArcGIS since ArcGIS 10.0. ArcGIS Desktop uses Python 2.7, which is currently up to release 2.7.18. ArcGIS Pro uses a new version of Python, Python 3. The most recent release of Python 3 as of publishing is Python 3.9.10. ArcGIS Desktop and ArcGIS Pro each will install a version of Python on your computer when you install them. Which version of ArcGIS Desktop or ArcGIS Pro you have installed will determine the version of Python you have. The most current install of ArcGIS Pro 2.8.0 contains Python 3.7.10.

To check what version of Python you have installed, follow these steps:

1. Open ArcGIS Pro.
2. Do not open a project. Click on **Settings** in the bottom-left corner:

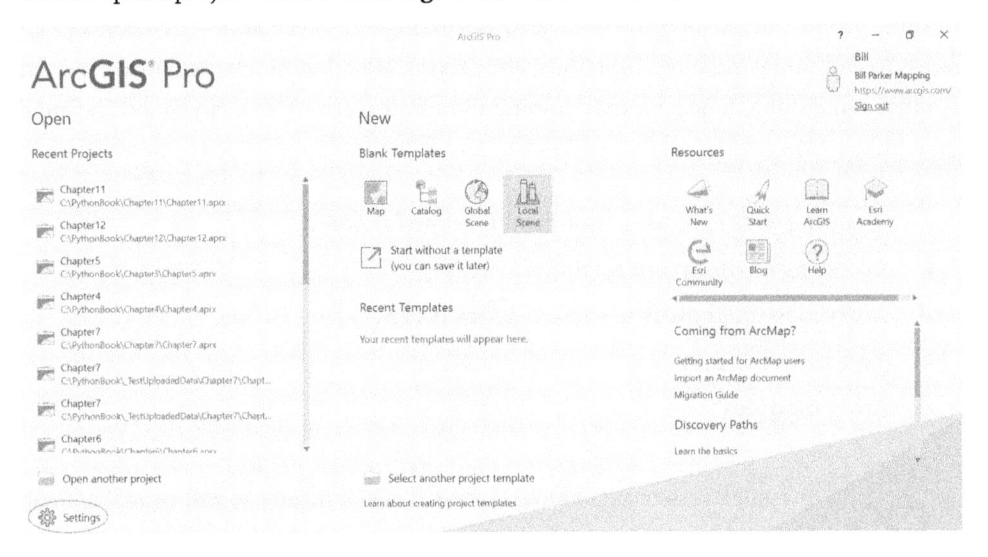

Figure 2.1: Settings button

3. Click on **Python** in the ribbon on the left:

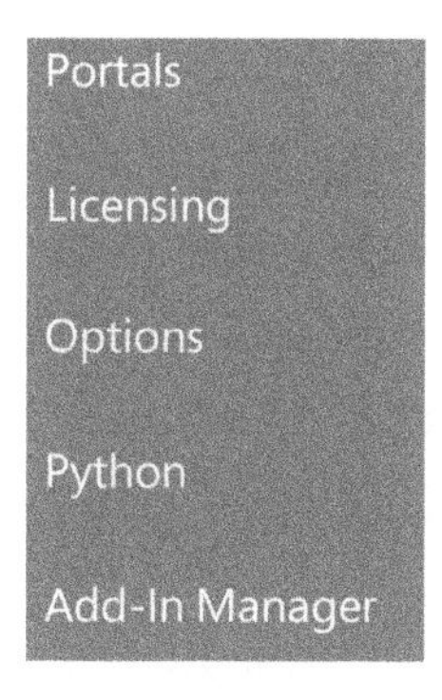

Figure 2.2: Clicking on Python

4. Scroll down **Installed Packages** to find **python**:

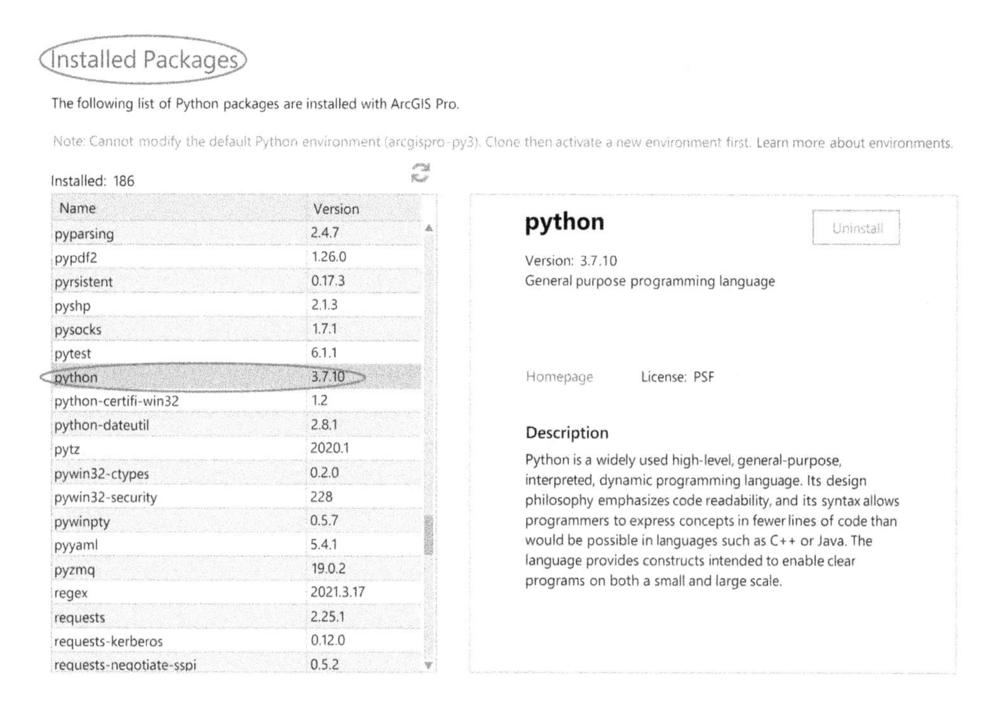

Figure 2.3: Installed Python package

To use the ArcPy package, it must be imported, as that will give you access to all the geoprocessing tools and modules included in it. Most scripts start with import statements to allow access to all the modules that are part of the package. To import ArcPy, use the following line of code:

```
import arcpy
```

Some of the more common modules within `arcpy` are:

- `arcpy.sa` (**Spatial Analyst**): This gives you access to the Spatial Analyst geoprocessing tools and specialized functions and classes for working with raster data.
- `arcpy.da` (**Data Access**): This helps you work with data by allowing control of edit sessions, cursor functions, and working with tables.
- `arcpy.mp` (**Mapping**): This allows you to automate cartographic tasks for map production.
- `arcpy.geocoding` (**Geocoding**): This allows you to set locator properties through the locator class and automate geocoding processes.
- `arcpy.na` (**Network Analyst**): This gives you access to the Network Analyst geoprocessing tools and specialized functions and classes for working with networks.

You will look at how to use the Spatial Analyst module later in this chapter. In *Chapter 4*, you will work with the Data Access module, and in *Chapter 7* and *Chapter 12*, you will work with the Mapping module.

Using the correct Python IDLE Shell

If you have both ArcGIS Desktop and ArcGIS Pro installed, you have multiple versions of Python installed. Because of this, you need to make sure when using the IDLE Shell that you are using the Python version associated with your install of ArcGIS Pro.

Unlike installs of ArcGIS Desktop, ArcGIS Pro does not install a desktop shortcut to its IDLE Shell. Most of the time, you will be writing scripts either directly into ArcGIS Pro's Python window, or ArcGIS Pro Notebooks in ArcGIS Pro. Using the IDLE Shell that comes with the ArcGIS Pro install is a convenient way to turn standalone scripts and Notebooks into script tools, as you can test portions of the script.

The easiest way to ensure you are accessing the Python IDLE that comes with the install of ArcGIS Pro is to create a shortcut, as one is not created on install.

When you want to use an IDLE to work with ArcGIS Pro, you need to use this shortcut, as it is associated with your install of ArcGIS Pro.

Follow these steps to do so in Windows:

1. Find the path to run IDLE. For a typical ArcGIS Pro installation, it is here: `C:\Program Files\ArcGIS\Pro\bin\Python\envs\arcgispro-py3\Lib\idlelib\idle.bat`. Dou-

ble-clicking on that will open IDLE.

2. To create a shortcut, right-click on your desktop, click **New** > **Shortcut**, and paste the full path of the `idle.bat` file:

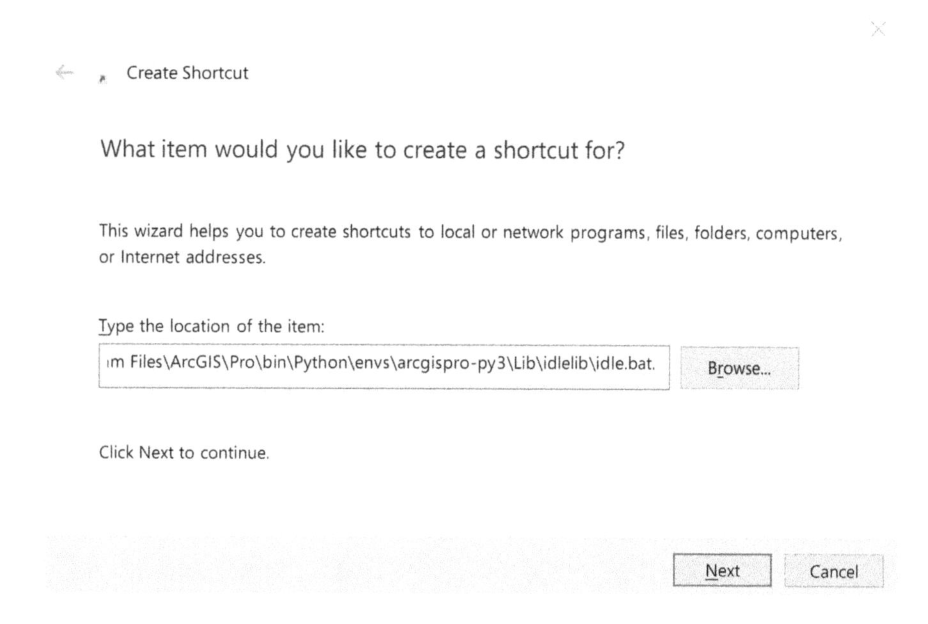

Figure 2.4: Creating a shortcut for IDLE

3. Click **Next**, give your shortcut a name, then click **Finish**:

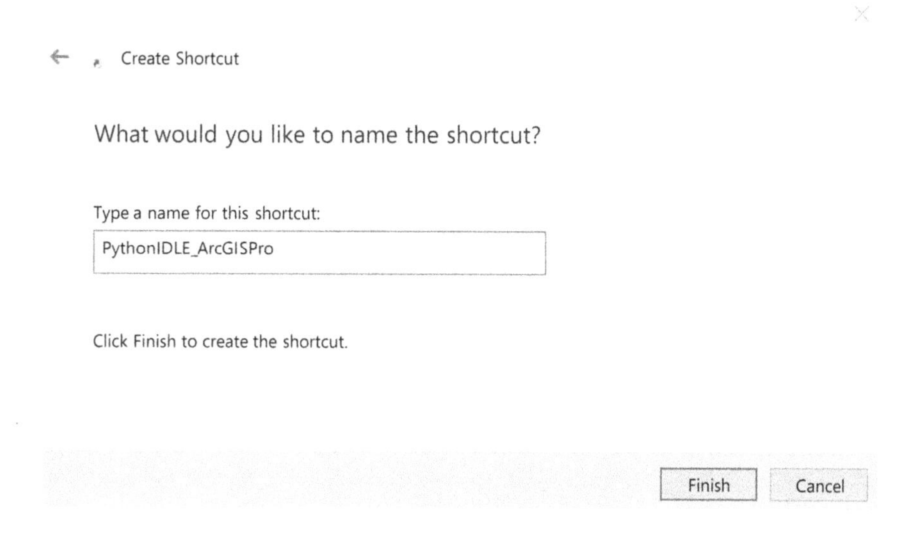

Figure 2.5: Naming the shortcut

I suggest using a name that will allow you to remember this is the Python environment that was installed with ArcGIS Pro.

The icon will be the default shortcut icon:

Figure 2.6: The IDLE shortcut

To change the icon to the standard Python IDLE icon, do the following:

1. Right-click on it and click **Properties**.
2. On the **Shortcut** tab, click on **Change Icon**. You may get an alert that there are no icons and that you need to choose an icon from a different file. If so, click **OK** so you can navigate to the icon location.
3. Navigate to the Python IDLE location, which should be here: `C:\Program Files\ArcGIS\Pro\bin\Python\envs\arcgispro-py3\Lib\idlelib\Icons`. Select the icon there and click **OK**.

A shortcut to access the IDLE install for ArcGIS Pro is now installed on the desktop:

Figure 2.7: The IDLE shortcut with IDLE icon

This install is a place to test snippets of code when working on complex script tools, something that will be explored in later chapters.

A good way to ensure the correct IDLE is being used is to import `arcpy`. If three carets (>>>) are displayed afterward, the install was successful.

```
Python 3.7.10 Shell
File Edit Shell Debug Options Window Help
Python 3.7.10 [MSC v.1927 64 bit (AMD64)] on win32
Type "help", "copyright", "credits" or "license()" for more information.
>>> import arcpy
>>> |
Ln: 4 Col: 4
```

Figure 2.8: The IDLE Shell with the associated ArcPy module

If you are using an IDLE Shell that is not associated with ArcPy, you will get an error when you attempt to import `arcpy`:

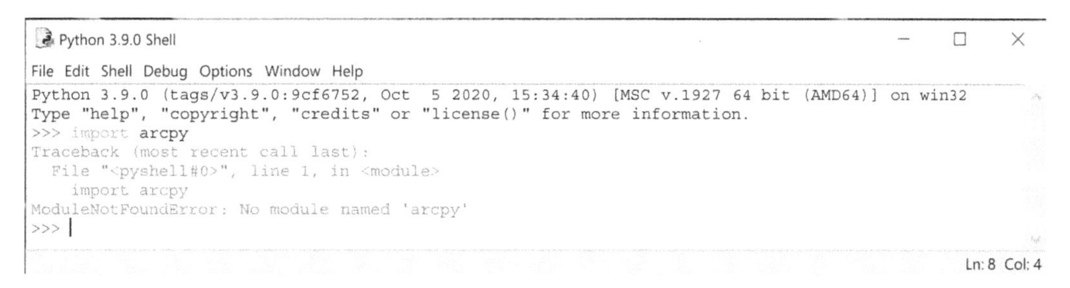

Figure 2.9: The IDLE Shell without the associated ArcPy module

Using the Python IDLE Shell

The Python IDLE Shell is a good place to try out code as it is interactive and will display immediate results:

```
Python 3.7.9 Shell
File Edit Shell Debug Options Window Help
Python 3.7.9 [MSC v.1922 64 bit (AMD64)] on win32
Type "help", "copyright", "credits" or "license()" for more information.
>>> x = 3
>>> y = 7
>>> x+y
10
>>> |
Ln: 7 Col: 4
```

Figure 2.10: The Python IDLE Shell

The IDLE Shell also shows how elements of the code will be interpreted using different colors. Strings are shown in green, functions are in purple, loops and condition statements are in orange, and results are in blue.

While it is useful to get immediate results from the IDLE Shell, it is not meant to be used for saving code. It is possible to copy code out if needed, but it is a better practice to write it into a script file for saving.

To start a script file, click in the menu bar of the IDLE Shell on **File** > **New File**. This opens a new window that is an empty Python script file called `untitled`. Unlike the IDLE Shell, it doesn't have a command prompt and the menu bar is different. Below is a comparison:

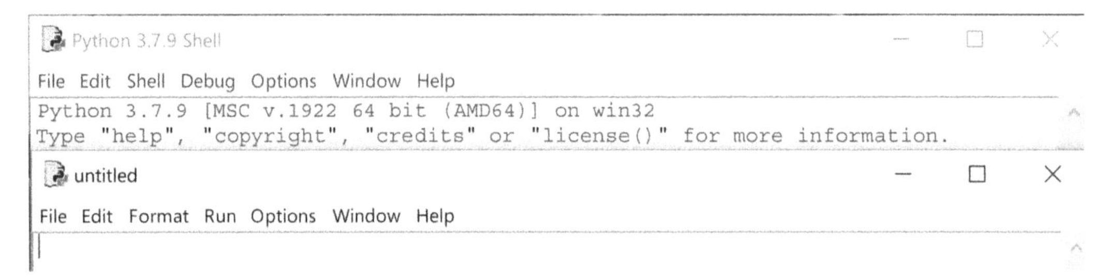

Figure 2.11: The IDLE Shell (top) and a new script file (bottom)

Let's first write some code into IDLE. IDLE knows when you are writing multiline connected code and won't run until you press *Enter* twice.

1. Type the following into IDLE:

   ```
   string = "Hello"
   ```

   Press *Enter*. Type:

   ```
   i = 1
   ```

   Press *Enter*. Type:

   ```
   while i < =5:
   ```

   Press *Enter*. Type:

   ```
   if i == 1:
   ```

   Press *Enter*. Type:

   ```
   print(string)
   ```

   Press *Enter*, and on the new line press *Backspace* to get the correct indentation. Type:

   ```
   else:
   ```

   Press *Enter*. Type:

   ```
   print(i)
   ```

   Press *Enter*, and on the new line press *Backspace* to get the correct indentation. Finally, type:

   ```
   i+=1
   ```

   Press *Enter* twice to run. The output will be the word `Hello` on one line, followed by `2`, `3`, `4`, `5`, each on a new line.

2. Copy the code you wrote in IDLE into a new script file:

```
*untitled*
File  Edit  Format  Run  Options  Window  Help
>>> string = "Hello"
>>> i = 1
>>> while i <= 5:
          if i == 1:
                    print(string)
          else:
                    print(i)
          i+=1

Ln: 2  Col: 4
```

Figure 2.12: IDLE code copied to a new file

3. Remove the carets (>>>).
4. Fix the indentation, remembering Python's 4-space indent convention we saw in the previous chapter:

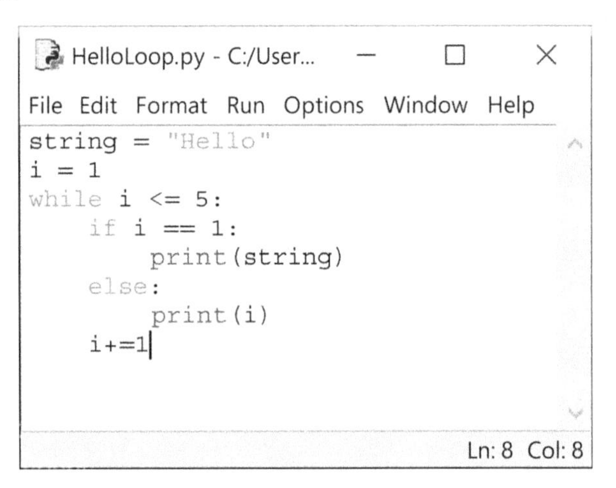

Figure 2.13: Python code with correct indentation

5. Save the file by clicking **File** > **Save** and naming it `HelloLoop`.

This new file, `HelloLoop.py`, has a `.py` extension, signifying that it is a Python file. It can be run by clicking **Run** > **Run Module**, which sends the results to the Python IDLE Shell. The results will look the same as when running the code from IDLE.

Now you should have an understanding of:

- How to ensure your Python environment is set up to work with ArcPy
- How to use the IDLE Shell
- How to start a new script file

You are ready to look at the Python window in ArcGIS Pro and what you can do there.

The Python window in ArcGIS Pro

ArcPy can be accessed not just through IDLE but also by using the Python window in ArcGIS Pro. This is accessed in the **Analysis** tab of the ribbon:

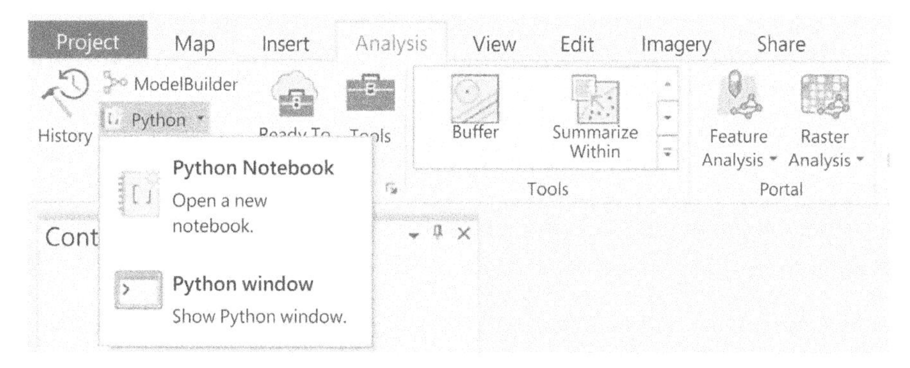

Figure 2.14: The Python window icon

The Python window allows you to write and run code directly in ArcGIS Pro and see the results of any geoprocessing tool as you run it. This can be an advantage when testing out new code to see what it is doing and how it is working. Code written in the Python window can then be copied or saved into large script tools. You will learn more about script tools in *Chapter 6, ArcToolBox Script Tools,* and *Chapter 12, Case Study: Advanced Map Production.* For now, let's look at the Python window in ArcGIS Pro and see how powerful it can be.

For this exercise, you will need to make sure you have downloaded the data from the GitHub site for *Chapter 2* and unzipped it.

1. Open ArcGIS Pro.
2. Click **Open another Project**, navigate to where you downloaded the *Chapter 2* data, and select `Chapter2.aprx` to load the `Chapter2` project.

3. Once the project is open, click the Python window icon to open the Python window. Usually, the first time, it will be docked at the bottom of your screen:

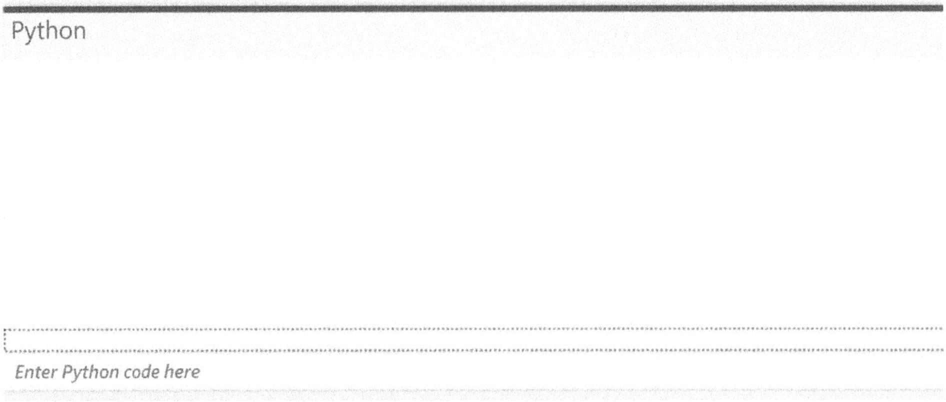

Python

Enter Python code here

Figure 2.15: Python window

Like all windows in ArcGIS Pro, the Python window can be docked anywhere or can be floating. You can move it to where it is best for you to work, using the same process of dragging and hiding as you would any ArcGIS Pro window.

The top part of the window is called the **transcript** and is where code you have previously written is located. The bottom, where it says **Enter Python code here**, is called the **prompt** and is where you type code. When you first open the Python window, the transcript is blank, as you haven't written any code yet.

4. Try some of the code you wrote in IDLE to see how it works in the Python window. Just like in IDLE, when you enter a line of code, you need to press *Enter*.

   a. Type in `x = 10` and press *Enter*.

   b. Type in `y = 3` and press *Enter*.

c. Type `x + y` and press *Enter*.

```
Python
x = 10
y = 3
x + y
13
```

Figure 2.16: Python window with results in the transcript window

You can see that this is working just like IDLE.

All of the standard Python functions and tools will work in the Python window the same as in the IDLE Shell.

The transcript can be cleared at any time by right-clicking in the transcript box and selecting **Clear Transcript**. This does not remove your code or your variables from memory.

5. Right-click in the transcript and select **Clear Transcript**:

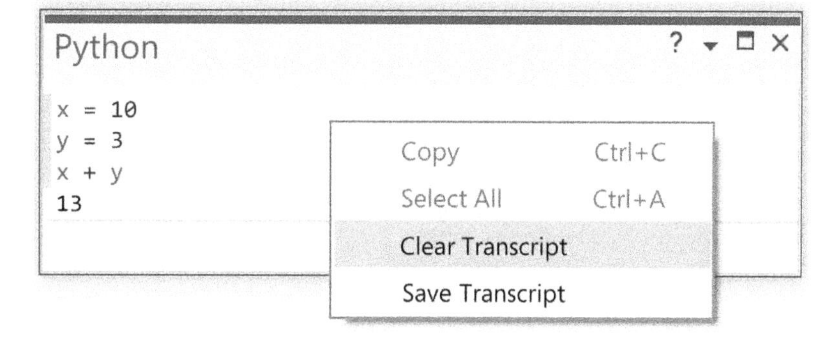

Figure 2.17: Clearing the transcript

6. Type `x + y` and press *Enter*:

Figure 2.18: Data is still held in memory

As you can see, the variables for `x` and `y` were saved in memory and are still usable even after clearing the transcript. These variables are even available if you save and close the project and open it again.

The variables are saved into the memory of the project so they can be used again later on the same project. This can be useful, but you will look at better ways to save code for reuse in the same and other projects in *Chapter 3, ArcGIS API for Python*, and *Chapter 6, ArcToolBox Script Tools.*

7. Just like in the IDLE Shell, the Python window understands when you are writing multiple lines of connected code. You can see this by writing in the code for the `HelloLoop.py` script you wrote earlier in this chapter.

   Do the following:

   a. Type in `String = "Hello"` and press *Enter*.
   b. Type in `i = 1` and press *Enter*.
   c. Type in `while i < 5:` and press *Enter*.

   Notice that the prompt window gets bigger and your cursor is indented. The Python window understands that the `while` statement is starting a block of code and is part of a multiline construct. As you continue to type in your code, the prompt window will get bigger as additional lines are needed. The `if` statement we will write next is also part of a multiline construct, so it too will get the additional lines and indent as the `while` statement does:

   d. Type in `if i == 1:` and press *Enter*.
   e. Type in `print(string)` and press *Enter*.

f. Type in `else:` and press *Enter*.

g. Type in `print(i)` and press *Enter*.

h. Type in `i+=1` and press *Enter*.

Your Python window should look like this:

```
Python                                  ? ▾ □ X

i = 1
while i <= 5:
    if i == 1:
        print(string)
    else:
        print(i)
    i+=1
```

Figure 2.19: HelloLoop in the Python window

When you are writing your multiple lines of code and hit *Enter*, you will just move down another line.

Press *Enter* again and the program will execute:

```
Python                                  ? ▾ □ X
    else:
        print(i)
    i+=1
Hello
2
3
4
5
```

Figure 2.20: HelloLoop output

What happens if you make a mistake in the Python window? It really depends on the mistake. If you forget the `i+=1` your code will run forever, you can click the **X** at the bottom of the Python window or type *Ctrl* + *C* in the prompt window to stop your running code. If you need to edit code already written, click where you need to edit and make the edits. Remember to follow the rules for indentation within your code to ensure it will still run.

ArcPy environment settings

ArcPy environment settings allow you access to general geoprocessing settings, as well as the geoprocessing settings of a specific tool. For tools, they act as parameters you can set to change the results of a tool. There are many that can be used, but there are some you will use more commonly than others.

In this section, we are going to look at two of the most common ones and see how to set them: `arcpy.env.workspace` and `arcpy.env.scratchWorkspace`. Setting your workspace and scratch workspace is a good idea, as it allows you to have a default location to send the data you are creating. They are also the workspaces that will be used when using the list functions you will explore below.

Using the workspace properties of the environment class, you can check and change your workspace or scratch workspace. Your **workspace** is the default location for any data you write and want to maintain. The **scratch workspace** is for data you do not want to maintain and is where intermediate steps in ModelBuilder will be written.

You can check your **workspace** by typing in:

```
arcpy.env.workspace
```

from the Python window and pressing *Enter*. The return value you see is your current workspace:

```
'C:\\PythonBook\\Chapter2\\Chapter2.gdb'
```

You can set your workspace by typing in:

```
arcpy.env.workspace =  r"C:\PythonBook\Project_2\Project_2.gdb"
```

and pressing *Enter*.

You can check your **scratch workspace** in a similar way; type in `arcpy.env.scratchWorkspace` and press *Enter*. You can set your scratch workspace by typing in `arcpy.env.scratchWorkspace = r"C:\PythonBook\Project_2\Project_2.gdb"` and pressing *Enter*.

In these examples, you have set your workspace and scratch workspace to geodatabases. However, you could set them to a folder or a feature dataset or any workspace you wanted.

The `workspace` and `scratchWorkspace` can only be called in IDLE if you have already set them. There is no default `workspace` or `scratchWorkspace` in IDLE and, if called without being set, they will return `None`.

What is the "r" in front of the path for the geodatabase?

Note the way you have typed in the locations, using an `r` followed by the location enclosed in double quotes. The `r` stands for **raw string** and means that Python will read everything within the quotes exactly as it is written. This is important because the \ character in Python is an escape character and can be used to insert characters otherwise not allowed in a string. Here, you don't want the escape character, so there are three options:

- Use an `r` in front of the quotes to create a raw string
- Change all the single backslashes (\) to double backslashes (\\)
- Change the single backslash (\) to a forward slash (/)

There are many other environment settings that could be useful to you depending on the process you are running. Most of the settings that you find for a tool in the properties of a tool can be set in the environment settings. Things like analysis extent can be set with `arcpy.env.extent`, or a snap raster when doing raster analysis with `arcpy.env.snapRaster`.

It is important to remember that once you set an environment setting, it stays set until you change it. In more advanced script tools, you will change it or have it set and reset throughout the code.

ArcPy tools: Geoprocessing using ArcPy

Now that you know some of the basics of how to use the Python window, it is time to look at how to use geoprocessing tools. In this hands-on section, you will learn how to use the following tools in the Python window:

- Select
- Buffer
- Make Feature Layer
- Select By Feature Layer
- Select Layer By Location
- Copy Features

Your task is to find all of the bus stops in Oakland that are within 1,000 feet of a park. You want the end result to be a feature class of all the bus stops that fall within 1,000 feet of any park.

To do some geoprocessing, you will need some data. If you do not have the `Chapter2.aprx` file open in ArcGIS Pro, do so now. You will be working first with the `CPAD_2020b_Units.shp` file that is already in the map. If it is not in the map, add the shapefile from where you downloaded the `Chapter2` folder. This is California Protected Areas Database data that shows parks and other protected areas throughout the state of California. For more information about the dataset, go to `https://www.calands.org/`.

A common GIS task consists of finding all the features within a distance of something else and creating a new feature from that selection. It can be the location of protected species within a proposed project, schools near a proposed new playground, or bus stops near community facilities. You will use the Python window to select the parks in Oakland, buffer them by 1,000 feet, select the bus stops within that 1,000-foot buffer, and create a new feature class. Let's begin:

1. Right-click the `CPAD_2020b_Units` file in the table of contents, select **Attribute table**, and examine the data. The `CPAD_2020b_Units` shapefile contains the name of the park, the agency responsible for the park, the type of agency, the city the park is in, a label for the park, and much more information about each park.
2. You are going to use the `AGNCY_NAME` field to run a Select tool to create a new feature class of just the protected areas in the City of Oakland. In the Python window, type in the following:

   ```
   arcpy.Se
   ```

   The Python window shows you some autocomplete options to help you find the tool you want. You are using the **Select** tool from the Analysis toolbox, so you want **Select() analysis**, the second option in the figure below:

   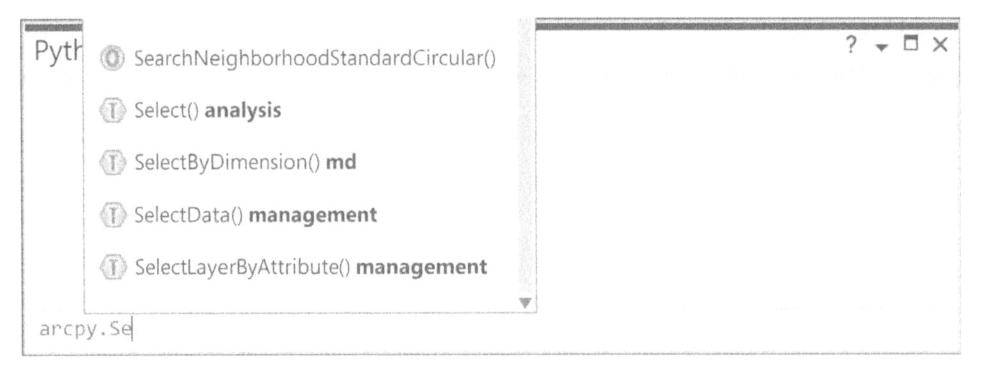

   Figure 2.21: The Python window autocompleting as you type

3. After selecting the tool, you can see what **parameters** the tool expects. Hover on the tool to get a help window popping up that shows you the tool parameters and what they mean. The Select tool takes the following mandatory parameters:

    1. `in_features`: The input feature class or shapefile.
    2. `out_features`**:** The output feature class or shapefile.

    It also takes the following optional parameter:

    3. `where_clause`: The `where` clause is in curly braces (`{}`) because it is optional. It is the SQL statement you will write to select features from `in_features`.

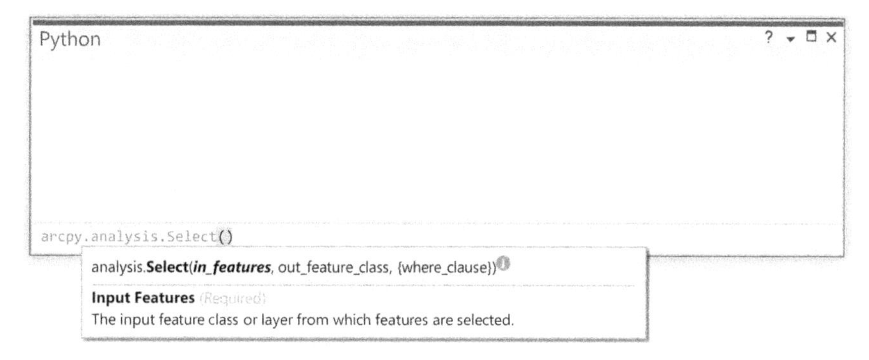

Figure 2.22: Hovering over the tool for more details

    Note that `in_features` is in bold because it is the parameter that the tool is currently expecting to be entered.

4. Complete the code as below to create the selection query:

```
arcpy.analysis.Select('CPAD_2020b_Units','CPAD_2020b_Units_
Oakland',"AGNCY_NAME = \'Oakland, City of\'")
```

    Press *Enter*.

How do we write the query in the where clause so it works? Using the escape character (\) properly

The backslash (\) marks are escape characters that are necessary when you need to use multiple single or double quotes. In this instance, since you are running a selection query on a shapefile, the attribute field being selected on needs double quotes and the string value needs single quotes. The entire `where` clause needs to be inside single or double quotes.

The easiest option is to wrap the entire query in a single quote and use escape clauses around the string being selected. If you wanted to wrap the entire `where` clause in double quotes, it would look like this: `"/"AGNCY_NAME" = 'Oakland, City of'"`. Both will work the same.

After running, it will look like the following:

```
<Result 'C:\\PythonBook\Chapter2\\Chapter2.gdb\\CPAD_2020b_Units_
Oakland'>
```

and you should have a new feature class that consists of just the protected areas in Oakland.

If you are working in an ArcGIS Pro project, then a new file will be created in that project's geodatabase, as that is the default workspace. If you have set a workspace through the environment settings, it will write to that workspace. If you have not set a workspace and are not working in an ArcGIS project, then it will be stored in a temp space and not written to disk.

How do you specify a different workspace if you don't want to use the default workspace?

To specify a different location, you need to write the full path when saving. To write out a shapefile to the `MyProject` folder, you would write the following:

```
arcpy.analysis.Select("CPAD_2020b_Units",r"C:\
Chapter2\Chapter2.gdb\CPAD_2020b_Units_Oakland.
shp",'"AGNCY_Name" = \'Oakland, City of\'')
```

5. Now you can take the selected parks and buffer them by 1,000 feet. The **Buffer** tool is in the **Analysis toolbox**, so to call it you type `arcpy.analysis.Buffer()`. You can see the parameters the Buffer tool takes by hovering over the parentheses. It takes the following mandatory parameters in this order:

    1. `in_features`: The input feature class or shapefile.
    2. `out_features`: The output feature class or shapefile to be written.
    3. `buffer_distance_or_field`: The buffer distance or attribute field from `in_features` to be used for buffering. It must include the units.

    It also takes the following optional parameters in this order:

    4. `line_side`: This is only used on line features and can be set to buffer the `LEFT` side, `RIGHT` side, or `BOTH`. `BOTH` is the default.
    5. `line_end_type`: This is only used on line features and will set the buffer end to `ROUND` or `FLAT`. `ROUND` is the default.
    6. `dissolve_option`: This is the type of dissolve for removing any buffer overlap. It can be set to `NONE`, `ALL`, or `LIST`. `NONE` is the default.
    7. `dissolve_field`: This is only used when `dissolve_option` is set to `LIST`. It is a list of the different fields to dissolve on. Even when one field is used, it must still be in square brackets as the tool requires a list.
    8. `method`: This is the distance method to use. It has `PLANAR` and `GEODESIC` as options. `PLANAR` is the default.

    You want to have a buffer of 1,000 feet for the parks. You will set the dissolve option to `LIST` and dissolve by the `UNIT_NAME` field. To get this, you will need to type in the `in_features`, `out_features`, `buffer_distance`, `dissolve_option`, and `dissolve_field` parameters. The `in_features`, `out_features`, and `buffer_distance` are the first three parameters, but `dissolve_option` and `dissolve_field` are the sixth and seventh parameters. To make sure they are in those positions, you will type a pair of single or double quotes in the fourth and fifth parameters. This signifies to the function that those optional parameters are blank, just as if they weren't entered, and allows you to enter parameters after them.

Type in:

```
arcpy.analysis.Buffer("CPAD_2020b_Units_Oakland","CPAD_2020b_Units_
Oakland_1000ft","1000 FEET", "","","LIST",["UNIT_NAME"])
```

Press *Enter*. The output will read:

```
<Result 'C:\\PythonBook\\Chapter2\\Chpater2.gdb\\CPAD_2020b_Units_
Oakland_100ft'>
```

The buffers should have been added to your map. You can explore them and see what they look like.

6. When you are ready, you are going to use the **Make Feature Layer** tool to make a feature layer of the bus stops feature class. This feature layer will be used for selecting the bus stops within the 1,000-foot buffer of the parks. Type in:

```
arcpy.management.MakeFeatureLayer()
```

You can see that the Make Feature Layer tool takes two mandatory parameters:

1. `in_features`: The input feature class or shapefile
2. `out_layer`: The name of the output feature layer

It also takes the following optional parameters:

3. `where_clause`: The SQL statement you will write to select features from `in_features`. If left blank, the feature layer contains all the data from `in_features`.
4. `workspace`: The input workspace used to validate the field names.
5. `field_info`: This can be used to hide some fields in the output.

The layer you are going to be making a feature layer from is already in your map and is in the default workspace you set earlier. Because of this, you do not have to use the entire path as the input and can just use the name of the layer. You will make a feature layer of all the bus stops by typing in the following:

```
arcpy.management.MakeFeatureLayer("UniqueStops_Summer21",
"AC_TransitStops_Summer21")
```

Press *Enter*. The output will read:

```
<Result 'AC_TransitStops_Summer21'>
```

7. The `AC_TransitStops_Summer21` feature layer will have been added to your map. You can explore it and see that it is just like the `UniqueStops_Summer21` feature class. However, because it is a feature layer, you can use the **Select Layer By Location** tool to select all of the bus stops within the buffer. Type in:

   ```
   arcpy.management.SelectLayerByLocation()
   ```

   You can see that the Select Layer By Location tool takes one mandatory parameter:

   1. `in_layer`: The input feature layer.

   It also takes the following optional parameters:

   2. `overlap_type`: The different overlaps that can be used for creating the selection layer. `INTERSECT` is the default, and the one you will use most of the time.
   3. `select_features`: The feature layer, feature class, or shapefile used to select `in_layer`.
   4. `search_distance`: The distance to search for additional features to be selected from `in_layer`. It is only valid when `overlap_type` is `WITHIN_A_DISTANCE`, `WITHIN_A_DISTANCE_GEODESIC`, `WITHIN_A_DISTANCE_3D`, `INTERSECT`, `INTERSECT_3D`, `HAVE_THEIR_CENTER_IN`, or `CONTAINS`.
   5. `selection_type`: This is how the selection will be applied to the `in_feature` layer. It can be `NEW_SELECTION` (default), `ADD_TO_SELECTION`, `REMOVE_FROM_SELECTION`, `SUBSET_SELECTION`, or `SWITCH_SELECTION`.
   6. `invert_spatial_relationship`: This will invert the selection so that the features that are not intersected or within a distance are those selected. It can be `NOT_INVERT` (default) or `INVERT`.

   To select the bus stops within the park buffers, type in the following:

   ```
   arcpy.management.SelectLayerByLocation("AC_TransitStops_
   Summer21","INTERSECT","CPAD_2020b_Units_Oakland_1000ft")
   ```

   Press *Enter*. The result will read:

   ```
   <Result 'AC_TransitStops_Summer21'>
   ```

   You should see the bus stops get selected. You can explore the data and see if this is what you were looking for. From here, you can export your data to a table, CSV, or feature class, or just use it for a map display.

8. Feature layers are temporary files, so you are going to export to a feature class. To do this, you will use the **Copy Features** tool. Type in:

```
arcpy.management.CopyFeatures()
```

The **Copy Features** tool takes two mandatory parameters:

1. `in_features`: The input feature class or shapefile, or feature layer.
2. `out_features`: The output feature class or shapefile.

It also takes one optional parameter:

3. `config_keyword`: A configuration keyword is only used if the output is a geodatabase. It does not need to be used to output to the geodatabase and is a rarely used parameter.

To copy the feature layer to a feature class, type in the following:

```
arcpy.management.CopyFeatures("AC_TransitStops_Summer21",
"AC_TransitStops_Within1000ft_OaklandPark")
```

Press *Enter*. The result will be:

```
<Result 'C:\\PythonBook\\Chapter2\\Chapter2.gdb\\AC_TransitStops_
Within1000ft_OaklandPark'>
```

Your resulting feature class is displayed in your map and written to your current workspace. The next steps with this data will be explored in *Chapter 4, Data Access Module and Using Cursors*. There, you will learn how to do this process all in memory and add the park names to the bus stops.

In the next section, you will look at some of the built-in ArcPy functions. These functions are things you do not have access to in Model Builder and are useful in automating the analysis process.

Built-in ArcPy functions

ArcPy has many built-in functions to help with geoprocessing. ArcPy functions look like geoprocessing tools in the way they are written. When you wrote the code to create a selection feature class in the previous exercise, you wrote `arcpy.analysis.Select(in_features, out_features, {where_clause})`. By encasing the input features, output features, and `where` clause in parentheses, you were calling the function and passing to it those parameters.

That is all a function is: a bundle of code that contains instructions for how to process the data you send it.

ArcPy has functions to assist with things such as the environment settings, describing data, licensing, ArcGIS Online, raster, listing data, along with functions for specific modules like the Spatial Analyst or Mapping modules. In this section, you will explore two of the more commonly used built-in functions:

- The `Describe` function
- `List` functions

These are common because they help you to set up and complete iterative processes, such as doing the same analysis on different feature classes in one location.

The Describe function

The `Describe` function will return different properties depending on what type of element it is being called on. It can be called on a variety of elements, including but not limited to shapefiles, geodatabases, feature classes, feature datasets, tables, LAS files, rasters, and map documents.

The `Describe` function returns an object with all the properties of the object, so you need to create a variable to hold those properties and then call them later. Let's try this on CPAD data in a new Python window:

1. Type in the following:

   ```
   desc = arcpy.Describe(r"C:\PythonBook\Chapter2\CPAD_2020b_Units.
   shp")
   ```

   Press *Enter*. It looks like nothing has happened, but now you can use that `desc` variable to get information about the shapefile.

2. You can see what data type `desc` is by typing in the following:

   ```
   desc.dataType
   ```

   Press *Enter*. The output will be `'Shapefile'`.

3. You can also see what type of geometry a feature class has by typing in the following:

   ```
   desc.shapeType
   ```

   Press *Enter*. The output will be `'Polygon'`.

You can see that if you don't know anything about a file, you can call a `Describe()` function on it and use the properties to find out information about the file. In the example above, you found out that the data is a shapefile. You can see how this information could be useful if you were searching through a folder and only wanted to run an analysis on the shapefiles.

List functions

Listing your data through data listing functions is a powerful tool. You create a list of all the data within a workspace, which you can then iterate over.

For these examples, you are going to use the data in the Chapter2 folder.

Data list functions take the current workspace you are in and will create a list of all the datasets or fields for that type of list function. There are the following list functions for listing data:

- `ListDatasets`
- `ListFeatureClasses`
- `ListFields`
- `ListFiles`
- `ListIndexes`
- `ListRasters`
- `ListTables`
- `ListVersions`
- `ListWorkspaces`

`ListDatasets`, `ListFeatureClasses`, `ListFiles`, `ListRasters`, `ListTables`, and `ListWorkspaces` need to have the workspace set before being run, as they will only run on the current workspace.

There are some additional list functions in ArcPy: `ListTools`, `ListToolboxes`, `ListSpatialReferences`, and `ListDataStoreItems`. These functions are designed to work with the specific objects they are referencing. Like the list functions above, they return a list that can be iterated through.

Often, you will want to run a similar process on all the data in a workspace. You can access all that data by creating a list of the data in the workspace. In this exercise, you will create a list of workspaces within a folder and then use that list to create a list of the data within the workspace. Let's get started:

1. Start by listing the workspaces that your `Chapter2.gdb` file is in. First, you need to set your workspace to the location of the geodatabase. Type the following line of code and press *Enter*:

   ```
   arcpy.env.workspace = r"C:\PythonBook\Chapter2"
   ```

2. Next, you will list the workspaces. The `ListWorkspaces` function has two optional parameters:

   1. `wild_card`: This can be used to limit the returned values to those that match the wildcard value. The wildcard will be encased in either single or double quotes, and an asterisk (*) can be used to select everything that starts with or ends with the wildcard. For example, if you wanted to select all the workspaces that start with `Project`, you would type in `"Project*"`; to select all the workspaces that end with `Project`, you would type in “`*Project"`; and to list all the workspaces that contain `Project` anywhere in the name, you would type in `"*Project*"`.
   2. `workspace`: This can be used to limit the type of workspace using the following:
      - `"Access"`: Limits to personal geodatabases.
      - `"Coverage"`: Limits to coverage workspaces.
      - `"FileGDB"`: Limits to file geodatabases.
      - `"Folder"`: Limits to shapefile workspaces.
      - `"SDE"`: Limits to enterprise databases.
      - `"All"`: All workspaces will be selected. This is the default.

   Assign the `ListWorkspaces` function to a variable called `wksp`. Type in the following and press *Enter*:

   ```
   wksp = arcpy.ListWorkspaces()
   ```

3. You can see what that looks like by just typing in `wksp` and pressing *Enter*. You can see it is all of the workspaces that are standard when creating a new project in ArcGIS Pro. They are a bit hard to read in this list:

```
['C:\\PythonBook\\Chapter2\\.backups', 'C:\\PythonBook\\Chapter2\\.
pyHistory', 'C:\\PythonBook\\Chapter2\\Chapter2.aprx', 'C:\\
PythonBook\\Chapter2\\Chapter2.gdb', 'C:\\PythonBook\\Chapter2\\
Chapter2.tbx', 'C:\\PythonBook\\Chapter2\\CPAD_2020b_Units.CPG',
'C:\\PythonBook\\Chapter2\\CPAD_2020b_Units.dbf', 'C:\\PythonBook\\
Chapter2\\CPAD_2020b_Units.prj', 'C:\\PythonBook\\Chapter2\\
CPAD_2020b_Units.sbn', 'C:\\PythonBook\\Chapter2\\CPAD_2020b_Units.
sbx', 'C:\\PythonBook\\Chapter2\\CPAD_2020b_Units.shp', 'C:\\
PythonBook\\Chapter2\\CPAD_2020b_Units.shp.BILL.26884.23180.
sr.lock', 'C:\\PythonBook\\Chapter2\\CPAD_2020b_Units.shp.
BILL.7612.23180.sr.lock', 'C:\\PythonBook\\Chapter2\\CPAD_2020b_
Units.shp.xml', 'C:\\PythonBook\\Chapter2\\CPAD_2020b_Units.shx',
'C:\\PythonBook\\Chapter2\\ImportLog', 'C:\\PythonBook\\Chapter2\\
Index']
```

To make them easier to read, let's iterate through the list, printing out each one. Type in:

```
for w in wksp:
    print(w)
```

Press *Enter*. Now you can really read what you have, as the workspaces are printed on one line each:

```
C:\PythonBook\Chapter2\.backups
C:\PythonBook\Chapter2\.pyHistory
C:\PythonBook\Chapter2\Chapter2.aprx
C:\PythonBook\Chapter2\Chapter2.gdb
C:\PythonBook\Chapter2\Chapter2.tbx
C:\PythonBook\Chapter2\CPAD_2020b_Units.CPG
C:\PythonBook\Chapter2\CPAD_2020b_Units.dbf
C:\PythonBook\Chapter2\CPAD_2020b_Units.prj
C:\PythonBook\Chapter2\CPAD_2020b_Units.sbn
C:\PythonBook\Chapter2\CPAD_2020b_Units.sbx
C:\PythonBook\Chapter2\CPAD_2020b_Units.shp
```

```
C:\PythonBook\Chapter2\CPAD_2020b_Units.shp.BILL.26884.23180.sr.lock
C:\PythonBook\Chapter2\CPAD_2020b_Units.shp.BILL.7612.23180.sr.lock
C:\PythonBook\Chapter2\CPAD_2020b_Units.shp.xml
C:\PythonBook\Chapter2\CPAD_2020b_Units.shx
C:\PythonBook\Chapter2\ImportLog
C:\PythonBook\Chapter2\Index
```

This is great, as you can see all the workspaces in the folder. But you only want to select the geodatabases in the folder. This is where the parameters come in. For this, you can use the workspace type parameter.

4. To select just the file geodatabase, you need to write the following:

```
wksp = arcpy.ListWorkspaces("","FileGDB")
```

Press *Enter*.

Why are there quote marks (““) followed by a comma (,)?

The first parameter is for the wildcard, and writing `""` will leave it blank. The quote marks need to be there, though, as functions take the parameters in the order they are written. If written as `wksp = arpcy.ListWorkspaces("FileGDB")`, the function will still run. But when you call it, you won't have any data in the list because there is no workspace called `"FileGDB"`.

5. If you call the `wksp` variable, you now have a list of just one value, `MyProject.gdb`. Type in:

```
wksp
```

Press *Enter*. The result is printed as:

```
['C:\\PythonBook\\Chapter2\\Chapter2.gdb']
```

While there is just the one value in the list, it is still a list and acts as such in Python. That means that functions in ArcPy that expect a string will fail if given a list. For example, you cannot update the workspace to this geodatabase location by using the `wksp` variable:

```
Python                                                                    ? ▾ □ ×
wksp
['C:\\PythonBook\\Chapter2\\Chapter2.gdb']
arcpy.env.workspace = wksp
Traceback (most recent call last):
  File "<string>", line 1, in <module>
  File "C:\Program Files\ArcGIS\Pro\Resources\ArcPy\arcpy\geoprocessing\_base.py", line 543, in set_
    self[env] = val
  File "C:\Program Files\ArcGIS\Pro\Resources\ArcPy\arcpy\geoprocessing\_base.py", line 605, in __setitem__
    ret_ = setattr(self._gp, item, value)
RuntimeError: Object: Error in accessing environment <workspace>
```

Figure 2.23: Error message when using the incorrect data type

6. To set the workspace, you need to use the list index to extract the workspace from the list. Since the list has just one value, it is at the 0 index of the list. To set the workspace, type in the following:

```
arcpy.env.workspace = wksp[0]
```

Press *Enter*.

What if you know you only have a single workspace that you are targeting?

In this example, you only have one item in the list as there was only one geodatabase in the folder. In these cases, you could just write `w = wksp[0]` to fetch the first (and only) element of the list. In fact, when you know you only have one item in your list you can just write the following to set your workspace:

```
arcpy.env.workspace = ListWorkspaces("","FileGDB")[0]
```

Be careful with using this notation, as if you have more than one workspace you will only be setting the workspace to the first one in the list.

7. Now that the workspace is set to your geodatabase, you can use the `ListFeatureClasses` function to get a list of all the feature classes in the geodatabase and assign it to a variable.

You are going to write the code to get a list of feature classes, and then write a `for` loop to iterate through the list so you can easily read what feature classes it contains. Enter the following code:

```
fcs = arcpy.ListFeatureClasses()
for fc in fcs:
    print(fc)
```

Press *Enter*. Here is the output we get:

```
tl_2019_06_prisecroads
UniqueStops_Summer21
Summer21RouteShape
tl_2019_06_tract
CPAD_2020b_Units_Oakland
CPAD_2020b_Units_Oakland_1000ft
AC_TransitStops_Within1000ft_OaklandPark
```

You now have a list that contains all the feature classes in your geodatabase. This list can be iterated through to give you a single feature class, which you can run through other ArcPy functions or geoprocessing tools. You could use the `Describe` function from above to find only the feature classes of a certain geometry to make sure you only run your analysis on that.

8. Starting with your list of feature classes stored in the variable `fcs`, you will iterate through it as in the previous step, when you just printed out the name. Then, you will use the `shapeType` property of feature classes to determine what the shape is of each feature class, and print out a statement saying that. To do this, write the following code in the Python window:

```
for fc in fcs:
    desc = arcpy.Describe(fc)
    fcName = desc.name
    if desc.shapeType == "Polygon":
        print("Shape Type for " + fcName + " is " +
            desc.ShapeType)
    elif desc.shapeType == "Polyline":
        print("Shape Type for " + fcName + " is " +
            desc.ShapeType)
    elif desc.ShapeType == "Point":
        print("Shape Type for " + fcName + " is " +
            desc.ShapeType)
    else:
        print(fcName + " is not a Point, Line, or Polygon")
```

You will need to press *Backspace* after each print statement line to ensure the indentation is correct, and press *Enter* twice after the last line for the code to run. The `for` loop will iterate through each feature class. For that feature class, you are creating a `desc` variable to hold the `Describe` properties of that feature class. You also create an `fcName` variable to hold the name of that feature class. Then, you write `if/elif/else` statements to test the `shapeType` property of the `Describe` object. The output statement will look like this:

```
Shape Type for tl_2019_06_prisecroads is Polyline
Shape Type for UniqueStops_Summer21 is Point
Shape Type for Summer21RouteShape is Polyline
Shape Type for tl_2019_06_tract is Polygon
Shape Type for CPAD_2020b_Units_Oakland is Polygon
Shape Type for CPAD_2020b_Units_Oakland_1000ft is Polygon
Shape Type for AC_TransitStops_Within1000ft_OaklandPark is Point
```

The wildcard parameter

Another way to select elements in the list functions before putting them into the list is to use the **wildcard parameter**. The wildcard limits what the function returns. It is not case sensitive and uses an asterisk (*) to include any number of characters before or after the asterisk. Let's look at some examples of how this works using our current geodatabase workspace.

The `ListFeatureClasses` function allows you to list all of the feature classes within a workspace. You will test different ways to use the wildcard to select data. First, you will use the wildcard to create a list of all of the CPAD data; next, you will create a list of all the data that ends with `Oakland`; and finally, you will create a list of all the 2019 census data. These are all examples of how to use the wildcard parameter to filter your lists to smaller datasets containing what you want.

Continue in the Python window with the workspace set to `C:\\PythonBook\\MyProject\\MyProject.gdb`:

1. Create a list of all the CPAD data. Type in the following and press *Enter*:

   ```
   cpad_fcs = arcpy.ListFeatureClasses("CPAD*")
   ```

2. View the data in the list using a `for` loop. Type in the following pressing *Enter* after the first line and *Enter* twice after the last line:

   ```
   for fc in cpad_fcs:
       print(fc)
   ```

The result printed out will be the feature classes that start with `CPAD`, and will look like the following:

```
CPAD_2020b_Units_Oakland
CPAD_2020b_Units_Oakland_1000ft
```

3. Create a list that contains just the CPAD units in Oakland feature classes. Type in the following and press *Enter*:

```
cpad_Oakland = arcpy.ListFeatureClass("*Oakland")
```

4. View the data using a `for` loop. Type in the following, pressing *Enter* after each line and *Enter* twice after the last line:

```
for fc in cpad_Oakland:
    print(fc)
```

The result printed out will be the feature classes that end with `Oakland`, and will look like the following:

```
CPAD_2020b_Units_Oakland
```

5. Create a list of the 2019 census feature classes. Type in the following and press *Enter*:

```
census_fcs = arcpy.ListFeatureClasses("*2019*")
```

6. View the data using a `for` loop. Type in the following, pressing *Enter* after each line and *Enter* twice after the last line:

```
for fc in census_fcs:
    print(fc)
```

The result printed out will be the feature classes with `2019` anywhere in their name, and will look like the following:

```
tl_2019_us_county
tl_2019_06_prisecroads
```

You have now seen how to use the * notation in the `wild_card` parameter to filter for different feature classes within the `ListFeatureClasses` function. It will work the same way on any data listing function that accepts `wild_card` parameters. In the next section, you will learn how to combine the `wild_card` parameter with feature types.

Combining wildcard and feature type parameters

The wildcard is one of the optional parameters in many of the list functions and can be used together with the other parameters. To illustrate, we'll look at the `ListFeatureClasses()` function.

You used the `ListFeatureClasses()` function in the previous section to illustrate how to use the `wild_card` parameter to create lists of specific feature classes. The `ListFeatureClasses()` function has a total of three optional parameters:

1. `wild_card`: This can be used to limit the returned values to those that match the `wild_card` value.
2. `feature_type`: This can be used to limit the returned values to specific feature classes. The valid parameters are `"Annotation"`, `"Arc"`, `"Dimension"`, `"Edge"`, `"Junction"`, `"Label"`, `"Line"`, `"Multipatch"`, `"Node"`, `"Point"`, `"Polygon"`, `"Region"`, `"Route"`, `"Tic"`, and `"All"` (default).

   The most common values you will use are `"Point"`, `"Polygon"`, and `"Polyline"`. Using one of those will limit the returned values to that type.
3. `feature_dataset`: This limits the feature classes to only those within the specified `feature_dataset`. If this is blank, only feature classes that are not within a feature dataset within the workspace will be returned to the list.

It is important to only return the data you will need to your list. This will ensure when doing analyses that you are only working on the correct datasets. In this exercise, you will use the `feature_type` parameter to further filter your feature classes returned using the `ListFeatureClasses()` function and return just the census data from 2019 that is a Polygon.

Continue in the Python window with the workspace set to `C:\\PythonBook\\Chapter2\\Chapter2.gdb`:

1. Create a list of just the 2019 census polygon data. Type in the following and press *Enter*:

```
census_fc_poly = arcpy.ListFeatureClasses("*2019*", "Polygon")
```

2. Verify the data by entering the variable and pressing *Enter*:

```
census_fc_poly
```

   The result printed out will be a list of the feature classes that correspond to those limits, and will look like the following:

```
['tl_2019_us_county']
```

3. Note that the feature class is stored within the square parentheses ([ ]), as it is in a list. To do any geoprocessing tasks on this, you either need to iterate through the list and do the tasks in the for loop, or extract the feature class using the list index to grab whichever list index you need. To select the individual feature class, type in the following and press *Enter*:

```
census_county = census_fc_poly[0]
```

4. Verify the data by entering the variable and pressing *Enter*:

```
census_county
```

The result printed out will be the single feature class and will look like the following:

```
'tl_2019_us_county'
```

Note that what the census_county variable returns is the name of the feature class. As long as your workspace is still the geodatabase, you can use just that name to do geoprocessing tasks. If you reset your workspace, ArcPy won't know where to find the feature class with that name.

So, it is good practice to use the **os library** to create a variable that contains the full path for your feature class. To use the os library, it needs to be imported like ArcPy when working in IDLE.

5. Continuing on from the previous step, type in the following and press *Enter* to import the os library:

```
import os
```

6. Create a variable with the census feature class full path. Type in the following and press *Enter*:

```
gdb = wksp[0]
```

7. You will use os.path.join() to create the full path. The os.path.join() method takes any number of arguments you need and joins them up with a backslash (\) between the arguments. This will give you the full path of the feature class. Type in the following and press *Enter*:

```
census_county_full = os.path.join(gdb,census_county)
```

8. Verify the data by entering the variable and pressing *Enter*:

```
census_county_full
```

The result printed out will be the full path of the feature class and will look like the following:

```
'C:\\PythonBook\\Chapter2\\Chapter2.gdb\\tl_2019_us_county'
```

Now you have the full path for the census county shapefile in a variable that you can use throughout any further code you may write. The above steps are common steps in creating automated analysis. You set a workspace, iterate through each dataset, set its full path, and do your analysis.

Introduction to ArcPy modules

ArcPy comes with a set of modules in addition to the geoprocessing tools and functions. As we've already seen, modules are just files that contain additional Python definitions and statements, including things like functions and variables. They are used to help organize code more logically.

ArcGIS Pro 2.8 comes with the following ArcPy modules:

- **Charts** module (`arcpy.charts`): Allows you to create charts of your data
- **Data Access** module (`arcpy.da`): Allows control of edit sessions and cursors for searching, inserting, and updating data
- **Geocoding** module (`arcpy.geocoding`): Allows you to set locators and automate geocoding
- **Image Analysis** module (`arcpy.ia`): Allows you to manage and process imagery
- **Mapping** module (`arcpy.mp`): Allows you to work with maps, layers, and layouts to automate map production
- **Metadata** module (`arcpy.metadata`): Allows you to access or manage an item's metadata
- **Network Analyst** module (`arcpy.na` or `arcpy.nax`): Allows you to work with the Network Analyst extension
- **Sharing** module (`arcpy.sharing`): Allows you to automate sharing data as web layers or map services
- **Spatial Analyst** module (`arcpy.sa`): Allows you to work with the Spatial Analyst extension
- **Workflow Manager** module (`arpcy.wmx`): Allows you to work with the Workflow Manager toolbox and automate business workflows

Some of the above modules do require specific licenses to use the functions and tools within them. For example, the Network Analyst and Spatial Analyst modules require you to have the Network Analyst and Spatial Analyst extensions available. The two you will look at in depth in later chapters, the Data Access module and the Mapping module, do not. The Data Access module can help you to streamline your data cleaning and analysis processes. The Mapping module can streamline mass map production and make creating hundreds of maps a simple process.

Spatial Analyst module

The Spatial Analyst module contains all of the geoprocessing tools associated with the Spatial Analyst extension. Because it uses the Spatial Analyst extension, you need to import the extension:

```
from arcpy.sa import *
```

In this exercise, you will learn how to write the code to run Spatial Analyst tools in the Python window using the FVEG data from CALFIRE. The FVEG data is a statewide raster land cover dataset. It has a raster attribute table showing different classification levels of the land cover at each raster grid square. The `Chapter2.gbd` file contains the CALFIRE FVEG data extracted to Alameda County, as the entire dataset is larger than GitHub will allow.

If you would like to download the full statewide dataset, directions on how to do this are in the `CalFireVegdownload.md` file in the `Chapter2` folder on GitHub. The data is also available for download here: `https://frap.fire.ca.gov/mapping/gis-data/`. The link will open a page to all the CALFIRE GIS data. To download the FVEG data, scroll down to find **FVEG** and click on it to expand the box. Click the **Download the FVEG geodatabase link** to download the data:

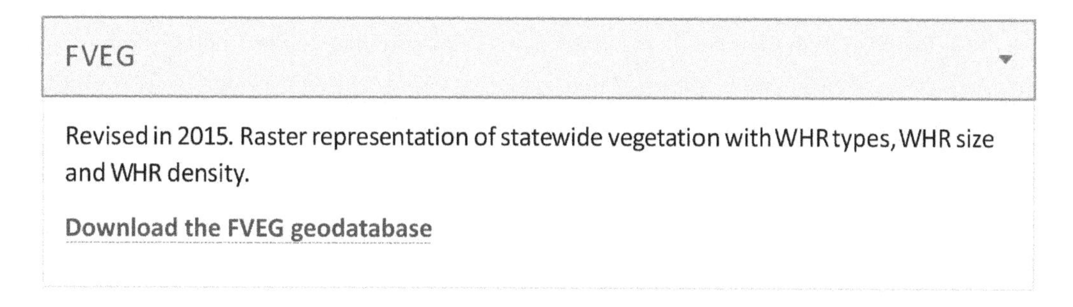

Figure 2.24: CALFIRE FVEG download

A common operation on raster data is extracting it to a study area. This is done using the `ExtractByMask` tool. The `ExtractByMask` tool is part of the Spatial Analyst toolset and, in ArcPy, is part of the Spatial Analyst module.

You will be extracting the CALFIRE FVEG data to the boundaries of the parks in the City of Oakland and running a `Con()` tool to find the areas that are not urban land cover. Let's get started. We have omitted the *Enter* instructions in each step, since you should be used to the interface by now:

1. In the Python window, type in the following:

```
from arcpy.sa import *
```

2. Check if a Spatial Analyst extension is available by typing in the following:

```
arcpy.CheckExtension("Spatial")
```

   It should return:

```
'Available'
```

   If not, you need to either enable your Spatial Analyst license or, if you are on a shared license network, have someone release theirs. You may need to contact your system administrator to ensure you have been given access to a Spatial Analyst license.

3. Once you have confirmed a license is available, you will need to **check out** the license to use it. Checking in and out an extension helps to manage floating licenses on systems where there are more users than extension licenses. Type in the following:

```
arcpy.CheckOutExtension("Spatial")
```

   It should return:

```
'Checked Out'
```

4. The CALFIRE FVEG data for Alameda County is already in the `Chapter2` map. If you need to add it in, it is in `Chapter2.gdb` as `C:\PythonBook\Chapter2\Chapter2.gdb\CalFireFVEG_AlamedaCounty_CO`.
5. Right-click on it and select **Symbology**.
6. Click the dropdown where it says **Stretch** and select **Unique Values**.
7. Click on the dropdown for **Field 1** and select **WHR10NAME**. You can play with the colors and color scheme if you want, or just select **Basic Random** to get each land cover symbolized by a different color.
8. If you did not create the `CPAD_2020b_Units_Oakland` feature class from earlier in the chapter, you will need to now. This consists of the parks that are just managed by the City of Oakland. If you have already created `CPAD_2020b_Units_Oakland`, you do not need to do this step. Type in the following:

```
arcpy.analysis.Select(r"C:\PythonBook\Chapter2\CPAD_202b_Units.
shp", r"C:\PythonBook\Chapter2\Chapter2.gdb\CPAD_2020b_Units_
Oakland",'"AGNCY_NAME" = \'Oakland, City of\'')
```

9. To extract the FVEG land cover data to the Oakland parks boundaries, you will use the `ExtractByMask()` tool. All the Spatial Analyst tools utilize a different syntax compared to the other tools you have been using. You still input the parameters of the tool, but there is no output parameter. The output parameter is created by setting the tool equal to a variable. That variable holds your newly created raster as a temporary file. To create the temporary file of the land cover within the Oakland parks, type in the following:

```
oaklandParksLandCover = ExtractByMask("CalFireFVEG_AlamedaCounty_
Co", r"C:\PythonBook\Chapter2\Chapter2.gdb\CPAD_2020b_Units_
Oakland")
```

The Python window will display no output, but a raster called `OaklandParksLandCover` will be added to your map and will contain data only within the boundary of the parks.

10. To save the temporary file, you will use the `save()` method on the variable that holds the temporary file. The `save()` method takes the full path of the raster you are saving. To save the raster, type in the following:

```
oaklandParksLandCover.save(r"C:\Chapter2\Chapter2.gdb\
OaklandParksLandCover")
```

The Python window will display no output, but if you right-click on `MyProject.gdb` in the **Catalog** window and click refresh, you will see that a new raster called `OaklandParksLandCover` was saved there.

11. You will now use the `Con()` tool to extract out the land cover within the park that is not urban. The `Con()` tool takes four parameters:

    1. `in_conditional_raster`: The input raster layer that is the true or false result of the condition
    2. `in_true_raster_or_constant`: Either a raster or constant value that will be used when `in_conditional_raster` evaluates as true
    3. `in_false_raster_or_constant`: Either a raster or constant value that will be used when `in_conditional_raster` evaluates as false
    4. `where_clause`: A SQL expression that determines if the values of `in_conditional_raster` are true or false

The first parameter is the raster to evaluate and will be the OaklandParksLandCover raster just created. The second parameter is what to return to the raster when the condition is true and that will be the same as the in_conditional_raster, since you want the land covers that are not urban. The third parameter is what is returned when the condition is false; you want that to be a NULL value, as you want to remove all the urban land from the parks land cover. The fourth parameter is the SQL clause that each cell in in_conditional_raster will be evaluated against as true or false. To create the new raster of all the non-urban parks land cover, type in the following:

```
oaklandParksNonUrban =
Con(oaklandParksLandCover,oaklandParksLandCover,"","WHR10NAME <>
'Urban'")
```

The Python window will display no output, but a new raster layer called oaklandParksNonUrban will be added to your map.

12. A problem with the Con() tool is that it does not carry through the raster attribute table from in_conditional_raster. All of the land cover values have been lost. To see this, open up the attribute table on the new raster in your table of contents. It only has a Value field and none of the rest of the attributes. This can be fixed by doing a join. The **Join Field** tool is in the management toolbox and works on rasters as well as feature classes. It creates a permanent join. The Join Field tool takes five parameters:

    1. in_data: The input dataset. It can be a feature class, table, or raster dataset with an attribute table.
    2. in_field: The field in the input table to be used for joining.
    3. join_table: The table to be joined to in_data. It can be a feature class, table, or raster dataset with an attribute table.
    4. join_field: The field in join_table to be used to join to in_data.
    5. fields: A list of the fields from join_table to be joined to in_data.

    You will want to join the oaklandParksLandCover data to the oaklandParksNonUrban dataset. You will use the Value field from each and list all the descriptive land cover fields as the fields. Type in the following:

    ```
    arcpy.management.
    JoinField(oaklandParksNonUrban,"VALUE",oaklandParksLandCover,
    "VALUE", ["WHRNAME","WHRTYPE","WHR10NAME","WHR13NAME"])
    ```

The following will be printed:

```
<Result 'C:\\Users\\William\\AppData\\Local\\Temp\\
ArcGISProTemp2356\\1d45ee13-5255-42fc-a41a-60f5727866ad\\Default.
gdb\\Con_oaklandP1'>
```

13. The result prints the location of the data that was joined. The `oaklandParksNonUrban` raster is still in the temp workspace. To save it, use the `save()` method on it by typing the following:

```
oaklandParksNonUrban.save(r"C:\Chapter2\Chapter2.gdb\
OaklandParksLandCover")
```

From the `CALFIRE FVEG` data, you have extracted just the land cover in the parks in Oakland. You further extracted just the non-urban lands within the parks using the Con tool. Finally, you joined back the land cover descriptions that were lost in the Con tool. This leaves you with a land cover dataset of just the non-urban data within the parks in Oakland.

Summary

In this chapter, we introduced ArcPy and showed you how to verify a proper install of ArcPy. A shortcut to the version of IDLE that connects to ArcPy for ArcGIS Pro was created and used to write Python code in both the IDLE Shell and in a standalone script file. You learned how to use the ArcGIS Pro Python window for listing data, and the wildcard parameter to filter lists. The data returned to the list was extracted and used for geoprocessing analysis. Raster analysis was completed using the Spatial Analysis module.

In the next chapter, you will be introduced to the ArcGIS API for Python for connecting ArcGIS Pro to ArcGIS Online. ArcGIS Notebooks will also be introduced as a way to write and store Python code.

3

ArcGIS API for Python

The **ArcGIS API for Python** is a Python package designed to work with web GIS. It allows you to work directly with data hosted on ArcGIS Online or ArcGIS Enterprise. Previously in this book you have been using ArcPy, which is excellent for desktop work, but has limited capabilities when working with hosted data. The ArcGIS API for Python provides tools to do many of the same functions that ArcPy does, such as creating maps, geocoding, managing data, and geoprocessing, but using data that is hosted within your organization. In addition to this, you can use it to manage your organization's data and ArcGIS Online account by managing users, groups, and items.

It is important to note that while all of the examples you will work through in this chapter are within ArcGIS Pro Notebooks, you don't have to work through ArcGIS Pro. You could install a standalone environment with **conda** and access everything through a Jupyter notebook. This book will not cover that, as it is focused on working with Python within ArcGIS Pro.

This chapter will cover:

- ArcGIS API for Python modules
- Managing virtual environments with the Python Package Manager
- ArcGIS Pro Notebooks
- Connecting to ArcGIS Online through the ArcGIS API for Python
- Searching for data

To complete the exercises in this chapter, please download and unzip the `Chapter3.zip` folder from the GitHub repository for this book: `https://github.com/PacktPublishing/Python-for-ArcGIS-Pro/tree/main/Chapter3`.

What is the ArcGIS API for Python?

The ArcGIS API for Python is like ArcPy, in that it is a Python package. It contains classes, modules, and functions. However, it is not just a Python package: it is also an **application programming interface (API)**. An API is code that allows different applications and software to talk to each other. It interacts primarily with the ArcGIS REST API. This means that you can use the module to make requests for data hosted on ArcGIS Online or ArcGIS Enterprise. This data is either in your own organization or is publicly available. It is a **Pythonic** API, in that it is designed to Python standards and best practices. As a Pythonic API, it allows Python programmers to easily use ArcGIS, and ArcGIS users familiar with Python to automate web GIS tasks.

ArcGIS API modules

The API is organized into different modules for your use. Each module has different functions and types to assist in your GIS.

These are the modules you are likely to use the most:

- `arcgis.gis`: This module is the one you will use the most. It allows entry to the GIS, connects you to your ArcGIS Online account, and provides the functionality to create, read, update, and delete GIS users, groups, and content.
- `arcgis.features`: This module contains spatial analysis functions for working with feature data, feature layers, collections of feature layers, and feature sets.
- `arcgis.geometry`: This module is for working with geometry types. It has functions that use geometry types as input and output, and for converting geometries to different representations.
- `arcgis.geocoding`: This module is for geocoding and reverse geocoding. It creates points of addresses with the output visualized on a map, or used as input data for spatial analysis.
- `arcgis.geoenrichment`: This module is for providing data about an area or location. Users can get information about people and places in an area or within a certain distance. It can assist by easily providing demographic data for models.
- `arcgis.env`: This module supplies a shared environment to be used by the different modules. It stores the currently active GIS and environment settings.

These are the modules that work with specific data types:

- `arcgis.raster`: This module contains classes and functions for working with raster and imagery data.

- `arcgis.realtime`: This module works with real-time data feeds. It is for use with streaming data to perform continuous analysis. It allows for Python scripts that can subscribe to data that is streamed and broadcast updates or alerts.
- `arcgis.network`: This module is for completing network analysis. It is for use on network layers and can be used to find best routes, closest facilities, and service areas, and calculate cost matrices.
- `arcgis.schematics`: This module is for working with schematics, which are simplified networks. It is for explaining the structure and way networks work.
- `arcgis.geoanalytics`: This module is for creating distributed analyses of large datasets, both feature and tabular. The tools are designed to work with big data as well as with feature layers.
- `arcgis.geoprocessing`: This module is for creating and sharing geoprocessing tools.

These are the modules most associated with visualization:

- `arcgis.mapping`: This module is for providing visualization capabilities for GIS data. It includes WebMap and WebScene to enable 2D and 3D visualization.
- `arcgis.widgets`: This module is for providing visualization of GIS data. It includes the Jupyter notebook MapView widget to assist with the display of maps and layers.
- `arcgis.apps`: This module provides the ability to manage the web-based applications available in ArcGIS.

In this chapter, the focus will be mostly on the `arcgis.gis` module to connect to your ArcGIS Online account and search for data and users both outside of and within your organization. In *Chapter 5, Publishing to ArcGIS Online*, you will learn how to publish and manage data, use the `features` module to query and edit data, and the `mapping` module to visualize your data.

What does it do and why use it?

The ArcGIS API for Python allows you to access your data in ArcGIS Online through the ArcGIS Pro interface. You can manage your ArcGIS Online or ArcGIS Enterprise organization, its users, and its data from either a Jupyter notebook or an ArcGIS Pro Notebook. By doing this in a Notebook and not through the ArcGIS Online web interface, you can use the full functionality of Python to iterate over data to run the same process multiple times, and schedule tasks to be run. The ArcGIS API for Python complements ArcPy, as it allows you to automate your organization's web GIS processes.

Just like using ArcPy to automate a process in ArcGIS Pro, you would use the ArcGIS API for Python when you need to automate a process on data that is stored in your ArcGIS Online account or ArcGIS Enterprise organization.

The Python Package Manager

Python comes in many versions and installs for different operating systems; each one is called a **Python distribution**. The Python distribution that you have is dependent on the version and the operating system you have. All of the different Python versions come with the standard library of built-in modules. You have already been introduced to the `sys` and `os` modules as two of the more important built-in modules. In addition to those, there are many third-party packages that can expand functionality. ArcPy is a third-party Python package that is installed with ArcGIS Desktop and ArcGIS Pro.

When you install ArcGIS Pro, you also install a custom Python distribution that works with ArcGIS Pro; for ArcGIS Pro 2.9, that is Python 3.7.11, at time of writing. This distribution includes all the standard libraries and packages, including ArcPy. To manage all of the different packages, ArcGIS Pro uses a package manager called **conda**. In addition to managing packages, conda also manages Python environments. Python environments are different collections of packages that can be switched between depending on the needs of a project.

Python environments

The **Python Package Manager** in ArcGIS Pro is where you can manage your different Python environments and packages. ArcGIS Pro comes with a default environment called `arcgispro-py3`. You can create additional environments, which are referred to as **virtual environments**, or sometimes **conda environments**, as conda is the package manager.

As we touched on in *Chapter 1, Introduction to Python for GIS*, a virtual environment is a Python installation with a unique set of packages that has been added. They are called **virtual environments** because each environment replicates a separate install of Python as if it were a different machine. Each virtual environment is isolated from the others. This allows you to have environments with additional packages or different versions of the same package, depending on what your project needs. The Python Package Manager is an alternative to managing your environments and packages through the command line.

How to create a new virtual environment

Although the Python Package Manager makes no reference to virtual environments or conda, it is a user interface designed for use in ArcGIS Pro to manage virtual environments. As mentioned, the default environment is called `arcigspro-py3` and can be viewed in the **Project Environment** field in the Python Package Manager:

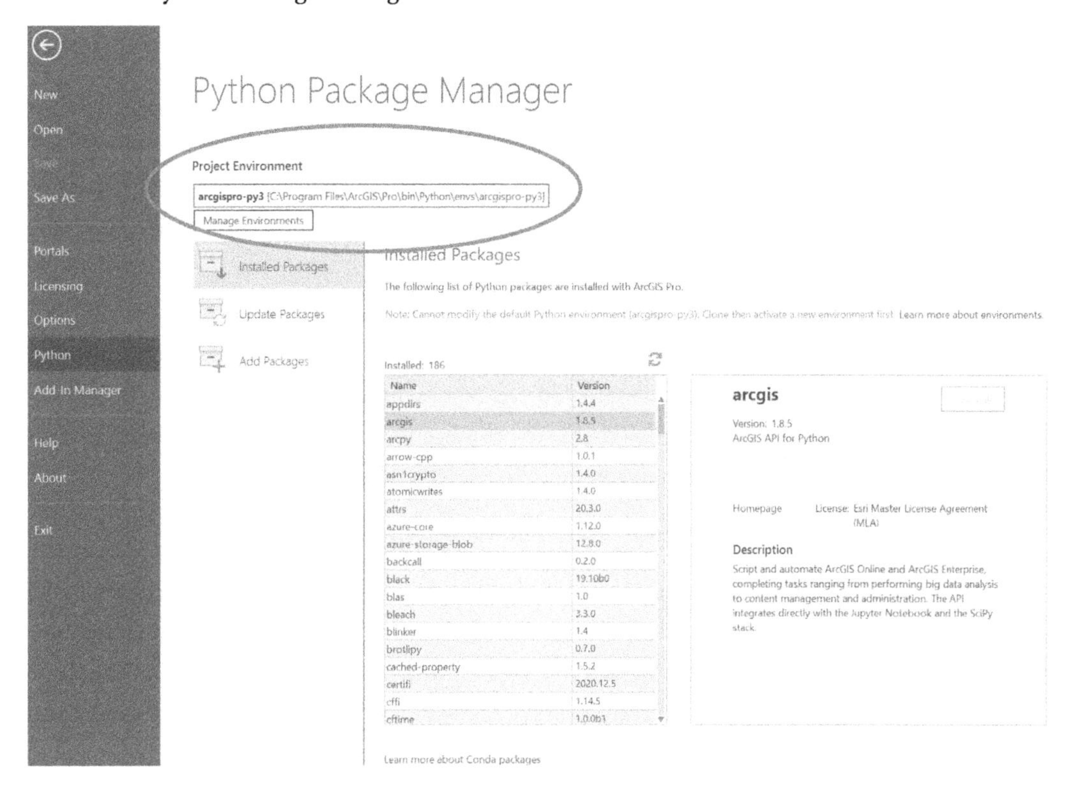

Figure 3.1: Python Package Manager with the default install

You cannot modify the default environment. It is kept this way so you always have a clean environment with default settings to switch back to if your other environments stop working. If you need to update any packages or install any new packages, you have to create a new virtual environment.

In this exercise, you will create a new virtual environment, upgrade all packages, and then add a package:

1. Open ArcGIS Pro and select **Open Another Project**.
2. Navigate to where you unzipped the `Chapter3.zip` folder, select `Chapter3.aprx`, and click **Open**.

3. Click the **Project** tab in the ribbon.
4. Click **Python** in the side ribbon. This will open the Python Package Manager.

   The `Chapter3` project is currently using the default environment. You will create a new environment.

5. Click the **Manage Environments** button. This will bring up the **Manage Environments** dialog window. `arcgispro-py` is the active environment. You need to clone the default environment first. There are two different ways to do that:

   - Click the **Clone Default** button. A new environment will be created called `arcgispro-py-clone`; if you have cloned `arcgispro-py` before, you may also see a number after `clone` depending on how many times you have cloned it. It will take some time for all of the packages to install.
   - Click on the two squares under the **Clone** header for the `arcgispro-py` environment. This will bring up the **Clone Environment** dialog window. You can choose a name and location to store your environment. It is suggested that you store it in the default location, as that is the conda environments folder that was created on your computer when you installed ArcGIS Pro.

   Choose one of these two ways to create a new environment.

6. When all the standard packages are installed for your environment, you can change the active environment to your new environment. Click on the radio button next to the environment under the **Active** header. An alert shows up at the bottom of the **Manage Environments** dialog box , stating **Restart ArcGIS Pro for your environment changes to take effect.**
7. Click **OK** to close the **Manage Environments** dialog window.

   You have now changed to a new virtual environment. The **Project Environment** is now set to `arcgispro-py3-clone`. For it to take effect, you need to restart ArcGIS Pro. Before doing that, though, you can make some changes to the packages in this virtual environment.

8. Click the **Update Packages** button in the Python Package Manager to update all of your packages. This will change the list of packages to those that have updates available and tell you how many have updates available. You can click on each package and update only the ones you want to update by clicking the **Update** button in the package description window. Alternatively, you can click the **Update All** button to update all of them.

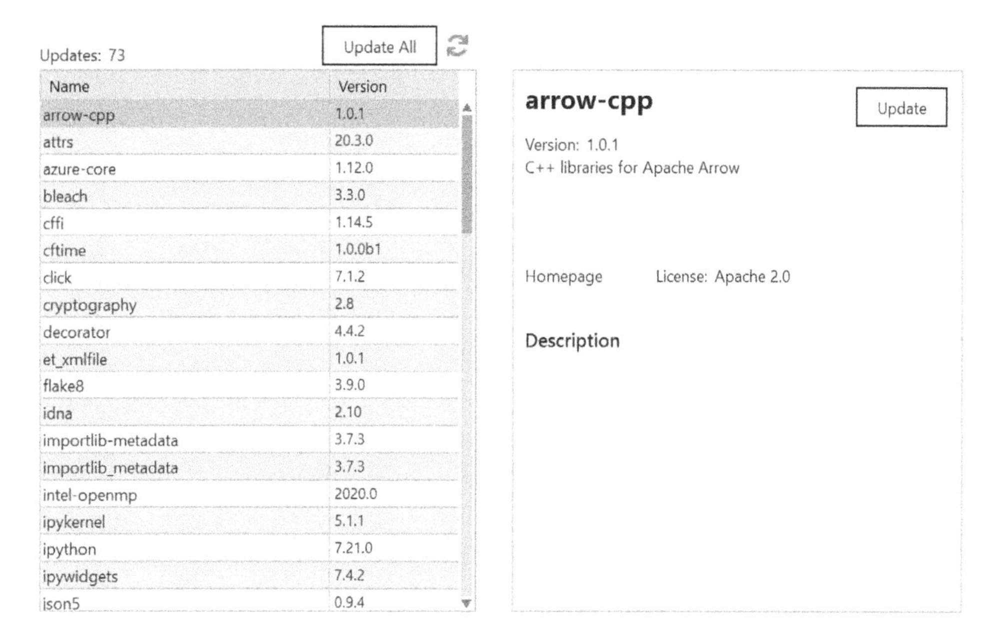

Figure 3.2: Update Packages

Click the **Update All** button to update all the packages.

You will now add a package. R is a free software that is useful in statistical analysis and graphics. There is a Python package called `rpy2` that will allow you to call R functions and methods in Python.

9. Add the `rpy2` package by clicking on **Add Packages** and typing `rpy2` in the search box. Select the `rpy2` package and click the **Install** button.

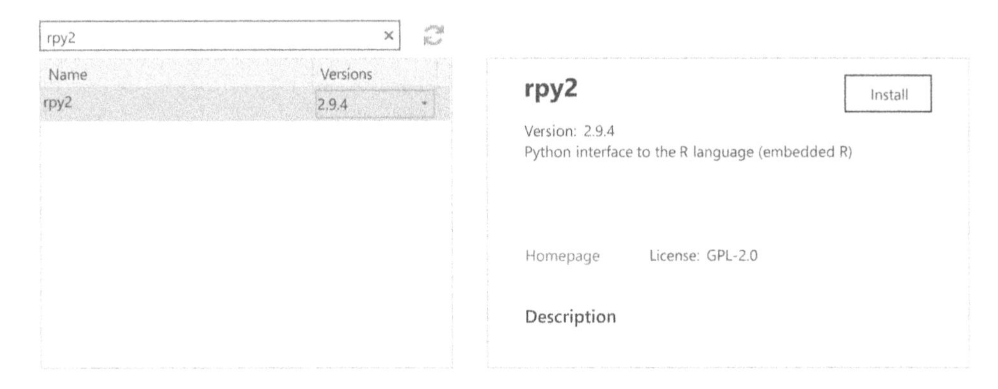

Figure 3.3: Add Packages window

10. An **Install Package** dialog window will open up. It displays the packages that the `rpy2` package depends on and a Terms and Conditions. Read through the Terms and Conditions, check **I agree to the terms and conditions**, and click **Install**.

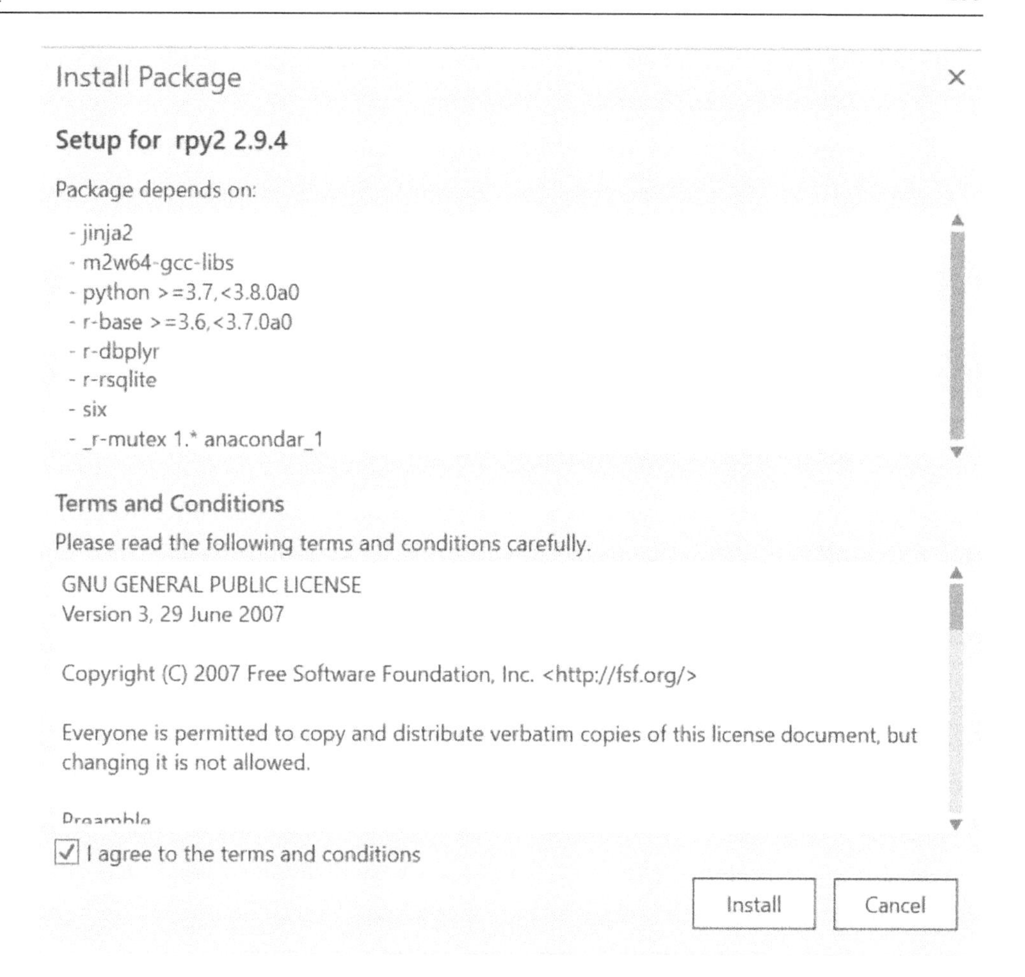

Figure 3.4: Install Package dialog window

The rpy2 package will now be installed. In addition to the rpy2 package, the packages that it depends on will also be installed. It may take a few minutes to complete the install.

11. For the change to the new virtual environment to be complete, you need to shut down ArcGIS Pro. Close ArcGIS Pro.
12. Open ArcGIS Pro back up. We will now follow the same steps from the start of this exercise.
13. Navigate to where you unzipped the `Chapter3.zip` folder, select `Chapter3.aprx`, and click **Open**.
14. Click the **Project** tab in the ribbon.
15. Click **Python** in the side ribbon. Observe that the Project Environment is now set to `arcgispro-py-clone`.
16. Click on **Installed Packages**. You will find the newly installed `rpy2` package and the dependent packages that were installed with it.

The virtual environment you created is available to use on any project and is the environment any project will be set to when you open it. To see this, close ArcGIS Pro and reopen it with any project. Then follow *steps 14-16* from above to see the Python package your project is using. It will be the package you were using in the last project you closed.

If you were to start a new blank project, it would use the `arcgispro-py-clone` virtual environment too.

You can also remove some packages from your cloned virtual environment. To do that, just click on the installed package and select **Uninstall** from the package description window. Some packages cannot be uninstalled; these include `arcpy`, `arcgis`, `numpy`, and `python`.

Now that you have created a new virtual environment, you will use ArcGIS Pro Notebooks to work with the ArcGIS API for Python.

ArcGIS Pro Notebooks

ArcGIS Pro Notebooks are a way to create, save, and share documents that contain Python code and visualizations. They are built on top of the open-source web application, **Jupyter notebook**. ArcGIS Pro Notebooks (referred to from here on as *Notebooks*) allow you to manage data, perform analysis, and view your results immediately.

With Notebooks, you can automate workflows and then easily share your automation by sharing your Notebook. In addition, they can be used as a sandbox to test your code and then save it when you have it working.

Notebooks have all the core Python functionality, in addition to ArcPy, the ArcGIS API for Python, and access to third-party libraries such as NumPy and pandas.

In *Chapter 2, Basics of ArcPy*, you used the Python window to write code in ArcGIS Pro; Notebooks are another way to write code in ArcGIS Pro. The following table shows you a comparison of the benefits of the Python window and an ArcGIS Pro Notebook:

Python window	ArcGIS Pro Notebook
Live code testing	Live code testing
Quick to start and test small code blocks	Save final finished code
Autocomplete while typing	Code can be shared
	Visualize output in the Notebook
	Tab completion – autocomplete when pushing tab
	Markdown code for commenting on your code

Creating a Notebook in ArcGIS Pro

In ArcGIS Pro, there are often different ways to complete a task. You have a few options for creating a new ArcGIS Notebook. When working in a Project, you can click on the **Insert** tab in the ribbon and click the **New Notebook** button:

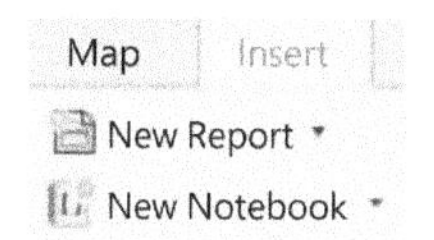

Figure 3.5: New Notebook in the Insert tab

Or you can click on the **Analysis** tab, click on the arrow next to **Python**, and select **Open a new notebook**:

Figure 3.6: New Notebook in the Analysis tab

Both of these will create a Notebook in your project. You could also open the **Catalog** pane and navigate to any folder, right-click that folder, and select **New > Notebook**:

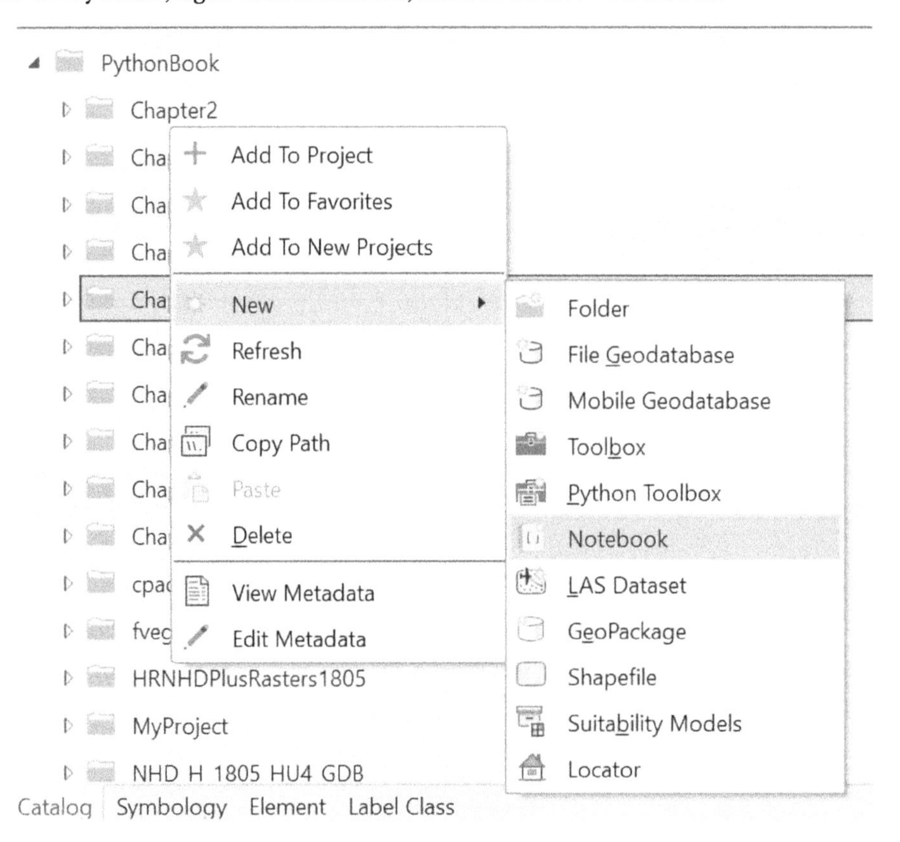

Figure 3.7: New Notebook in the Catalog pane

This will create the Notebook wherever you choose. This approach allows you to create Notebooks outside of projects, which can be helpful for Notebooks that might be needed across many projects.

Creating your first Notebook

This section will walk you through creating a Notebook. Your first Notebook will be a simple one that connects to ArcGIS Online and displays a map of Oakland, California. This will also test out your virtual environment and installation of the `arcgis` module:

1. If you closed ArcGIS Pro after the last section, open it back up and open up the `Chapter3.aprx` file.

2. Create a new Notebook using one of methods mentioned previously:
    - Click on the **Insert** tab and then **New Notebook.**
    - Click on the **Analysis** tab and then **Python** > **New Notebook.**
    - Right-click on your project folder and select **New** > **Notebook.**
3. The Notebook created is named `New Notebook`. Rename it to `Chapter3_FirstNotebook` by right-clicking on it in the **Catalog** pane and selecting **Rename.** Your Project should look like this now:

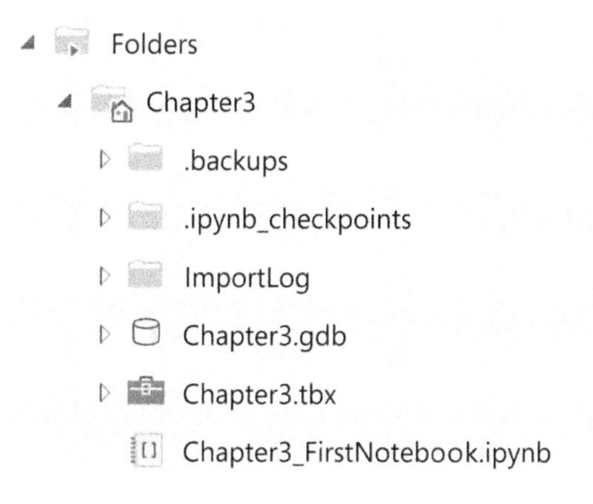

Figure 3.8: New Project with a Notebook

ArcGIS Notebook structure

Before writing code in the Notebook, take a look at the structure and what you can see in it:

Figure 3.9: New Notebook

Cells

This is where you write the code in a Notebook. You can write as much code as you want in a cell, but it is good coding practice to keep related things together as much as possible in single cells. This is because you can run your code one cell at a time.

By doing this, you can test out portions of code and make sure you are getting the output you want before running the entire Notebook.

Figure 3.10: Empty cell

Edit tab

The **Edit** tab is where you can edit your Notebook. It is where you find the tools to cut, paste, delete, split, merge, and move cells.

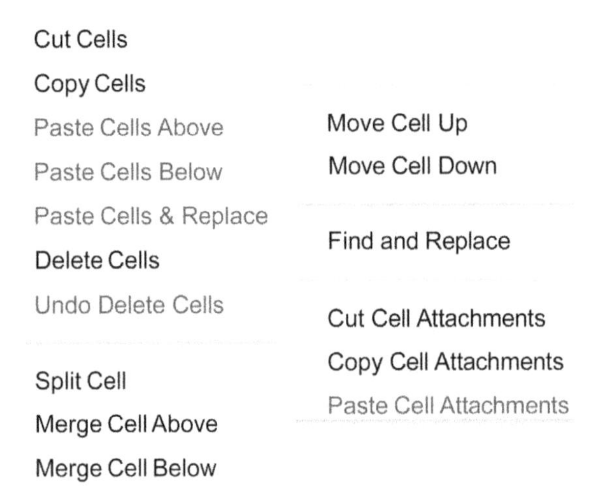

Figure 3.11: The Edit tab

View tab

The **View** tab is where you can change the different view properties of your Notebook. You can toggle on/off the toolbar, toggle on/off line numbers, and add different toolbars to the cell.

Toggle Toolbar
Toggle Line Numbers
Cell Toolbar ▸

Figure 3.12: View tab

It is useful to have line numbers inside your cells when you have cells that will be more than one line. To add line numbers, click **View > Toggle Line Numbers**. You should now see a line number in your cell.

Insert tab

The **Insert** tab is where you can insert new code cells and heading cells. Heading cells are a type of **Markdown** cell. Markdown is a markup language that allows you to create formatted text using plain text. When a Markdown cell is run, the plain text will be turned into rich text with the Markdown syntax.

For more information on Markdown syntax, see the Markdown Guide: `https://www.markdownguide.org/`.

Markdown cells are the cells you use to add comments to code. Like comments in other coding languages, comments you add are not read by the computer. They are there to give the human reading the code direction on what the code is doing.

Cell tab

The **Cell** tab is where you can run single cells, groups of cells, or the entire code. It is also where you can modify the setting for **Output** cells. Output cells are the output from anything you run in a cell. This can be as simple as print statements, or data frames when using `pandas`, or maps when using the ArcGIS API for Python. The **Cell** tab is also where you can change a cell from code to Markdown.

Help tab

The **Help** tab is where you can get help. It has a tour of the user interface and keyboard shortcuts. It also has links to documents on GitHub for general help with Notebooks, Markdown, and Jupyter extensions.

Toolbar

The toolbar contains the most common tools you will use:

Figure 3.13: Notebook toolbar

On the toolbar, you see the following:

- The + icon, which will add in a new cell below your selected cell.
- The **scissors** icon, which will cut your selected cell.
- The **two pages** icon, which will copy the selected cell.
- The **single page**, which will paste what you copied below the selected cell.
- The **up arrow**, which will move a cell up.
- The **down arrow**, which will move a cell down.
- The **Run** button, which will run a selected cell.
- The **Code** dropdown, which lets you toggle cells between code and Markdown.
- The **keyboard** icon. This will open up the command palette, which you can use to run different commands for cells. It will also show you keyboard shortcuts.

The toolbar gives you quick access to many of tools you will use the most. There are also keyboard shortcuts for many of the same functions.

Keyboard shortcuts

Depending on how you work, you may want to use keyboard shortcuts instead of the mouse to point and click. There are two different modes when working in an ArcGIS Pro Notebook, **command mode** and **edit mode**. Command mode is whenever your cursor is not active in a cell and edit mode is whenever your cursor is active in the cell. You can tell which mode you are in based on the color of the outline around a cell. A blue outline means you are in command mode and can use command mode keyboard shortcuts:

Figure 3.14: Command mode

A green outline means you are in edit mode and can use edit mode keyboard shortcuts:

Figure 3.15: Edit mode

Below are some of the command mode keyboard shortcuts you may find useful:

- *F* – Find and replace
- *Alt + Enter* – Run cell and insert cell below
- *Y* – Change cell to code
- *M* – Change cell to Markdown
- *A* – Insert cell above
- *B* – Insert cell below
- *X* – Cut selected cell
- *C* – Copy selected cell
- *V* – Paste cells below
- *Z* – Undo cell deletion

Below are some edit mode keyboard shortcuts you may find useful:

- *Tab* – Code completion or indent
- *Ctrl +]* – Indent
- *Ctrl + [* – Dedent
- *Ctrl + Z* – Undo
- *Ctrl + Y* – Redo
- *Esc* – Enter command mode
- *Alt + Enter* – Run cell and insert cell below

There are many keyboard shortcuts. To see a full list, go to **Help** > **Keyboard Shortcuts** in the Notebook menu.

Connecting to ArcGIS Online or ArcGIS Enterprise

There are different ways to connect to ArcGIS Online or ArcGIS Enterprise when using the ArcGIS API for Python. You can connect as an anonymous user, through your ArcGIS Pro connection, through built-in users, or with a URL, username, and password for an ArcGIS Online account. All connections are made by constructing a GIS object by typing in `gis = GIS()`. What varies is the parameters that are entered inside the parentheses, as that will determine the connection type.

Before connecting, you will need to import the GIS class from the gis module. You do that by typing in the following in the first cell:

```
from arcgis.gis import GIS
```

Anonymous users

Connecting to ArcGIS Online as an anonymous user allows only for limited tasks. You have the ability to query and view publicly available data; however, you cannot create or modify any of the data you see, and you cannot perform any analysis. To connect as an anonymous user, type in the following in the cell:

```
gis = GIS()
```

ArcGIS Pro connection

You can connect through ArcGIS Pro using the Pro authentication scheme. This will connect your Notebook to your ArcGIS Online portal using the credentials used to sign in to ArcGIS Pro. This is the most common way you will connect when you are in Pro, as it gives you access to all of your data on your ArcGIS Online Portal. To connect to ArcGIS Online with the Pro authentication scheme, type in the following in the cell:

```
gis = GIS("Pro")
```

Built-in users

ArcGIS Online and ArcGIS Enterprise come with a built-in identity store that allows you to create accounts and manage them. Connecting to ArcGIS Online using a built-in account is similar to connecting through a Pro authentication scheme when in Pro. The connection will use the credentials you are logged in to ArcGIS Pro with to make a connection to your ArcGIS Online Portal. To connect to ArcGIS Online using a built-in account, type in the following:

```
gis = GIS("home")
```

The difference between the built-in account and Pro authentication is that the Pro authentication scheme only works when ArcGIS Pro is installed locally and is running concurrently.

Connecting to an ArcGIS Enterprise account using a built-in account requires you to enter the portal URL, username, and password as parameters. This will connect you to your Enterprise portal and allow you access to your data stored there. To connect to ArcGIS Enterprise using a built-in account, type in the following.

```
gis = GIS("https://portalname.domain.com", "username", "password")
```

Most of the time, you will be connecting using either `gis = GIS('home')` or `gis = GIS('Pro')`.

Creating a Notebook

Now that you are familiar with structure of a Notebook and how to connect to ArcGIS Online, you will create a map. This exercise will help you become familiar with some of the steps you will always take to connect to ArcGIS Online when using the ArcGIS API for Python, and ensure your `arcgis` package and virtual environment are correctly installed.

1. If you closed ArcGIS Pro after the last section, open it back up and open up the `Chapter3.aprx` file.
2. Find the `Chapter3_FirstNotebook` you created above and open it by double-clicking on it.
3. In the first cell, you will import the `GIS` class from the `arcgis` module. This will be in the first cell of all of your ArcGIS API for Python Notebooks. The `GIS` class is what you use to create the connection to either your ArcGIS Online or ArcGIS Enterprise account. Type in the following:

   ```
   from arcgis.gis import GIS
   ```

   Click **Run** to run the cell.

   When you run a cell with no cells below it by clicking **Run** or using the *Alt + Enter* keyboard shortcut, a new cell is created below it.

4. In the newly created cell below your first cell, you will create the connection to ArcGIS Online. You will connect through an anonymous connection, as you just want to create a test map using publicly available data. Type in the following:

   ```
   gis = GIS()
   ```

   Run the cell.

5. In the next cell, you will create the map variable. When creating a map, you can pass many different things to the parameter in the map widget to set the view. These different options will be explored in more depth in *Chapter 5, Publishing to ArcGIS Online*, when creating maps of your data.

You will be using a city name to center your map on right now, as you just want to test your setup. Type in the following:

```
map1 = gis.map("Oakland, California")
```

Run the cell.

6. The map is not displayed, as you have just created the variable for it. To display the map in the Notebook, you just have to call the variable. In the next empty cell, type in the following:

```
map1
```

Run the cell.

The results will look like the figure below:

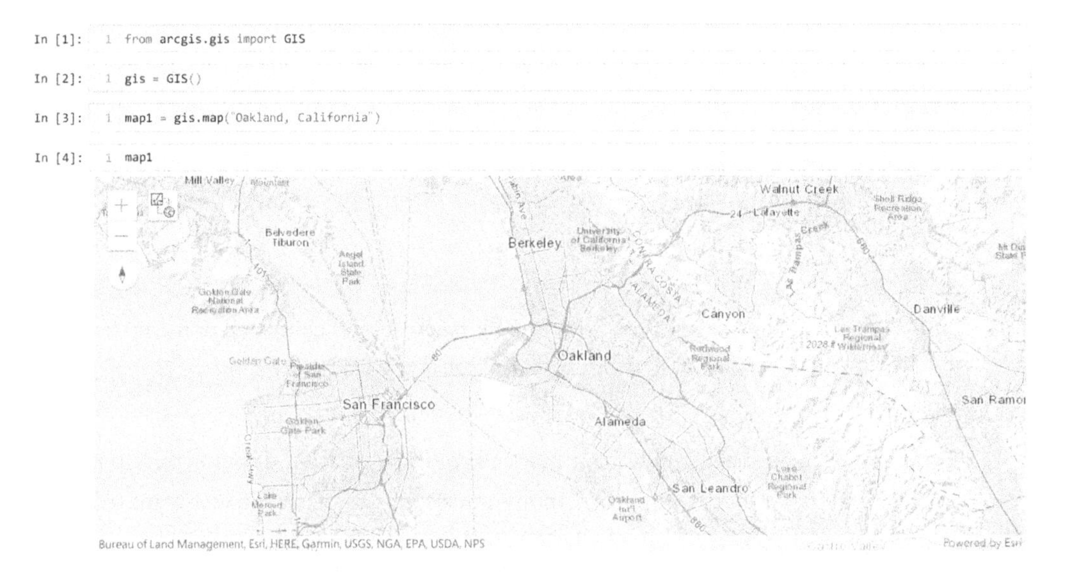

Figure 3.16: First Notebook results

Why is the variable map1 and not just map?

You cannot use `map` as a variable in Python, as `map` is reserved for the `map()` function. You will most commonly see `map1` or `m` used as a variable when creating a map in the ArcGIS API for Python.

You have just created your first map in a Notebook and used the ArcGIS API for Python. You created an anonymous connection to ArcGIS Online and displayed a map in your Notebook. The results show that you have a proper installation and virtual environment set up to work with the ArcGIS API for Python in an ArcGIS Notebook. In the next section, you will continue to explore the `gis` module and how it can be used to search for data.

Using the gis module to manage your GIS

Using the `gis` module, you can access and manage your folders, content, groups, and users. If you have any repetitive tasks and workflows, you can automate them as scripts. In this section, you are going to see how to search for data, access and manage groups, and access and manage users.

Searching for data, users, or groups

Creating a GIS object allows you access to many of the different classes and properties of the GIS object. To search for users, groups, or content, you will use the `UserManager`, `GroupManager`, or `ContentManager` classes through the `users`, `groups`, or `content` properties of the GIS object. This means that when searching for users, groups, or content, you will use similar syntax within the `search()` method. A search for data will use the `ContentManager` class as the `content` property of the GIS object, with the following syntax: `gis.content.search()`.

The `search()` method will return a list of items based on the arguments given. It has several arguments that it can take. The only argument that is mandatory is a query argument. In this section, you will look at how to query based on an item's title or owner.

Searching for public data as an anonymous user

In the previous example, you were connecting anonymously to ArcGIS Online. For this search example, you will still be connecting anonymously, as it is easier to find public data this way. You will see later how to find public data when connected to your organizational account; it takes more arguments.

Connecting anonymously in these examples also allows you to get the same data returned as we do here. If you were searching with your own organizational account, you would not see the same data.

You are going to search for publicly available feature layers for Oakland.

1. If you closed ArcGIS Pro after the last section, open it back up and open up the `Chapter3.aprx` file.
2. Right-click on the `Chapter3` folder and select **New** > **Notebook**. Rename the Notebook to `SearchForDataSample`.
3. You are going to create your GIS using an anonymous login, and import a `display` module that will make viewing the data returned easier. Type in the following:

   ```
   from arcgis.gis import GIS
   from IPython.display import display
   gis = GIS()
   ```

   Run the cell.

4. In the next cell, you will search for feature layers associated with Oakland, limited to just five items, and display the results:

   ```
   oaklandResults = gis.content.search(query="Oakland",
   item_type="Feature Layer",max_items=5)
   for item in oaklandResults:
       print(item)
   for result in oaklandResults:
       display(result)
   ```

   The two `for` loops will return the data details to you in two different ways. The first simply prints out the results, while the second uses the `display` module to show more details.

   Run the cell, and you should see something like the following:

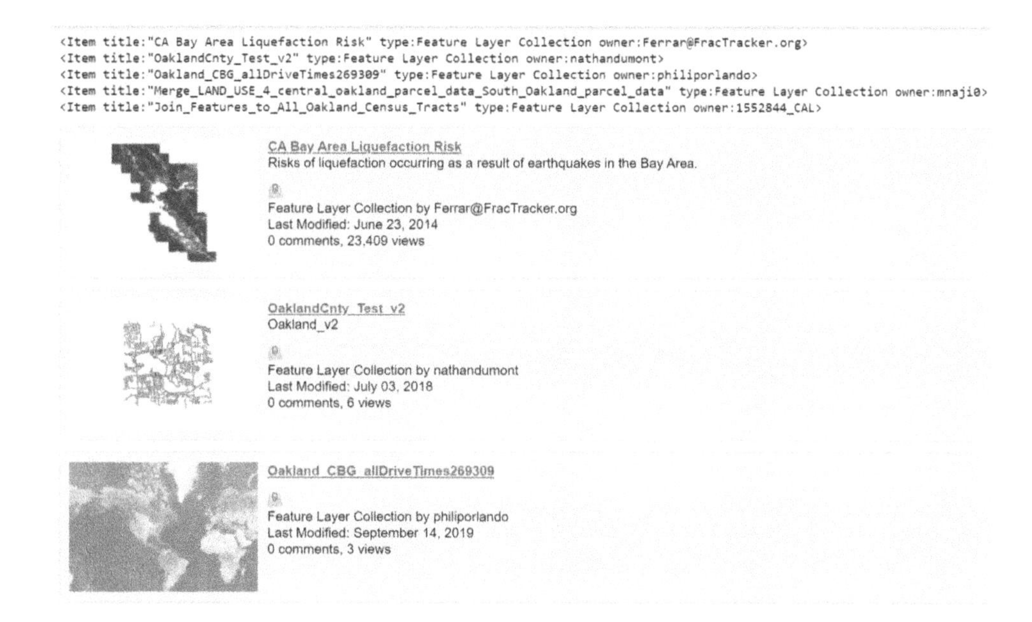

Figure 3.17: Results of the search for Oakland data

The results are just the first five feature layers returned that are associated with Oakland. Your results may vary, as content on ArcGIS Online changes over time.

There are many more arguments that can be used to find different data. You can apply any or all of the following arguments:

- `query`: This can be used to query for title or owner and can use wildcards.
- `item_type`: This can be used to query any type of item that can be on an ArcGIS Online portal. It can find shapefiles, feature layers, feature collections, CSVs, tables, maps, web scenes, and more. It also can take wildcards.
- `sort_field`: This can be used to sort the data on a field, such as the title, owner, or number of views.

- `sort_order`: This can be used with `sort_field` to sort in ascending or descending order.
- `outside_org`: This can be used when logged in to your organization to search for data outside of it.

The `outside_org` parameter is important to remember when you are logged in to your ArcGIS Online account and are searching for public data. If you do not set this to `True` when you search, you will only be searching within your ArcGIS Online account.

You are going to test some of these arguments to see how you can get different results from the `search()` method.

First, you will take the last search and modify it to find feature layers or collections that have Oakland in the title, sorting them by the number of views:

5. You are now searching for data with `Oakland` in the title and any item type that begins with `feature`. You are also sorting this by descending number of views to get the most viewed items, and returning only the first five of them. In the same Notebook, type in the following:

```
oaklandResults2 =
gis.content.search(query="title:Oakland",item_type="Feature *",
sort_field="numViews",sort_order="desc",max_items=5)
for item in oaklandResults2:
    print(item)
for result in oaklandResults2:
    display(result)
```

Run the cell. The output should look something like this:

```
<Item title:"Oakland City Limit Line" type:Feature Layer Collection owner:davidlok>
<Item title:"Oakland Transit Stations" type:Feature Layer Collection owner:kfong88>
<Item title:"Oakland_Demographics" type:Feature Layer Collection owner:antievictionmapdev>
<Item title:"Oakland" type:Feature Collection owner:achuman>
<Item title:"Oakland_UD_Eviction" type:Feature Layer Collection owner:antievictionmapdev>
```

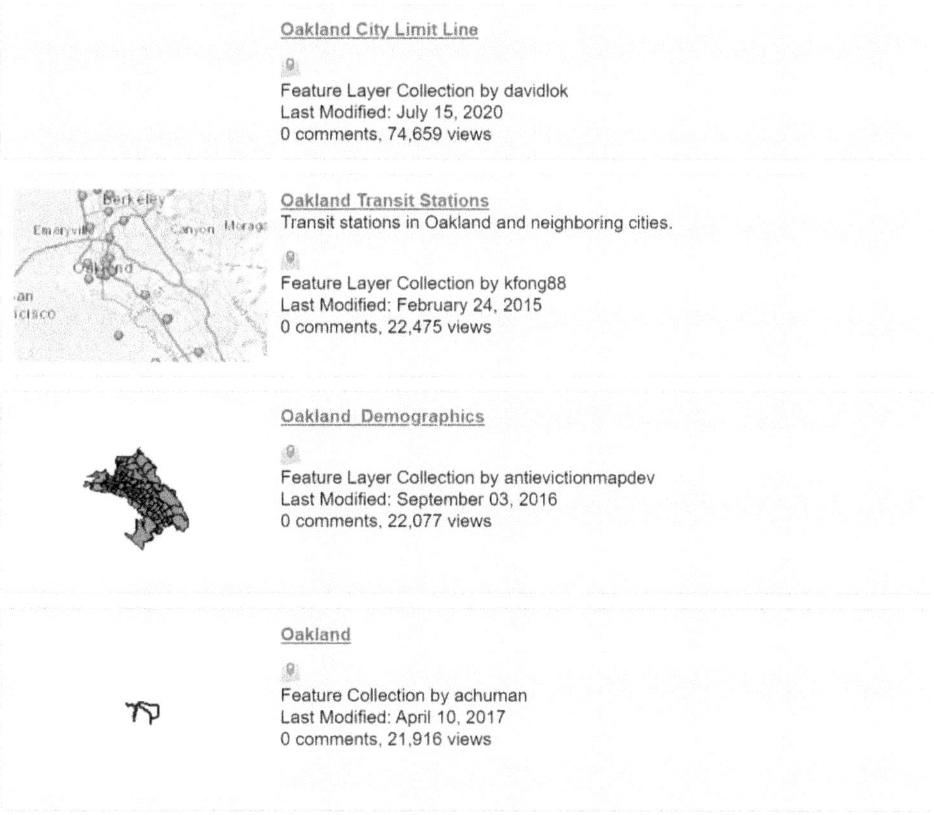
Oakland City Limit Line
Feature Layer Collection by davidlok
Last Modified: July 15, 2020
0 comments, 74,659 views
Berkeley
Emeryville
Canyon
Oakland
Oakland Transit Stations
Transit stations in Oakland and neighboring cities.
Feature Layer Collection by kfong88
Last Modified: February 24, 2015
0 comments, 22,475 views
Oakland Demographics
Feature Layer Collection by antievictionmapdev
Last Modified: September 03, 2016
0 comments, 22,077 views
Oakland
Feature Collection by achuman
Last Modified: April 10, 2017
0 comments, 21,916 views

Figure 3.18: Results of search query for Oakland with different search arguments

When using `title` for a query, you do not need to use wildcard characters. The search looks for any instance of the word "Oakland" in the title and returns those feature classes. This is not true for other queries; note that you had to use the wildcard character in the `item_type` query.

You can also search for data by the owner of the data. Your query argument is structured like this: `query="owner:username"`. This will only return data that the owner has made publicly available. You will see this now by using the owner of one of the datasets in the above example and searching for all data owned by them.

6. In the same Notebook in the next cell, type the following:

```
oaklandResults3 = gis.content.
search(query="owner:antievictionmapdev", item_type="Feature *")
print(len(oaklandResults3))
```

   Run the cell and see that they own 10 feature layers.

7. Now that you know there are 10 feature layers or feature collections, in the next cell you can type the following to display them all:

```
for result in oaklandResults3:
    display(result)
```

   Run the cell to see the layers be displayed:

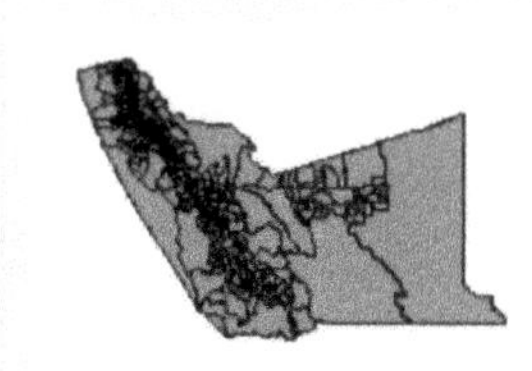
alameda co foreclosures non ud 2005 15
Feature Layer Collection by antievictionmapdev
Last Modified: August 23, 2016
0 comments, 51 views

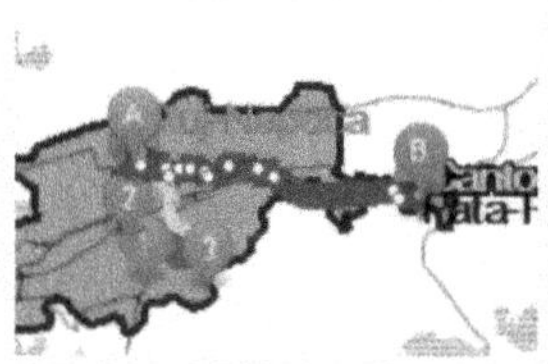
Ionica 4th
Route and directions for Ionica 4th
Feature Collection by antievictionmapdev
Last Modified: October 02, 2016
0 comments, 35 views

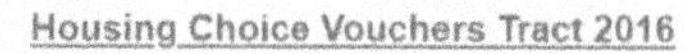
Housing Choice Vouchers Tract 2016
Feature Layer Collection by antievictionmapdev
Last Modified: June 14, 2017
0 comments, 195 views

Images
Feature Layer Collection by antievictionmapdev
Last Modified: June 27, 2018
0 comments, 2 views

transit routes bay area 2008
Feature Layer Collection by antievictionmapdev
Last Modified: August 17, 2016
0 comments, 54 views

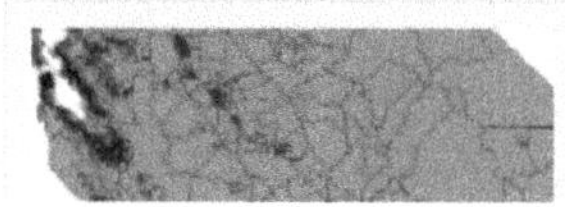
CalEnviroScreen
Feature Layer Collection by antievictionmapdev

Figure 3.19: List of layers from the owner query

You can click on the layer name in the **Output** cell and a browser will open, showing you the overview page for the item you clicked on.

You have seen how to use the `search()` method to search for data when anonymously connected to ArcGIS Online. Next, you will see how to search when connected to your organization.

Searching for data when connected to your organization

So far, you have seen how to use the `search()` operation to search for public data as an anonymous user. As you have seen, there are different ways to connect to your organization, depending on how you are using the ArcGIS API for Python, and whether you are using ArcGIS Online or ArcGIS Enterprise. Because you will be connected to your organization in this exercise, there will be a limited number of figures showing the output cells as these will depend on what data you have in your organization.

1. To search for data in your organization, you are going to continue to use the `SearchForDataSample` Notebook from the exercises above. If you have closed ArcGIS Pro, open it up and open up the `Chapter3.aprx` project.
2. When `Chapter3.aprx` is open, right-click on **SearchForDataSample** in the **Project** tab of the **Catalog** pane to open the Notebook.
3. Go to the bottom of the Notebook and, if you need to, create a new blank cell. In this cell, you are going to create another connection to ArcGIS Online using the account you are logged into in ArcGIS Pro, by typing in the following:

   ```
   gis2 = GIS('home')
   ```

   This will create a GIS object under `gis2` that you can use to access and manage content and users in your ArcGIS Online instance. If you have an ArcGIS Enterprise portal, you need to type the following:

   ```
   gis2 = GIS("https://portal/domain.com/
   webadapter","username","password")
   ```

   The address above is for your organization's ArcGIS Enterprise portal, and the username and password are your username and password for accessing that portal.

   Run the cell.

4. You can see properties of the user you are signed in under by typing the following:

```
gis2.properties.user
```

   Run the cell. The result will be a data dictionary containing all the information about the user. The data dictionary contains not just the username, full name, and email of the user, but also information about credits and privileges.

5. All of this data can be further accessed and assigned to variables if needed. To store the first name and then display it, type in:

```
firstName = gis2.properties.user.firstName
firstName
```

   Run the cell. The output will be the first name of the account you are logged in with.

6. You can also access your user information by using `users.me`. This will display the picture for the user, along with their full name, bio, username, and date joined. It is not displayed as a data dictionary, but as a display card. To see this, type in the following:

```
gis2.users.me
```

   Run the cell. The output will be a card like below, but with your user information displayed:

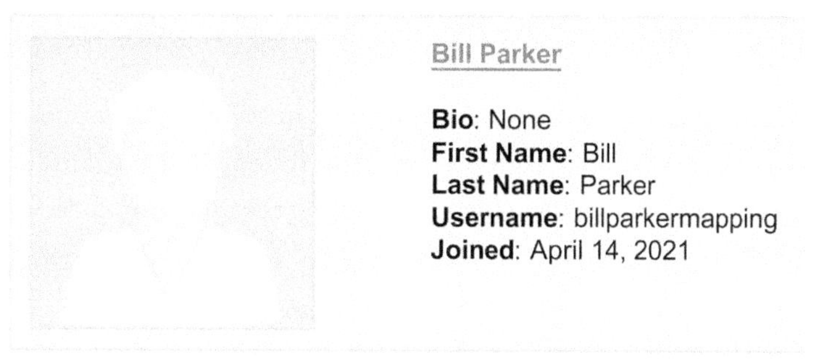

Figure 3.20: A users.me card

7. You can set the user information to a variable. You will create a variable to hold the username so that you can use that later for searching through your data. Type in the following:

```
myUsername = gis2.users.me.username
```

   Run the cell. There will be no output displayed, but you can now use the `myUsername` variable to search for data owned by you.

8. Searching through your content is the same as when you are logged in anonymously. The only difference is you are searching through data that is within your organization. In the next cell, type in the following:

```
searchResults = gis2.content.search(query="*",
item_type="Feature Layer")
for result in searchResults:
    display(result)
```

Run the cell. It will display all of the feature layers in your organization.

9. To search for just the items owned by you, type in the following:

```
searchResults = gis2.content.search(query="owner:"+myUsername, item_
type="Feature Layer")
for result in searchResults:
    display(result)
```

Run the cell.

The only argument required in the `search()` function is the query. Because wildcards can be used, you can search for everything by just writing `query="*"`. But be careful – if you have a lot of layers, the search can be slow.

10. When connected to your organization, you can still search for publicly available data by setting the `outside_org` argument to `True`. You can find the same Oakland datasets we saw in the previous section in `gis2` by writing the following code:

```
oaklandResultsHome =
gis2.content.search(query="title:Oakland",
item_type="Feature *",
sort_field="numViews", sort_order="desc",
max_items=5,outside_org=True)
for result in oaklandResultsHome:
    display(result)
```

Run the cell. The results should be the same as when connected anonymously:

Oakland City Limit Line

Feature Layer Collection by davidlok
Last Modified: July 15, 2020
0 comments, 74,659 views

Oakland Transit Stations
Transit stations in Oakland and neighboring cities.

Feature Layer Collection by kfong88
Last Modified: February 24, 2015
0 comments, 22,475 views

Oakland Demographics

Feature Layer Collection by antievictionmapdev
Last Modified: September 03, 2016
0 comments, 22,077 views

Oakland

Feature Collection by achuman
Last Modified: April 10, 2017
0 comments, 21,916 views

Oakland UD Eviction

Feature Layer Collection by antievictionmapdev
Last Modified: September 02, 2016
0 comments, 21,038 views

Figure 3.21: Results from searching outside your organization

In this section, you have seen how to search for data both anonymously and when you are connected to your organization. Now that you can find data, you will learn how to search for groups within your organization.

Searching for groups

Searching for groups is very similar to searching for data. You can search for groups that are open to all when you are logged in anonymously, or search for groups within your organization when logged in to your organization.

You are going to first search for groups anonymously and access the properties of the groups you found. Then, you will search for groups within your organization.

1. If you have closed ArcGIS Pro, open it up and open up the `Chapter3.aprx` project.
2. Right-click on the `Chapter3` folder and select **New** > **Notebook**. Rename the Notebook to `SearchForGroups`.
3. In the first cell, type in your import statements and create your GIS object. You are going to create the GIS object anonymously:

   ```
   from arcgis.gis import GIS
   from IPython.display import display
   gis = GIS()
   ```

   Run the cell.

4. In the next cell, you will create your search and then display the results. Just like with the feature layers, you are going to limit your data search to the first five records. You will also be using the `display` module to better show the group information. Type in the following:

   ```
   oaklandGroups = gis.groups.search('title:Oakland', max_groups=5)
   for group in oaklandGroups:
       display(group)
   ```

Run the cell. You should have results that look like the figure below:

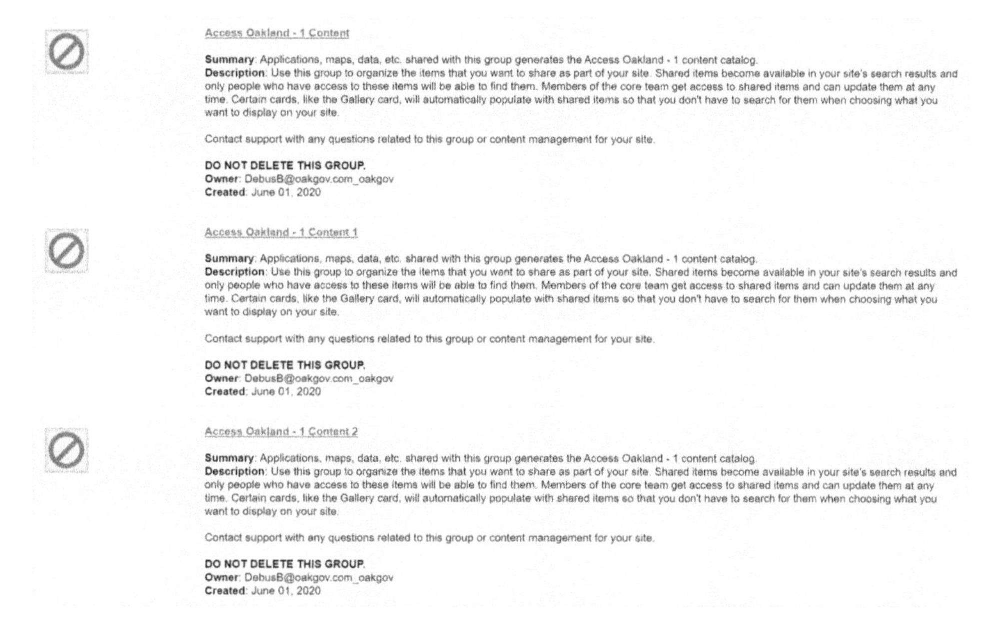

Figure 3.22: Results of group search for Oakland groups

5. Just like with items, you can search for groups by owner instead of title. You will use one of the group owners from the search results. Type in the following code:

```
oaklandGroups2 =
gis.groups.search('owner:DebusB@oakgov.com_oakgov', max_groups=5)
for group in oaklandGroups2:
    display(group)
```

Run the cell. You should see results that look like the figure below:

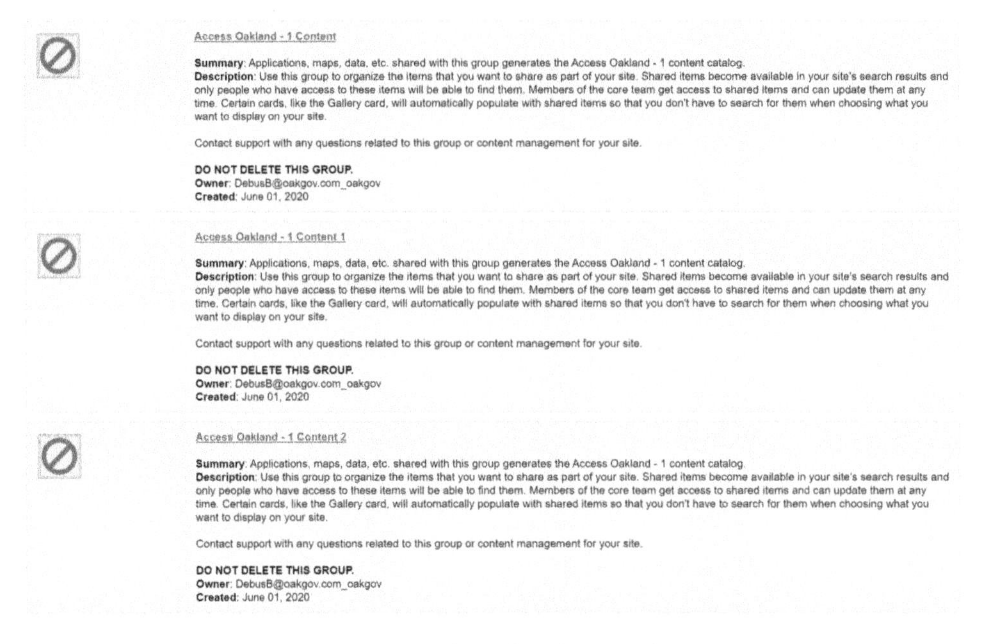

Figure 3.23: Results of searching for group by owner

6. Just like with searching for items, the group search returns a list. To look further at a group's properties, you need to select it using the list index. You are going to select the first group (index 0) from the first search to look at its properties. Type in the following code:

```
oaklandGroup1 = oaklandGroups[0]
oaklandGroup1
```

Run the cell. You should have results that look like the figure below:

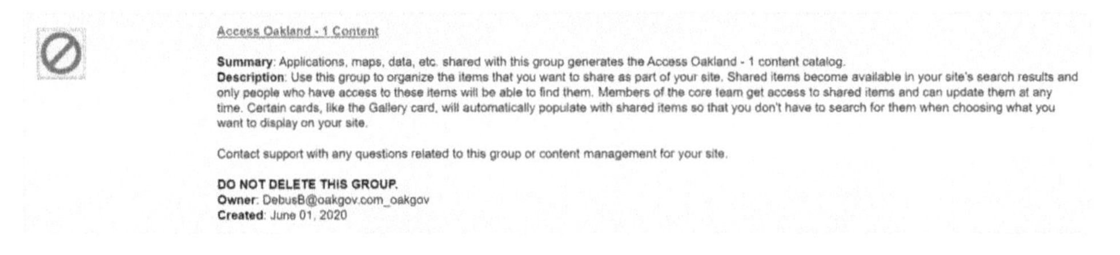

Figure 3.24: Result of selecting a group from the group list

7. Now you can see some the properties of the group. You are going to print out the property values, using `.format()` to add some context to them:

   ```
   print("Group Access is: {}".format(oaklandGroup1.access))
   print("Group id is: {}".format(oaklandGroup1.id))
   print("Group Tags are: {}".format(", ".join(oaklandGroup1.tags)))
   print("Group is Invitation only: {}".format(oaklandGroup1.
   isInvitationOnly))
   ```

   Run the cell. You should have the following results:

   ```
   Group Access is: public
   Group id is: ae0252ba6fd64ab8bb2b8e507d659c51
   Group Tags are: Hub Group, Hub Content Group, Hub Site Group
   Group is Invitation only: False
   ```

   There are more properties that you can access for a group. A full list of them is here: `https://developers.arcgis.com/rest/users-groups-and-items/group-search.htm`.

8. To search for groups within your organization, you need to be logged in to your GIS. Create a new GIS object in this workbook by typing the following in the next cell:

   ```
   gis2 = GIS('home')
   ```

   Run the cell.

9. In the next cell, you will run a search for all the groups in your organization that you have access to. Type in the following:

   ```
   myGroups = gis2.groups.search(query="*", max_groups=5)
   for group in myGroups:
       display(group)
   ```

   Run the cell. You should see at most five groups. If you are not a member of five groups, you will only see those that you are a member of. If you want to see all the groups you are a member of, remove `max_groups=5`.

You have now seen how to search for data and groups within and outside of your organization. Next, you will see how to manage users.

Managing users

Managing users in your organization through the ArcGIS API for Python can be a timesaver, as you can have Notebooks that can be run to quickly create new users, access user data, reassign user content, and delete users. The first step is understanding the `users` class to see what information you can see about users.

User properties

In order to learn more about the properties of users, you are going to look at yourself and explore the different user properties.

1. If you have closed ArcGIS Pro, open it up and open up the `Chapter3.aprx` project.
2. Right-click on the `Chapter3` folder and select **New** > **Notebook**. Rename the Notebook to `UserProperties`.
3. You are going to log in to your organization's GIS through the user you are currently logged in to ArcGIS Pro with, by typing in the following:

```
from arcgis.gis import GIS
from IPython.display import display
gis = GIS('home')
```

   Run the cell.

4. In the next cell, you are going to view your own account by using the `me` property, as you did in *step 6* in the *Searching for data when connected to your organization* exercise. This time, you will view the different properties for a user in the steps below. Type in the following:

```
me = gis.users.me
me
```

   Run the cell. You should see the same output as *Figure 3.19*.

5. You can identify many different aspects of a user's profile, such as the user's first and last name, email address, when they last accessed their account, what groups they are a member of, and how much storage they are using. You are going to extract and write out all of this information. You will need to import the `time` module to convert the time returned to month/date/year format.

   The `groups` property returns a list of all the groups with each group's information stored in a data dictionary. The group's name is stored in the `"title"` key of each group's data dictionary. To access the group name, you will create an empty list.

Then, you'll iterate through the groups list and access the "title" key of each data dictionary to get the group name. You will append() it to the list you created for storing group names. Finally, you will use the join() function to write the contents of the list of the groups to a string. In the next cell, type the following:

```
import time
firstName = me.firstName
lastName = me.lastName
email = me.email
accessedLast = time.localtime(me.lastLogin/1000)
groups = me.groups
myGroupsList =[]
for group in groups:
    groupName = group["title"]
    myGroupsList.append(groupName)
groupsName = ", ".join(myGroupsList)
storageAssigned = me.storageQuota
storageUsed = me.storageUsage
prctStorage = round((storageUsed/storageAssigned)*100,4)

print("First Name:             {0}".format(firstName))
print("Last Name:              {0}".format(lastName))
print("email:                  {0}".format(email))
print("Last Accessed:
{0}/{1}/{2}".format(accessedLast[1],accessedLast[2],accessedLast[0]))
print("Groups:                 {0}".format(groupsName))
print("Storage Assigned:       {0}.".format(storageAssigned))
print("Storage Used:           {0}.".format(storageUsed))
print("Percent Storage Used: {0}".format(prctStorage))
```

In a similar manner to before, you print out the user information, using format() to help you.

Run the cell. You should see the following returned, but with your name and user information:

```
First Name:            Bill
Last Name:             Parker
email:
Last Accessed:         10/24/2021
Groups:                Census Demographic Data, Alameda County
Farmers Markets, City of Oakland Buses And Parks
Storage Assigned:      2199023255552.
Storage Used:          5827358.
Percent Storage Used: 0.0003
```

Why divide lastLogin by 1000, and what is round() doing?

You divide `lastLogin` by 1000 because the time returned is the time from the beginning of the *epoch* in milliseconds. On Windows and most Unix systems, the epoch begins on January 1, 1970. When you divide it by 1000, you get seconds. The `time.localtime()` function will convert the seconds since the start of the epoch to the year, month, day, hour, minute, second, and day of the week.

`round()` is there to round to a number of digits after the decimal point. In this case, you are using 4 in the second argument to round the data to four digits after the decimal point.

You have just been searching for information about yourself. If you have administrator privileges in your organization, you can search and display this information for the other users in your organization.

Searching for users

You can search for users just like you would with items or groups. You can set a query to look for a user by username or find users by email address. In this exercise, you will set up a Notebook with examples of both that you can use to search for users within your organization.

1. If you have closed ArcGIS Pro, open it up and open up the `Chapter3.aprx` project.
2. Right-click on the `Chapter3` folder and select **New > Notebook**. Rename the Notebook to `SearchForUsers`.
3. You are going to log in to your organization's GIS through the user you are currently logged in to ArcGIS Pro with, by typing in the following:

```
from arcgis.gis import GIS
from IPython.display import display
gis = GIS('home')
```

Run the cell.

First, you will search for users by username. The search for users works just like all the previous searches, in that it returns a list of values. Since you are searching for a specific username, you should get back a list of one item. To be sure of this, you will run a test to print out the length of the list. Type in the following, replacing `{userName}` with the username you want to search for:

```
userNameSearch = gis.users.search(query="username:{userName}")
len(userNameSearch)
```

Run the cell. You should see a `1` returned, as you have created a list of users containing just one user.

4. To access the user returned, you need to use the list index to extract the first user, then display those results. Type in the following:

```
userNameSelect = userNameSearch[0]
userNameSelect
```

Run the cell. You should see something like the following:

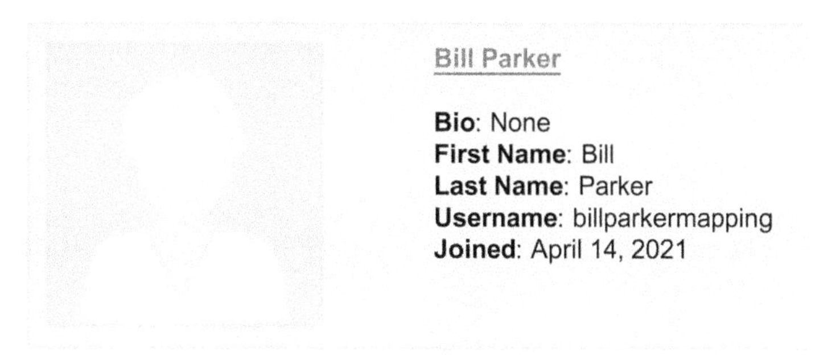

Figure 3.25: Result of selecting a single user

5. You can also search by email using the wildcard *. This allows you to search for all email addresses from the same email provider. The code is the same as for searching by username, except for the query. In this example, you are again going to find the length of the list returned to you, before extracting a user from it.

Type in the following in the next cell, replacing `{@email.com}` with your own email provider:

```
emailSearch = gis.users.search(query="email: *{@email.com}")
len(emailSearch)
```

Run the cell.

Depending on how many people in your organization have that email host, you may have a large number of search results. You are going to iterate through them using a `for` loop and print out the user information from the last Notebook. Type in the following:

```
import time
for user in emailSearch:
    firstName = user.firstName
    lastName = user.lastName
    email = user.email
    accessedLast = time.localtime(user.lastLogin/1000)
    groups = user.groups
    myGroupsList =[]
    for group in groups:
        groupName = group["title"]
        myGroupsList.append(groupName)
    groupsName = ", ".join(myGroupsList)
    storageAssigned = user.storageQuota
    storageUsed = user.storageUsage
    prctStorage = round((storageUsed/storageAssigned)*100,4)

    print("-------------------------------------------")
    print("First Name:              {0}".format(firstName))
    print("Last Name:               {0}".format(lastName))
    print("email:                   {0}".format(email))
    print("Last Accessed:
{0}/{1}/{2}".format(accessedLast[1],accessedLast[2],accessedLast[0]))
    print("Groups:                  {0}".format(groupsName))
    print("Storage Assigned:        {0}.".format(storageAssigned))
    print("Storage Used:            {0}.".format(storageUsed))
    print("Percent Storage Used: {0}".format(prctStorage))
```

Run the cell. It will print out the user information for each user in that search separated by `----------`, with each user looking like this:

```
--------------------------------------------
First Name:            Bill
Last Name:             Parker
email:
Last Accessed:         10/24/2021
Groups:                Census Demographic Data, Alameda County
Farmers Markets, City of Oakland Buses And Parks
Storage Assigned:      2199023255552.
Storage Used:          5827358.
Percent Storage Used: 0.0003
```

You have seen how to search for users within your organization and print out user information. The Notebook created here can help you easily identify the usage levels for each user to help you manage access and credits.

Summary

In this chapter, we showed you how to set up a virtual environment, introduced ArcGIS Pro Notebooks, and the ArcGIS API for Python. You learned how to add additional modules to your virtual environment to allow you to extend your Python analyses. ArcGIS Pro Notebooks were used to begin exploring the ArcGIS API for Python. You created some sample Notebooks that you can use to search for content inside and outside of your organization and groups within your organization, as well as searching for and displaying user information.

In *Chapter 5, Publishing to ArcGIS Online,* you will learn more about the ArcGIS API for Python and see how to add, move, and share data to your ArcGIS Online account. In the next chapter, we will return to ArcPy and explore the Data Access module.

Part II

Applying Python Modules to Common GIS Tasks

4

The Data Access Module and Cursors

The Data Access module is for working with data. You have already seen some functions from the Data Access module (the `Describe` function from *Chapter 2, Basics of ArcPy*) that assist you in finding different data types. In addition to that, the Data Access module can be used to walk through directories to find data; it contains cursors to assist in finding and updating data and an editor class that allows you to edit data on enterprise systems when other users are accessing the data.

In this chapter, you will learn how to walk through a directory to extract all the ZIP files present and move them to an organized geodatabase structure. You will also create a Notebook using cursors that will insert census demographic data into a census geography feature class.

This chapter will cover:

- `arcpy.da.Walk` to walk through a directory and find data
- **Search, insert**, and **update cursors** to search, write, and update data

To complete the exercises in this chapter, you'll need data from the `Chapter4` folder in the GitHub repository for this book: `https://github.com/PacktPublishing/Python-for-ArcGIS-Pro/tree/main/Chapter4`.

Walking through a directory to find data

So far, you have been working with single datasets and using ArcPy to do many things that you could do as single tools. The benefit of ArcPy has been in helping you track your tasks and in being able to use Notebooks, which make sharing your analysis easy. However, what about when you have a lot of datasets and need to search through them and organize them, or do analysis on them? This is where the Data Access module comes in.

The first thing to look at in the Data Access module is the `walk` function, which will allow you to walk through directories.

arcpy.da.Walk

The Python `os` module is one that you have seen in previous chapters. You have used `os.path.join` to create a full path of a file from the directory and filename. It also has a `walk()` function that will walk through a directory tree and find data. This means that you can run it on a folder and be able to walk through all the data, not just within that folder but within the subfolders as well.

The problem with `os.walk()` is that it does not recognize database contents. That means it will not find feature classes, tables, or rasters within geodatabases. This is where `arcpy.da.Walk` comes in. It can see the data within databases. This makes it very useful, as you can use it to find data within a geodatabase that is within a folder or subfolder.

In the following exercise, you are going to use `arcpy.da.Walk` to walk through a folder that contains multiple unzipped folders and copy the data from each folder to the correct geodatabase for that data.

arcpy.da.Walk exercise

The US Census Bureau TIGER program creates many useful shapefiles of different geographies throughout the US. They are updated each year and have polygon geographies from the state level all the way down to block groups. They also have line data such as roads and railroads, and point data such as landmarks. In this exercise, you will write code in a Notebook that will use `os.walk` to walk through the download folder and unzip each ZIP file. Then, it will use `arcpy.da.Walk` to walk through the folders with the unzipped shapefiles and copy each shapefile to the correct geodatabase.

Unzipping files using os.walk

For this exercise, you should have downloaded the `Chapter4CensusDownloads` folder from the GitHub repository. Take a look inside and see that there are 26 different ZIP files. They are a collection of different data types from different state and national files. You could manually unzip each one, manually create the needed geodatabases, and then manually import each shapefile into the correct geodatabase. However, ArcPy and the Data Access module can do that for us:

1. Create a new Notebook in the `Chapter4` project folder by right-clicking on the `Chapter4` project folder and selecting **New Notebook**.
2. Rename the Notebook `ExtractAndCopyCensusData`.
3. Insert a heading cell by clicking **Insert** > **Insert Heading Above**.
4. Type in the following for the heading:

   ```
   # Extract Multiple Zip Files and Move to Geodatabases
   ```

   Run the cell.

5. Since this is an ArcGIS Notebook, you do not need to import `arcpy`; it is already loaded. You will be using the os module to start with, so you need to import it. In the next cell, type in the following:

   ```
   import os
   ```

6. You also need to import the **ZipFile** module. After `import os`, press *Enter* to get a new line, and type in the following:

   ```
   from zipfile import ZipFile
   ```

   Run the cell. You now have access to the `os` module and the `ZipFile` module, which will allow you to create the walk to find and extract the ZIP files.

7. You are going to create a variable for the workspace that contains the ZIP files, and another one where you will create the geodatabases to store the data. Type in the following:

   ```
   zipWksp = r"C:\PythonBook\Chapter4\Chapter4CensusDownloads"
   gdbWksp = r"C:\PythonBook\Chapter4\Chapter4"
   ```

   Run the cell.

8. Next, you need to create the walk object to walk through the downloads folder with all of the ZIP files. In the next cell, type in the following:

   ```
   zipWalk = os.walk(zipWksp)
   ```

   Run the cell.

9. Now that you have the walk created, you can use a `for` loop to walk through the directory names, directory paths, and filenames within that workspace. Within the first loop, you will create another loop to find each filename. Within that loop, you will create a conditional to test for filenames that end in `.zip`. For those that do, you will create a new path name for the extracted folder that is the same as the ZIP folder name without the `.zip` part. You will use `os.path.isdir()` to test whether the path exists; if it does not, you will use `os.mkdir()` to create the new folder. You will then use the `ZipFile` module to open the ZIP file in read mode, and extract all to the newly-created directory. To do this, in the next cell, type in the following:

   ```
   for dirpath, dirnames, filenames in zipWalk:
       for filename in filenames:
           if filename[-4:] == ".zip":
               path = os.path.join(dirpath,filename[:-4])
               if os.path.isdir(path) == False:
                   os.mkdir(path)
               with ZipFile(os.path.join(dirpath,filename),"r")
   as zipObj:
                   zipObj.extractall(path)
   ```

   Run the cell. Since you haven't put any `print` statements in the code, the only way to see it is running is to watch the * in the cell number. You can also look at the folder where the new folders containing the unzipped data are being created, to see the folders created and the contents being unzipped.

When it is done, you should have all 26 files unzipped. Now we can move on to using the `arcpy.da.Walk` function to walk through them, create geodatabases, and import the shapefiles into them as feature classes.

Copying shapefiles to feature classes using arcpy.da.Walk

Taking a closer look at all the census folders that were unzipped, you see that each folder has a single shapefile in it. The folders and shapefiles all have a similar name structure of `tl_YYYY_XX_Name`.

That uniformity is something you can use to create geodatabases for data and move the correct data to the correct geodatabase. In this exercise, you will continue in the same Notebook as above, to move the data for each state or country into a geodatabase for that state or country.

This can be done because the two-digit number or two letters after the year tell you what state the data is for, or if it is country-wide data. For example, `tl_2019_06_tract` is tract data for the entire state with a **Federal Information Processing Standard (FIPS)** code of 06. There are many sites that have the state FIPS code lookup, but I prefer this one from the United States Department of Agriculture (USDA) Natural Resource Conservation Service (NRCS): `https://www.nrcs.usda.gov/wps/portal/nrcs/detail/?cid=nrcs143_013696`. It gives a list of states, their zip code, and their FIPS code.

With this, continue on in the `ExtractAndCopyCensusData` Notebook from above:

1. The first step is creating a data dictionary to look up the different FIPS codes and find the state they are associated with. This will allow you to create geodatabases with state names instead of FIPS codes, which will be much more useful as not everyone knows every state's FIPS code. The data dictionary will be a simple key/value pair of strings, with the key being the FIPS code and the value being the state name. You have the following FIPS codes downloaded: `04`, `06`, `16`, `32`, `41`, `53`, `56`, `us`. These correspond to the following states: Arizona, California, Idaho, Nevada, Oregon, Washington, Wyoming, and United States (US datasets cover the entire country). Click in the code box for the next cell and type in the following code:

```
stateCountry_dict = {
    "04" : "Arizona",
    "06" : "California",
    "16" : "Idaho",
    "32" : "Nevada",
    "41" : "Oregon",
    "53" : "Washington",
    "56" : "Wyoming",
    "us" : "US_Full",
}
```

Run the cell. The data dictionary can now be used later in your Notebook by calling it.

2. Now create your `arcpy.da.Walk` by typing in the following in the next cell:

```
shpWalk = arcpy.da.Walk(zipWksp, datatype = "FeatureClass")
```

Run the cell. Note that the `arcpy.da.Walk()` function takes one mandatory parameter: the workspace. It also has the following five optional parameters:

- `topdown`: A Boolean with `True` being the default. When it is set to `True`, the tuple generated for the directory is generated before the workspace. Much of the time, you will leave this set to `True`.
- `onerror`: A function that can be set to report any errors. It is set to `None` by default and errors are ignored. Much of the time, you will leave this as `None`.
- `followlinks`: A Boolean with `False` as the default. When set to `False`, the walk will not walk into connection links. A **connection link** is the same as a symbolic link: a file that contains a reference to another file or directory. Much of the time, you will leave this set to `False`.
- `datatype`: A string of the data type to limit the returns to. This can be set to a number of different data types to ensure you are only finding the type you want. The default is `Any`, which will find all the data types in each directory. For this task, you are setting it to `FeatureClass` as you only want to find the shapefiles. It can be set to select only data types like tables, rasters, tools, maps, and many more. Multiple data types are allowed in a list or tuple.
- `type`: A string that can be used to further select within rasters or feature classes. This allows you to limit your feature classes to point, polyline, polygon, multipatch, or multipoint. Raster data can be limited to different raster types. Multiple types are allowed in a list or tuple.

Why is datatype = written as a parameter in the function argument?

The `datatype` is the fourth parameter, but instead of writing `arcpy.da.Walk(zipWksp, "", "", "", "FeatureClass")` to get to it, you can just type the parameter name and what it is equal to. Both ways of writing the code are correct.

3. In the next cell, you will walk through the walk to get each shapefile. You will create a variable that is just the shapefile name without the `.shp` part, and extract the FIPS code from that by using the `split()` method and list index locations.

The FIPS code will be used as a key to look up the state in the `stateCountry_dict` data dictionary. Once you have the state or country name, you will use that to determine whether there is a geodatabase for it, create it if not, then copy the feature class to the geodatabase. You will include print statements throughout to track your progress. Write the following code into the cell:

```
for dirpath, dirnames, filenames in shpWalk:
    for filename in filenames:
        fcName = filename[:-4]
        censusType = filename.split("_")[2]
        fileFullPath = os.path.join(dirpath,filename)
        stateCountry = stateCountry_dict[censusType]
        print(fileFullPath)
        print(censusType)
        print(stateCountry)
        gdb = os.path.join(gdbWksp,stateCountry+".gdb")
        if arcpy.Exists(gdb) == False:
            arcpy.management.CreateFileGDB(gdbWksp,stateCountry)
        fcFullPath = os.path.join(gdb,fcName)
        arcpy.management.CopyFeatures(fileFullPath,fcFullPath)
        print(fcFullPath + " was copied to " + gdb)
```

Print statements can be very valuable in error checking. They allow you to see what your code is actually doing. If you are getting results you don't understand or errors you don't understand, putting in print statements that return your variables can help you see what the error might be.

Run the cell. The code should run for a bit, as there are 26 shapefiles that need to be copied to feature classes in the new geodatabases. Since you have `print` statements written in, you can watch them run. You can also see the different geodatabases being created in the **Catalog** pane and the feature classes written to them.

Why do you check to see if something exists before creating it?

It is always a good idea to check and see if something exists before creating it. If you do not, your code could throw an error. Or worse, it will work and overwrite what you are creating. In this case, if it did that, you would end up with a geodatabase that only has one feature class in it.

In this section, you have learned how to use both the `os.walk()` and `arcpy.da.Walk()` functions to walk through folders looking for different data types. The `os.walk()` function was used to find the ZIP folder and the `arcpy.da.Walk()` function was used to find shapefiles. You also learned how to use a data dictionary as a lookup tool to find and decode FIPS codes. The end product is a Notebook that will find all the ZIP files in a directory and its subdirectories, extract them, and then copy any feature classes/shapefiles to a geodatabase that corresponds to the name of that census geography. You were able to extract and copy 26 unorganized shapefiles to organized geodatabases through using ArcPy.

Cursors

A **cursor** is used to access data. It is an object created using the Data Access module and can be used to iterate over the rows in a table or insert data into the table. Cursors created in the Data Access module have access to a feature class's geometry and can read, write, and update geometries. There are three different cursors in the data access module: **search**, **insert**, and **update**. In this section, you will see examples of each and how they can be applied to automate your workflows.

Search cursor

`arcpy.da.SearchCursor` will search through your feature class, shapefile, or table line by line and return the data to you. It can return the shape and attribute data, but the data is returned as a tuple so it is immutable. The search cursor has the following two required parameters:

- `in_table`: The feature class, layer, table, or table view to be searched.
- `field_names`: A list or tuple of the field names to be returned. When using a single field, you can use a string instead of a list; when needing all the fields, you can use an * instead of listing them all.

It has the following five optional parameters:

- `where_clause`: A SQL statement used to query the records returned and limit them.
- `spatial_reference`: This will transform the spatial reference of the input feature class to this spatial reference.
- `explode_to_points`: This will transform a feature into individual points or vertices. Each point or vertex will be returned to the cursor object as an individual row.
- `sql_clause`: A tuple of SQL prefix and postfix clauses to organize the records returned. Prefix clauses can be `None` (default), `DISTINCT`, or `TOP`. Postfix clauses can be `None` (default), `ORDER BY`, or `GROUP BY`.

For more information on how to use the `sql_clause` parameter, visit the following page: `https://pro.arcgis.com/en/pro-app/latest/arcpy/data-access/searchcursor-class.htm`.

- `datum_transformation`: A string used with the `spatial_reference` parameter for when the projection from one reference to another requires a datum transformation.

For more information on this, visit the following page: `https://pro.arcgis.com/en/pro-app/latest/help/mapping/properties/geographic-coordinate- system-transformation.htm`.

When a cursor is created, it returns a **cursor object**. This object can be iterated over using a `for` loop. In each loop, the cursor returns a single row from the dataset. The data returned can be accessed using list indexing. The first item in the tuple is 0, the next is 1, and so on. The order of the items returned corresponds to the order in the field list.

The fields passed into the search cursor must match the name of those in the attribute table, not the aliases.

In addition to the attribute table data, `arcpy.da.SearchCursor` has access to the shape field in the form of **geometry tokens**. A geometry token is a shortcut that allows access to specific properties of the geometry. They are a time-saving option compared to accessing the full geometry when you only need specific properties of the geometry. A full list of the geometry tokens available can be found here: `https://pro.arcgis.com/en/pro-app/latest/arcpy/get-started/reading-geometries.htm`. Most of the time, you will use the following, as they are the most common:

- `SHAPE@XY`: Returns a tuple of the feature's centroid *(x, y)* coordinates
- `SHAPE@X`: Returns a double of the feature's *x* coordinate
- `SHAPE@Y`: Returns a double of the feature's *y* coordinate

Let's take a look at a simple way of using the search cursor to access the attribute data of a dataset and create a list of the unique values that we can use later for analysis.

Accessing the geometry of a feature class

In this exercise, you will create a CSV of all of the transbay bus stops and the *(x, y)* coordinates of each. This data could then be uploaded to ArcGIS Online to display the data or used for further spatial analysis. You are simplifying the data from the shapefile and extracting the geometries of it for use by other analysts. By putting it in a CSV file, you are using a format that people without access to ArcGIS can also utilize.

A search cursor will work well for this, as it can access the attributes to create a list of transbay bus routes in the feature class. An additional search cursor will then be used with a SQL query to select the different stops for a bus route and export the information, along with the XY data of the stops, to a CSV for each bus route.

1. In the `Chapter4` project folder, right-click and select **New > Notebook**. Rename the Notebook to `AC_TransitTransbayStops`.
2. You will import two modules, `os` and `csv`. You have used the `os` module before to create paths and directories. The `csv` module will allow you to open, read, write, and append to CSV files. In the first cell, type in:

   ```
   import os, csv
   ```

   Run the cell.
3. To output to a CSV, you are going to use the `csv.writer()` and `writerow()` functions in the `csv` module. Instead of calling them multiple times when you need them, it is more efficient to create a function so you can call it whenever you have data to write to a CSV.

   You will create a function called `createCSV` with three parameters: the input data, the name of the CSV, and the mode set to write (`'w'`). Within the function, you will open the CSV in a `with` statement, pass in the mode from the parameter, and set the `newline` parameter to `''`.

   The newline parameter has to be set to `''` or your CSV output will have a blank line between each line.

   In the next cell, type in the following:

```
def createCSV(data, csvName, mode ='w'):
    with open(csvName, mode, newline='') as csvfile:
        csvwriter = csv.writer(csvfile)
        csvwriter.writerow(data)
```

Run the cell. This function allows you to write one line of code when you need to write data to a CSV.

Functions are very useful when you have code that you will need multiple times throughout a script. They are usually written at the top of a script so that you can call them later. In Python, all the functions you write will be named functions, meaning that you give them a name to call them later. To declare a function, you use `def FunctionName(... , ...)` and you put your parameters in the parentheses. These parameters are the values you will pass through to the function when you run it later.

4. In the `Chapter4` folder, find the `UniqueStops_Summer21.shp` shapefile and add it to your map. Create a variable in your Notebook by typing in:

   ```
   AC_TransitStops = r"C:\PythonBook\Chapter4\UniqueStops_Summer21.shp"
   ```

5. Open up the attribute table and look at the **ROUTE** field. This is how you are going to identify which routes are transbay routes. For AC Transit, all routes that start with a letter are transbay lines. You can use this knowledge and how the table is set up to create your query for each transbay line. The problem is that a stop can have multiple lines and the letter you are looking for will not always be the first letter. This means that you need to be careful about structuring your query, and that you want to set up your script to be able to go through each bus line individually.

   By looking at the current list of AC Transit transbay routes (`https://www.actransit.org/maps-schedules#transbay`), you can see that you will need to look for the following bus lines: F, G, J, L, LA, NL, NX, O, P, U, V, W. You want to create a list of these that you can iterate through. In the same cell as the previous step, type in:

   ```
   transbayRoutes =
   ["F","G","J","L","LA","NL","NX","O","P","U","V","W"]
   ```

6. You need headers for your CSV file so you know what the data is that you are extracting. You are going to extract the `511 Stop ID`, `Stop Description`, `Route`, `X`, and `Y` from each point. Since the headers are going to be written to a CSV file, you want to put them in a list so they will be comma-separated. In the same cell as the previous step, type in:

```
csvHeader = ["511 Stop ID","Stop Description","Route","X","Y"]
```

7. The last variable you need is a folder to write all of the CSVs to. You will be using the name of the bus route for each CSV name, so you just need a location to put the CSVs. A folder by the name `TransbayStops` needs to already exist in this location for you to write your CSVs to it. In the same cell as above, type:

```
csvFolder = r"C:\PythonBook\Chapter4\Chapter4\TransbayStops"
```

Run the cell.

8. Now you are going to create a `for` loop to loop through that list. Within that loop, you will create the SQL query for each bus route, create your search cursor to find, and write the information about each stop that you want. **This code will be written into the next cell over the next two steps.**

In this step, you will loop through the routes in the `transbayRoutes` list and create a SQL statement to select any stop that contains that route. You will create a CSV name, the full path for the CSV, and then call the `createCSV` function to write the headers to that CSV. Within the loop, you will also write some print statements to track your output. In the next cell, type in the following:

```
for route in transbayRoutes:
    sql = '"ROUTE" LIKE \'%{0}%\''.format(route)
    print(sql)
    csvName = "TransbayStopsRoute_{0}.csv".format(route)
    print(csvName)
    csvFullPath = os.path.join(csvFolder,csvName)
    print(csvFullPath)
    createCSV(csvHeader,csvFullPath)
```

What is the {0} and the .format() in the sql variable?

The `.format` method allows you to insert data from a variable into a string. The `{}` is a placeholder within your string, and the variable within the `()` is inserted into the string in that location. You can include as many placeholders and parameters as you need. The order in the `()` is 0-based, meaning the first parameter is 0, the next is 1, and so on.

9. You have now created a SQL statement that will find all of the transbay stops for a particular route, and you also have a CSV file ready for the output. You are ready to create a search cursor.

   The search cursor will use a `with...as...` statement. The advantage of using this is that it closes and deletes the cursor each time through, so you don't have to remember to delete the cursor at the end of your process. This helps reduce accidental schema locks on your data. The search cursor will take the bus stops feature class and the list of fields as the required parameters. It will take the `sql` variable as the `where_clause` parameter to limit the records returned. You will create a `for` loop to iterate through each row returned in the cursor object. This creates a list of data for each row and you will extract the data from the list using list indexing on the row variable. You will create a list of this data in the order of your headers from above, print this to the **Out** cell to track your results, and write it to the CSV using the `createCSV` function. You will change the mode to `'a'` in the `createCSV` function to append each row to the CSV. In the same cell as above, type in the following:

```
with arcpy.da.SearchCursor(AC_TransitStops,
["STP_511_ID","STP_DESCRI","ROUTE","SHAPE@XY"],sql) as cursor:
        for row in cursor:
            stopID = row[0]
            stopDesc = row[1]
            route = row[2]
            locX = row[3][0]
            locY = row[3][1]
            csvData = [stopID,stopDesc,route,locX,locY]
            print(csvData)
            createCSV(csvData,csvFullPath,mode='a')
```

The SHAPE@XY data returns a tuple of the *x* and *y* value. In order to extract the *x* value, you need to get the value at the first position of the tuple. The *y* value is in the second position.

Run the cell.

When the Notebook runs, you should see print statements telling you the route that is being run, the CSV that is being created, and each row being written to the CSV. When the first one is done, you can open it up and see what the data looks like. Navigate to the folder you created them in and double-click on the CSV. If you have Excel installed, it will default to opening them up in Excel. If not, you can open them in Notepad or any other text editor:

	A	B	C	D	E
1	511 Stop ID	Stop Description	Route	X	Y
2	52252	Adeline St & Alcatraz Av	12 F	-122.2710909	37.8491153
3	52525	Adeline St & Alcatraz Av	12 F	-122.2717477	37.8484501
4	53327	Adeline St & Ashby Av	F	-122.2686188	37.855424

Figure 4.1: Sample of output data to CSV opened in Excel

```
511 Stop ID,Stop Description,Route,X,Y
52252,Adeline St & Alcatraz Av,12 F,-122.2710909,37.8491153
52525,Adeline St & Alcatraz Av,12 F,-122.2717477,37.8484501
53327,Adeline St & Ashby Av,F,-122.2686188,37.8554239999999
```

Figure 4.2: Sample of output data to CSV opened in Notepad

Using a search cursor with a data dictionary as a lookup value

In addition to accessing the shape of a feature class or shapefile, search cursors can be useful for creating data dictionaries for lookup values for later use in code. This is important when you don't know how many different values a table may have. It can also save you from making errors in attempting to create a lookup dictionary yourself. Instead of you typing in the values, Python extracts them from the feature class or table.

Let's consider an example. You want to be able to extract just the census tracts for a county and you know that the tract FIPS code contains the county FIPS code within it. However, you don't know what the FIPS code is for any counties in the state. You do have the county feature class for the state and the tract feature class for the state. You could do this using the intersect tool, select by location, or possibly spatial join tools.

However, those can cause slivers, and it would be quicker just to select the tracts by the FIPS code. `arcpy.da.SearchCursor` can help by creating a lookup data dictionary and then using that information to select the tracts you need.

In this next exercise, you are going to create a lookup table of county FIPS IDs for California. Then you will use those lookup values to extract the census tracts of just one county based on the FIPS code of the census tracts. Let's get started:

1. In the `Chapter4` project folder, right-click and select **New > Notebook**. Rename the Notebook to `CensusCountyExtractTract`.
2. You will need the os module later to write data out. In the first cell, type:

   ```
   import os
   ```

   Run the cell.
3. In the next cell, you are going to declare the two variables to hold the county data and the tract data. Type in:

   ```
   usCounty = r"C:\PythonBook\Chapter4\Chapter4\US_Full.gdb\tl_2019_us_
   county"
   caTract = r"C:\PythonBook\Chapter4\Chapter4\California.gdb\
   tl_2019_06_tract"
   ```

   Run the cell.
4. In the next cell, you are going to create an empty data dictionary to hold all of the county FIPS code and county name pairs. Type in:

   ```
   countyLookUp = {}
   ```

   Run the cell.
5. In the next cell, you will create a SQL statement to limit counties returned to those from a single state. County FIPS codes are specific to a state, so two states could have the same FIPS code for a county. If you wanted to do this for multiple states, you could create a data dictionary for each state. You are just working in California for now, so you want to limit the county data to only counties in California. The county feature class has an attribute called `STATEFP` that contains the state FIPS code. You know from the `arcpy.da.Walk` example that the FIPS code for California is 06. Type in:

   ```
   sql = "STATEFP = '06'"
   ```

Run the cell.

It is a good idea to check the type of field you are writing your query on. Since the FIPS code is 06, it is a string; if it was an integer, it would be 6. However, don't always assume a number is stored as a number in the attribute table. Your SQL query will look different, as strings are encased in '' and numbers are not.

6. In the next cell, you create the search cursor and add key/value pairs to the data dictionary. The search cursor will take the `usCounty` feature class and a list of the fields as the required parameters. It will take the `sql` variable as the `where_clause` parameter to limit the results returned. You will create a `for` loop to iterate through each row returned in the cursor object. Within the loop, you will test if the county name is in the `countyLookUp` dictionary, and if not, you will add the name as the key and the county FIPS code as the value. You will also add a print statement to track your results. Type in the following code:

```
with arcpy.da.SearchCursor(usCounty,
['STATEFP','COUNTYFP','NAMELSAD'],sql) as cursor:
    for row in cursor:
        if row[2] not in countyLookUp:
            countyLookUp[row[2]] = row[1]
            print("Adding key: {0} and value: {1} to countyLookUp".
format(row[2],row[1]))
```

Run the cell. You should see in the output cell the print statement when each county is added. The first few lines will look similar to this:

```
Adding key: Sierra County and value: 091 to countyLookUp
Adding key: Sacramento County and value: 067 to countyLookUp
Adding key: Santa Barbara County and value: 083 to countyLookUp
```

7. You can check and see if all the counties were added. California has 58 counties; you can check the length of the dictionary to see if it has 58 entries. In the next cell, type in the following:

```
len(countyLookUp)
```

Run the cell. The result should indeed be 58.

8. Now you can use that `countyLookUp` data dictionary to extract the tracts for a single county. You need to create a variable for the feature class you are going to create. Since this will be California data, you want to put it into the California geodatabase. You already have a geodatabase for California created in the Notebook above. You are going to use that for this data. Type in:

```
gdb = r'C:\PythonBook\Chapter4\Chapter4\California.gdb'
```

   Run the cell. You should not see any results in the **Out** cell.

9. Next, you will select the single county you want. You will put it into a variable to make it easy to change later if you need to extract the tracts for another county.

If you needed to extract data for multiple counties, you could create a list and iterate through that list.

   In the next cell, type in the following:

```
countyName = "Alameda County"
```

   Run the cell. You should not see any results in the **Out** cell.

10. Now you can use the `countyLookUp` data dictionary to find the county FIPS code for Alameda County. In the next cell, type in the following:

```
countyFips = countyLookUp[countyName]
```

   Run the cell. You should not see any results in the **Out** cell.

11. You can take the `countyFIPS` variable and create a SQL statement from it. You will use the `GEOID` field in the tract feature class. The `GEOID` field is set up to have the state FIPS code, county FIPS code, and then the tract FIPS code. In order to select all the tracts within the county, you have to structure your SQL statement properly. You do that by using the `LIKE` and `%` values. In the next cell, type in the following:

```
sql = "GEOID LIKE '06{0}%'".format(countyFIPS)
print(sql)
```

   Run the cell. You should see the following in the **Out** cell as your SQL statement:

```
GEOID LIKE '06001%'
```

12. The next step is creating a feature class to write to. In the next cell, you will type two lines of code. The first will create a variable for the county name with the spaces removed. The second will create a variable for the new feature class you will create, of just the tracts in Alameda County. In the next cell, type in the following:

    ```
    tractCounty = countyName.replace(" ","")
    tractCountyFull = os.path.join(gdb,"Tracts_"+tractCounty)
    ```

    Run the cell.

13. The last step is using the query to select the tracts. As we saw in *Chapter 2, Basics of ArcPy*, the `Select` function takes three parameters: an input feature class, output feature class, and where clause. All three of these parameters are variables you have declared above. The input feature class is the state-wide tract data, the output feature class is the new feature class you just declared, and the `where` clause is the SQL statement. Type in the following:

    ```
    arcpy.analysis.Select(caTract,tractCountyFull,sql)
    ```

    Run the cell. You should get an output message with the name of the new feature class:

    ```
    Out[168]:
    Output
    C:\PythonBook\Chapter4\Chapter4\California.gdb\Tracts_AlamedaCounty

    Messages
    Start Time: Tuesday, July 20, 2021 10:49:42 PM
    Succeeded at Tuesday, July 20, 2021 10:49:43 PM (Elapsed Time: 1.23 seconds)
    ```

    Figure 4.3: Output message

    The feature class should have been added to your map as well. Click over to your map and take a look at the data:

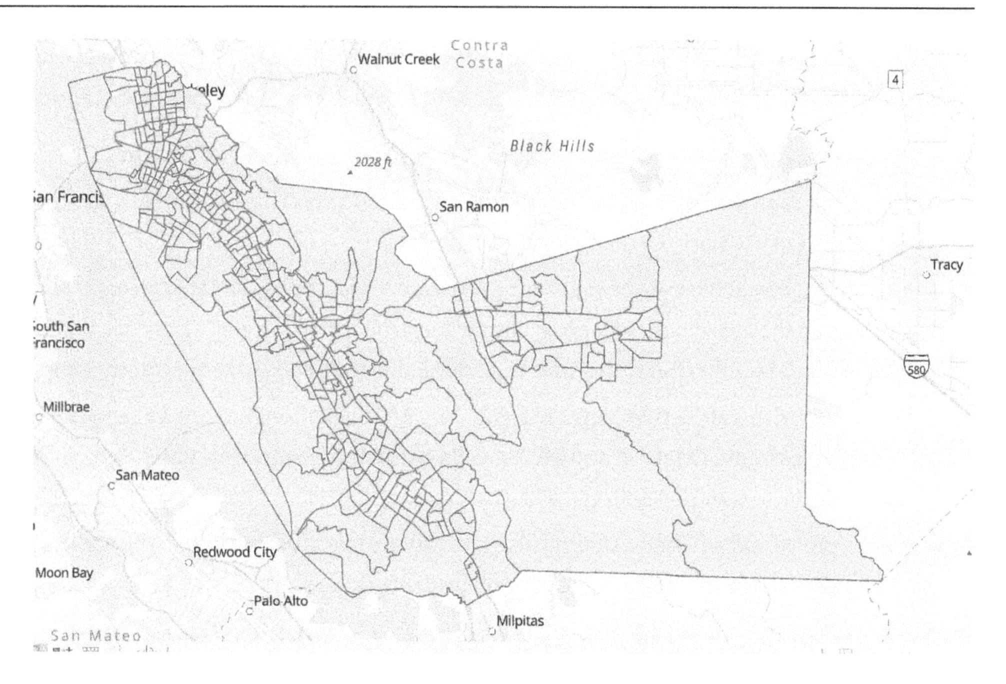

Figure 4.4: Alameda County tracts

You now have a feature class of the tracts from Alameda County. You also have a Notebook you can utilize again to select out the tracts from any county in the US. Next, you will learn how to not just select the tracts but add and calculate a field using the update cursor.

Update cursor

While it is nice to be able to use the county dataset to extract just the tract data in the county and not have to worry about any slivers or extra tracts, it would be really nice to add the county name to the statewide tract data. This would allow easier SQL querying of the statewide tract data by county, as you and your team will not have to remember or look up the FIPS codes when you need to extract county data. The **update cursor** can help you do that.

The update cursor gives you read and write access to the attributes in a feature class or a table. It is set up in a similar way to the search cursor and has the same parameters, but the biggest difference is that the update cursor returns a **list** and not a tuple. Since it gives you a list, you can make changes to the data.

In this exercise, you are going to take the code you were working on for the California census tracts in the previous exercise and add an update cursor to calculate a new county field. Let's begin:

1. Right-click on the `CensusCountyExtractTract.ipynb` file and click **Copy**.
2. Right-click on the `Chapter4` project folder and click **Paste**. This will create a new Notebook named `CensusCountyExtractTract_1.ipynb`.
3. Rename `CensusCountyExtractTract_1.ipynb` to `AddCountyToStateCensusTract.ipynb` and open the Notebook. You are going to keep the first five cells. These are the cells that create variables for the county and tract data, the state SQL, and the county lookup table.
4. Starting with cell 6 that contains `gdb = r"C:\PythonBook\Chapter4\Chapter4\California.gdb`, delete it and all the cells below it by selecting them and clicking the scissors button, or pressing the *d* key twice.
5. You need to add a field for the county name to the tract feature class. The `AddField` function takes the following three required parameters:

    - `in_table`: The feature class, shapefile, coverage, table, or raster with attribute table to add a field to.
    - `field_name`: A string for the name of the field being added.
    - `field_type`: A string for the type of field to be added. It can be any of the following values: `"STRING"`, `"LONG"`, `"SHORT"`, `"DOUBLE"`, `"FLOAT"`, `"DATE"`, `"BLOB"`, `"RASTER"`, `"GUID"`.

    It also has the following seven optional parameters:

    - `field_precision`: A long value that is the number of digits stored in the field.
    - `field_scale`: A long value that is the number of decimal places stored in the field.
    - `field_length`: A long value that is the limit for the number of characters in the string field.
    - `field_alias`: The alias name for the field.
    - `field_is_nullable`: A Boolean value that is either `NON_NULLABLE` for fields that cannot be set to null or `NULLABLE` (default) for fields that can be set to null.
    - `field_is_required`: A Boolean value that is either `NON_REQUIRED` (default) for fields that are not required or `REQUIRED` for fields that are required and cannot be deleted.
    - `field_domain`: The existing domain in the geodatabase to be applied to the field.

To add the field, in cell 5 press *Enter* after the last line and type in the following:

```
arcpy.management.AddField(tract,"CountyName","STRING")
```

6. In the next cell, you are going to create a `for` loop to loop through the `countyLookUp` dictionary and then use the update cursor to calculate the county name for each tract. **The code will be written into this cell over the next two steps.** In this step, you will create the `for` loop to loop through the `countyLookUp` dictionary and create variables for the county name, FIPS code, and SQL statement to select all the tracts within a county. You will include print statements to track your progress. Type in the following:

```
for key in countyLookUp:
    countyName = key
    countyFIPS = countyLookUp[key]
    print(countyName)
    print(countyFIPS)
    sqlTract = "GEOID LIKE '06{0}%'".format(countyFIPS)
    print(sqlTract)
```

7. Continuing the same cell, you will create a `with...as...` statement to create the cursor object. The update cursor will take the `tract` feature class and a list of the fields as the required parameters. It will take the `sqlTract` variable as the `where_clause` parameter to limit the results returned. Then, you will write a `for` loop to loop through each row from the cursor. Within the loop, you will set the row with the `CountyName` field equal to the `countyName` variable. You will then call the `updateRow` property on the cursor with `row` as an argument to write the value to each row:

```
with arcpy.da.UpdateCursor(tract,
["GEOID","CountyName"],sqlTract) as cursor:
        for row in cursor:
            row[1] = countyName
            cursor.updateRow(row)
```

8. When you are ready, click **Cell > Run All** to run the Notebook.

You must remember to use the `updateRow` property on your cursor with your row as an argument. If you do not, the row you just calculated will not be written.

After the code has run, switch over to the map to explore your data. Make sure the `tl_2019_06_tract` feature class is loaded in and open up its attribute table. Scroll to see the `CountyName` field and see if it has all the county values. Open up the attribute table and try a few Select By Attributes to see how you can now select all the tracts that are within a county. You should now see how to use both the search and update cursors to easily add value to your attribute table.

Insert cursor

The **insert cursor** is used to add in new rows of data to tables or feature classes. Like the search and update cursors, it can act on the geometry of a feature class as well as the attribute tables. While you can still use the `with...as...` format for insert cursors, the more common syntax seen in the ArcGIS documentation involves creating a cursor, looping through it, and deleting the cursor. You will write the code below using this convention.

To better understand why using an insert cursor to insert data into a table from the census CSV is necessary, we should look at the census CSV file. Open up the `ACSDT5Y2019.B03002_data_with_overlays_2021-07-22T010002.csv` file in the `CensusCSV` folder. This is the *Hispanic or Latino Origin By Race for the American Community Survey (ACS) 5-Year Estimates Detailed Tables* for years 2014-2019, Table B03002 for Alameda County at the tract level. It shows the population totals for Hispanic or Latino and non-Hispanic or Latino broken down by race. Here is a section of the data:

	A	B	C	D
1	**GEO_ID**	NAME	B03002_001E	B03002_001M
2	id	Geographic Area Name	Estimate!!Total:	Margin of Error!!Total:
3	1400000US06001400100	Census Tract 4001, Alameda County, California	3120	208
4	1400000US06001400200	Census Tract 4002, Alameda County, California	2007	120

Figure 4.5: Census CSV file

There are some important things to notice about this table:

- It has two rows for headers:
    a. The first row contains coded values.
    b. The second row contains values that would not work well as field names for an attribute table, as they have spaces and special characters.
- It contains a lot of data. In addition to the estimates for each race being split in terms of non-Hispanic or Latino and Hispanic-Latino, it has a margin of error for each. You want to get the totals for each race that is non-Hispanic or Latino, and just the total Hispanic or Latino. This will allow you to calculate the total and percent minority of each tract.

- The GEO_ID field is different from the one in the tract feature class. The GEO_ID field in the CSV is prefixed with 1400000US before getting to the values in the GEOID field in the feature class.

For these reasons, it is a good idea to create a table from this CSV containing just the data you need. That table can then be used to join to the feature class so you can map the data. You are going to do this by reading the CSV in a similar way to using a search cursor. Then, you are going to use an insert cursor to insert those values into an empty table. Finally, you will join that table to the feature class so you can display the data on a map:

1. Right-click on the Chapter4 project folder and select **New** > **Notebook**. Rename the Notebook to CreateCensusTableInsertRows.
2. In the first cell, you will import the modules you will need, the csv and os modules. Type in the following:

   ```
   import csv, os
   ```

3. In the next cell, you are going to set variables that you will need throughout the Notebook. They are:

   a. The geodatabase you are working in
   b. The tract feature class for Alameda County
   c. The CSV file with the census data in it
   d. The table name
   e. The table full path
   f. The new census polygon full path

   To do so, type in the following:

   ```
   gdb = r"C:\PythonBook\Chapter4\Chapter4\California.gdb"
   tract = r"C:\PythonBook\Chapter4\Chapter4\California.gdb\Tracts_
   AlamedaCounty"
   csvFile = r"C:\PythonBook\Chapter4\CensusCSV\ACSDT5Y2019.
   B03002_2021-07-22T010004\ACSDT5Y2019.B03002_data_with_overlays_2021-
   07-22T010002.csv"
   table = "AlamedaCounty_RaceHispanic"
   tablePath = os.path.join(gdb,table)
   censusPoly = os.path.join(gdb,table+"_Tract")
   ```

4. In the next cell, you are going to create a data dictionary of the field names, field aliases, and field types. By using a data dictionary, you can use a loop to add all the fields and their aliases. The data dictionary will not only contain the different Hispanic/race types from the census data but also fields for the total minority and percent minority. For this example, you will assume that all Hispanic/races that are not white are a minority.

Why do we take all non-white races to be minority?

In most environmental documents, it is common to consider all non-white to be minority; occasionally you will see them not include two or more races as minority. If you want to change that definition, you just have to change the math later in the code that adds up the values for minority.

The data dictionary will have a key that is the field name, and a list of values that are the field alias and the field type. This will allow you to loop through the dictionary and use the key and data from the value to create a field. Type in the following:

```
fields = {"geoid_census":["GeoID_Join","STRING"],
          "total_pop":["Total Population","LONG"],
          "white":["White","LONG"],
          "prct_white":["Percent White","FLOAT"],
          "black":["Black","LONG"],
          "prct_black":["Percent Black","FLOAT"],
          "am_indian_nat_alaska":["American Indian/Native
Alaskan","LONG"],
          "prct_am_indian_nat_alaska":["Percent American Indian/
Native Alaskan","FLOAT"],
          "asian":["Asian","LONG"],
          "prct_asian":["Percent Asian","FLOAT"],
          "nat_hawaiian_pac_island":["Native Hawaiian/Pacific
Islander","LONG"],
          "prct_nat_hawiian_pac_island":["Percent Native Hawaiian/
Pacific Islander","FLOAT"],
          "some_other":["Some Other Race","LONG"],
          "prct_some_other":["Percent Some Other Race","FLOAT"],
          "two_or_more":["Two Or More Races","LONG"],
          "prct_two_or_more":["Percent Two Or More Races","FLOAT"],
          "hispanic_latino":["Hispanic/Latino","LONG"],
```

```
            "prct_hispanic_latino":["Percent Hispanic/
Latino","FLOAT"],
            "total_minority":["Total Minority","LONG"],
            "percent_minority":["Percent Minority","FLOAT"],
            }
```

5. In the next cell, you are going to create a table. You will be writing the demographic data you extract from the CSV to this table. The `CreateTable` tool takes the following two mandatory parameters:

    - `out_path`: The workspace to create the table in
    - `out_name`: The name of the table

    It also takes the following three optional parameters:

    - `template`: A table with an attribute schema that will be applied to the new table.
    - `config_keyword`: The configuration keyword that will determine where and in what format to store a table in an enterprise geodatabase. You will use the `Default` much of the time.

For more information on configuration keywords, see the documentation here: `https://desktop.arcgis.com/en/arcmap/latest/manage-data/geodatabases/what-are-configuration-keywords.htm`.

    - `out_alias`: The alias of the table.

    You will just be using the mandatory parameters. Type in the following:

```
arcpy.management.CreateTable(gdb, table)
```

6. In the next cell, you are going to iterate through the data dictionary to add all the fields to the new empty table. You will start by creating an empty list called `tableFields` that you will add the field names to for future use. You will then loop through each field in the data dictionary. Within the loop, you will create a variable to hold the name of the field, which is the key from the data dictionary and what is returned when looping through a data dictionary. You will then use the key to create variables for the alias and data type.

Remember that the values in the data dictionary are a list, and when using the key to access them you will be returned a list. You can use list indexing to return the specific values from the list.

You will then append the field name to the `tableFields` list. Finally, you will use the `AddField` tool to add the field to the table passing in the table, field name, data type, and alias variables you created. Type the following code:

```
tableFields = []
for field in fields:
    name = field
    alias = fields[field][0]
    dataType = fields[field][1]
    print(name)
    print(alias)
    tableFields.append(field)
    arcpy.management.AddField(tablePath,name,dataType,
field_alias = alias)
```

7. In the next cell, you are going to open the CSV and create a `csv.reader` object to read in the data from the CSV row by row. Type in the following:

```
fileRef = open(csvFile)
csvRef = csv.reader(fileRef)
```

8. In the next cell, you are going to read in each line of the CSV, extract just the values you need, use them to calculate the percentage of each Hispanic/race group, the total minority, and percentage minority, then write that data to a row in the table you just created using the insert cursor. **The code will be written into the next cell over the next six steps.**

   In this step, you will create the `for` loop. Within the `for` loop, you will first create a conditional to check the line number of the CSV using the `line_num` property of the CSV reader object. You will check if the line number is less than or equal to 2. If so, the `continue` keyword will be called. If not, the rest of the code will run. Type the following in the cell:

```
for row in csvRef:
    if csvRef.line_num <= 2:
        continue
```

What does continue do?

The `continue` keyword tells the code not to run any of the code beneath it and go back to the top of the `for` or `while` loop, starting on the next value. In your code here, it is run to skip the first two lines of the CSV, but it could be used to skip any line if you knew which one you wanted to skip. `continue` is often compared to `break`. The difference is that `continue` resets to the top of the loop, going to the next value, while `break` stops the loop and breaks out of it.

9. In the same cell, you will now create a variable to join the feature class and table on. To create the variable, you need to remove the first nine characters from the value in the CSV, so it will match the value in the feature class. This will allow you to join the table to the feature class.

1400000US06001400100

Figure 4.6: First column value in the CSV

06001400100

Figure 4.7: GEOID value in the tract feature class

Type the following in the cell:

```
geoJoin = row[0][9:]
```

10. In the same cell, you will now create a variable that holds the total minority of the tract. Remember from above that the total minority is the total Hispanic or Latino and all of the non-white non-Hispanic or Latino races. Type in the following:

```
totMinority =
int(row[8])+int(row[10])+int(row[12])+int(row[14])+int(row[16])
+int(row[18])+int(row[24])
```

In order to keep all the indexes straight, you could open up the CSV in Excel, add a row at the top, write 0 in cell A1, write 1 in cell B1, and then autocomplete the rest. Now when you go back to the CSV to see what index you need, it will be written there. Just be sure not to save it, or if you do save it, change your `if` statement to be `<=3`, to account for the extra row you've added.

11. In the same cell, you will now create the percentages of each race and of the Hispanic or Latino population in the tract. Since some tracts have a total population of 0 and dividing by 0 will cause an error, you will create a conditional to check if the total population is equal to 0. If it is not, you will divide the race or Hispanic Latino population by 0 and multiply the result by 100, using the `round()` function to round the results to two decimal places. If the population is 0, you will assign the percentage to `-999`. Type in the following:

```
if int(row[2]) != 0:
    prctWht = round((int(row[6])/float(row[2]))*100,2)
    prctBlk = round((int(row[8])/float(row[2]))*100,2)
    prctAmIn = round((int(row[10])/float(row[2]))*100,2)
    prctAsi = round((int(row[12])/float(row[2]))*100,2)
    prctNatHaw = round((int(row[14])/float(row[2]))*100,2)
    prctSmOth = round((int(row[16])/float(row[2]))*100,2)
    prctTwoMr = round((int(row[18])/float(row[2]))*100,2)
    prctHispLat = round((int(row[24])/float(row[2]))*100,2)
    prctMinority = round((totMinority/float(row[2]))*100,2)
else:
    prctWht = -999
    prctBlk = -999
    prctAmIn = -999
    prctAsi = -999
    prctNatHaw = -999
    prctSmOth = -999
    prctTwoMr = -999
    prctHispLat = -999
    prctMinority = -999
    prctMinority = -999
    prctMaj = -999
```

12. In the same cell, you will now create a list to hold all of the population totals and percentages of each race or Hispanic or Latino group. The order they are in in the list is the order they will be written to the table in. You need to check against the data dictionary you created above to make sure you are using the same order. Type in the following:

```
value =
[geoJoin,int(row[2]),int(row[6]),prctWht,int(row[8]),prctBlk,
int(row[10]),prctAmIn,int(row[12]),prctAsi,int(row[14]),prctNatHaw,
```

```
int(row[16]),prctSmOth,int(row[18]),
prctTwoMr,int(row[24]),prctHispLat,totMinority,prctMinority]
```

13. In the same cell, you will now create the insert cursor with the table path and list of fields as the arguments. You will add a print statement to track the values being inserted. Then, you will call the `insertRow` property on your cursor to insert the list of the values you created above into a new row. Finally, you will delete the cursor using the `del` command. Deleting the cursor will remove any locks on the table so you can continue to write to it when your loop continues through each row. Type in the following:

```
cursor = arcpy.da.InsertCursor(tablePath,tableFields)
    print(value)
    cursor.insertRow(value)
    del cursor
```

14. In the *next* cell, you will create a new feature class and then join the data from the table up to it. You use the `CopyFeatures` tool to create the new tract polygon you stored in the `censusPoly` variable. You will also use the `JoinField` function to join the new table of data to the new tract polygon. The `JoinField` tool takes the following four mandatory parameters:

    - `in_data`: The input feature class, shapefile, table, or raster with attribute table for the data to be joined to.
    - `in_field`: The field from the `in_data` feature class, shapefile, table, or raster with attribute table to join on.
    - `join_table`: The feature class, shapefile, table, or raster with attribute table to be joined to `in_data`.
    - `join_field`: The field from `join_table` to be joined to `in_field` on.

    The `JoinField` tool also takes the following optional parameter:

    - `fields`: A list of fields to be transferred from the join table to the input table

    You are going to use the newly created census polygon as your `in_data`, the `GEOID` field as `in_field`, the table you have created as `join_table`, the `geoid_census` field as `join_field`, and the `tableFields` list created as the optional `fields`. Type in the following:

```
arcpy.management.CopyFeatures(tract,censusPoly)
arcpy.management.JoinField(censusPoly,"GEOID",tablePath,"geoid_
census",tableFields)
```

Why create a copy of a feature class before joining the data?

Creating a new feature class before the join means you keep your original tract data containing no demographic data for future use, so you will be able to add different demographic data to it too.

15. When you have entered all the code, click on **Cell > Run All** to run the entire Notebook.

When it is finished, you can open up the attribute table of the `AlamedaCounty_RaceHispanic_Tract` feature class and the `AlamedaCounty_RaceHispanic` table. You will see that the values from the CSV were written to the table, and then joined to the feature class:

AlamedaCounty_RaceHispanic_Tract

Field: Add Calculate Selection: Select By Attributes Zoom To Switch Clear Delete Copy

	Total Population	White	Percent White	Black	Percent Black	American Indian/Native Alaskan	Percent American Indian/Native Alaskan	Asian	Percent Asian
1	5407	910	16.83	130	2.4	15	0.28	3648	67.47
2	6546	830	12.68	156	2.38	19	0.29	4528	69.17
3	6175	1000	16.19	98	1.59	0	0	4324	70.02
4	5233	697	13.32	349	6.67	13	0.25	3394	64.86
5	4982	1372	27.54	450	9.03	16	0.32	1953	39.2
6	8113	1767	21.78	144	1.77	0	0	4237	52.22
7	7543	2600	34.47	141	1.87	18	0.24	3765	49.91
8	3909	639	16.35	262	6.7	74	1.89	1623	41.52
9	6250	924	14.78	179	2.86	31	0.5	4209	67.34
10	5462	731	13.38	78	1.43	29	0.53	4399	80.54

Figure 4.8: AlamedaCounty_RaceHispanic_Tract attribute table (truncated)

In this section, you have learned how to use a data dictionary with a list of lookup values to add multiple fields to a table. You have learned how to read in data from a CSV row by row and insert that data into a table using the insert cursor. Now that you have this in a Notebook, you can use it for multiple different geographies, or you can take the base and use different tables and different demographic data.

Summary

In this chapter, you have learned about `arcpy.Walk` and cursors in the Data Access module. You have seen how the Data Access module allows you to access and modify your data. By using the `arcpy.Walk` module in the Data Access module, you are able to walk through directories and sub-directories and find geospatial data that the `os.walk` module would miss.

This allows you to extract and transfer data programmatically. You explored how to use data dictionaries as lookup tables and applied that in multiple examples. Finally, you used the search, update, and insert cursors to find data, update it, and insert new data.

In the next chapter, you will use the ArcGIS API for Python to publish, organize, and manage access to data on your ArcGIS Online account.

5

Publishing to ArcGIS Online

In *Chapter 3*, you were introduced to the ArcGIS API for Python and used it to search for data, groups, and users in your organization. The data you were searching for was published to ArcGIS Online. This can be done by publishing maps and service definitions from within ArcGIS Pro, but you can also add and publish CSVs, shapefiles, and geodatabases by using the **ArcGIS API for Python**. By creating Notebooks or script tools to publish data to your organization, you can automate repetitive tasks for data that needs to be updated regularly and reduce the number of clicks you have to make.

In this chapter, you will cover:

- Using the `ContentManager` class to publish new content and organize it in folders
- Using the `GroupManager` class to create new groups and share content with them
- Using the `features` module to work with feature layers
- Using the `mapping` module to visualize your data

To complete the exercises in this chapter, please download and unzip the `Chapter5.zip` folder from the GitHub repository for this book: `https://github.com/PacktPublishing/Python-for-ArcGIS-Pro/tree/main/Chapter5`

Using ContentManager for publishing and organizing data

You have already seen how the `ContentManager` class can be used through the `content` property of the GIS object to search for data. In *Chapter 3, ArcGIS API for Python*, you searched for data both within and outside of your organization.

The `content` property can also be used to add data and publish it as a feature layer, and organize your data into folders within ArcGIS Online.

In this section, you will add data from a CSV, publish it, and move it to a folder by using the ArcGIS API for Python in an ArcGIS Pro Notebook.

Publishing data

When you publish data to ArcGIS Online or ArcGIS Enterprise, much of this is achieved within ArcGIS Pro. This is useful and convenient when you are publishing a map that contains all of your layers. However, it is less convenient when you want to publish a CSV, shapefile, or geodatabase, as you have to create the map in ArcGIS Pro to publish it. Using the ArcGIS API for Python, you can take a CSV, shapefile, or geodatabase, add it to your organization, and publish it with a few lines of code.

Adding data from a CSV

To add data to your ArcGIS Online account, you are going to use the `add` method. Like the `search()` method, the `add()` method is part of the `ContentManager` class of the GIS object. The `add()` method takes the following parameters, with only `item_properties` being required:

- `item_properties`: A data dictionary with a set of key/value pairs. The list below contains the most common key/value pairs you will use.
- `data`: A path or URL to the data.
- `thumbnail`: A path or URL to a thumbnail image.
- `metadata`: A path or URL to the metadata.
- `owner`: A string that defaults to the logged-in user.
- `folder`: The name of the folder in your ArcGIS Online account to place the data in.

The following are the most common `item_properties` values you will see. You will use most of these for every item you create:

- type: The type of item being added. You will mostly use CSVs, shapefiles, and file geodatabases.

A comprehensive list of acceptable types can be found here: `https://developers.arcgis.com/rest/users-groups-and-items/items-and-item-types.htm`

- `title`: The title of the item being added.
- `tags`: The tags for the item being added. They are listed as comma-separated values, or a list of strings.
- `description`: The description for the item being added.
- `snippet`: A short description of fewer than 250 characters for the item being added.

The `add()` method will simply add the CSV to your ArcGIS Online account. To make it viewable on a map, you need to publish the data to a hosted web layer. To do this, you use the `publish()` method, which will create a hosted feature layer that can be displayed on maps and shared in groups. The `publish()` method can be used to create hosted feature services from many file types, including CSVs, file geodatabases, shapefiles, and service definitions. The `publish()` method does not have any required parameters and often you will not need to set any. Some of the optional parameters you may need are the following:

- `publish_parameters`: A data dictionary with publish instructions and customizations. The different parameters available depend on the item type being published.

Each item type has an extensive list of publish instructions that you can set. Below you will see an example of how to set some for a CSV. To see all of the customizations available for each item, visit this site: `https://developers.arcgis.com/rest/users-groups-and-items/publish-item.htm`

- `address_fields`: A data dictionary that maps the columns of your input data to address fields. It is used when geocoding data to ArcGIS Online.
- `geocode_service`: A geocoder that can be set. If it is not set, the default ArcGIS Online geocoder will be used.
- `file_type`: A string of the file type being published. It can be used when the file type is not being automatically detected or you want to make sure it is detecting the correct file type. The `file_type` values you can specify are `"serviceDefinition"`, `"shapefile"`, `"csv"`, `"tilePackage"`, `"featureService"`, `"featureCollection"`, `"fileGeodatabase"`, `"geojson"`, `"scenepackage"`, `"vectortilepackage"`, `"imageCollection"`, `"mapService"`, and `"sqliteGeodatabase"`.

You should have data for the farmers' markets in Oakland and Berkeley in a CSV called `AlamedaCountyFarmersMarket.csv`. It is included in the `Chapter5.zip` file you downloaded at the start of this chapter.

Open up the `AlamedaCountyFarmersMarket.csv` file to see the data you will be adding. It is a basic CSV containing market names, opening days, opening times, location, city, latitude, and longitude:

	A	B	C	D	E	F	G
1	MarketName	Days	Time	Location	City	Latitude	Longitude
2	Downtown Berkeley	Saturday	10 am - 3 pm	Center Street and Martin Luther King Jr. Way	Berkeley	37.869336	-122.272118
3	North Berkeley	Thursday	3 pm - 7 pm	Shattuck Avenue and Vine Street	Berkeley	37.881804	-122.269392
4	South Berkeley	Tuesday	2 pm - 6:30 pm	Adeline Street and 63rd Street	Berkeley	37.847751	-122.27194
5	Grand Lake	Saturday	9 am - 2 pm	Splash Pad Park	Oakland	37.810721	-122.247899

Figure 5.1: Farmers' market CSV

In this exercise, you will create a Notebook to add this data in and publish it as a feature layer.

1. Open up ArcGIS Pro, navigate to where you unzipped the `Chapter5.zip` folder, and open up `Chapter5.aprx`.
2. Right-click on the `Chapter5` folder and select **New** > **Notebook**. Rename the Notebook to `AddPublishData`.
3. In the first cell, type in your import statements and create a GIS object that will be logged into the same ArcGIS Online account you are using in ArcGIS Pro:

```
from arcgis.gis import GIS
from IPython.display import display
gis = GIS('home')
```

4. In the next cell, you are going to create a variable for the CSV. Type in the following:

```
csvFM = r"C:\PythonBook\Chapter5\AlamedaCountyFarmersMarket.csv"
```

If your CSV is saved to a different location, make sure you are using that instead.

5. In the next cell, you are going to create the data dictionary of the CSV properties. You will fill in the properties for the title, description, and tags as the keys, with their properties as the values. Type the following:

```
csvProperties = {
    "title": "Farmers Markets in Alameda County",
    "description": "Location, days, and hours of Farmers Markets in
Alameda County",
    "tags": "Farmers Market, Alameda County, ArcGIS API for Python"
}
```

6. In the next cell, you are going to create a variable to hold the CSV item being added. You will use the `add()` method from the `content` property. The arguments passed are the properties dictionary and the variable you created above with the path to the CSV. Type the following:

```
addCsvFM = gis.content.add(item_properties=csvProperties,data=csvFM)
```

7. In the next cell, you will publish the CSV item you just added by calling the `publish()` method. Type the following:

```
farmersMarketFL = addCsvFM.publish()
farmersMarketFL
```

By assigning the `publish` method to a variable, that variable will contain the feature layer. You can call that variable to display the feature layer's properties.

8. In the next cell, you are going to create a quick map to visualize your data and verify that the feature layer was created. Type the following:

```
map1 = gis.map("Oakland, California")
map1.add_layer(farmersMarketFL)
map1
```

9. Click **Cell** > **Run All** to run all the cells. The output map should look like this:

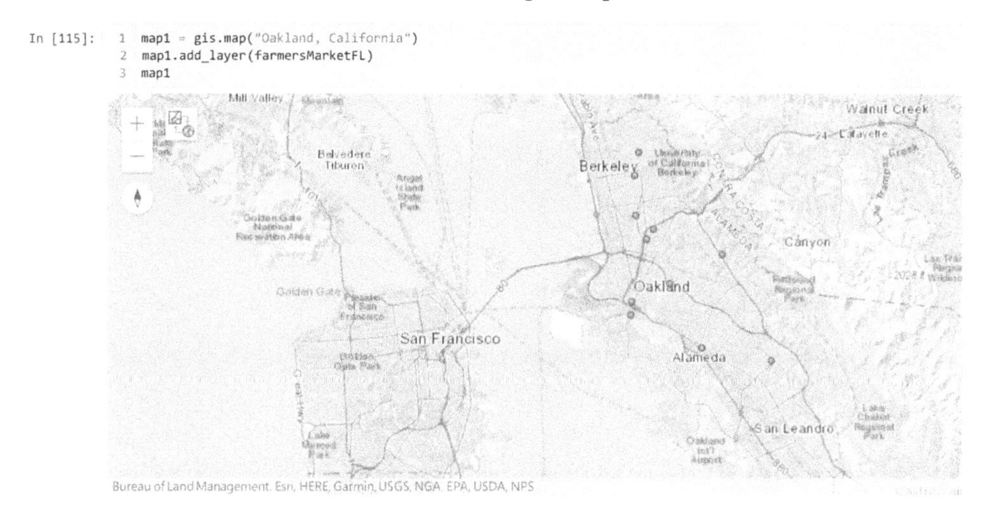

Figure 5.2: Map widget showing the farmers' market feature layer

You have now published a CSV of address locations to your ArcGIS Online account. Any time you have a CSV of points that needs to be published to ArcGIS Pro, you just need to update the path to the CSV and the `item_properties` in the data dictionary and then you can add and publish that CSV.

Adding and publishing tips

You have seen how to publish a CSV that has latitude and longitude columns for point data. This process could be turned into an iterative process using a loop to publish multiple CSVs; all you would need to write is a properties data dictionary for each CSV. However, your data is not always going to be a CSV of point locations with latitude and longitude. Below are some tips for publishing other types of data:

- **When publishing a CSV with latitude and longitude fields, make sure they are named "latitude" and "longitude".** The module is looking for those field names. If they are not found, it will not locate the points correctly. You can specify the field names to use by creating a `publish_parameters` dictionary in the `publish()` method. The `publish_parameters` dictionary will be used to set the `latitudeFieldName` and `longitudeFieldName` values. To set those field names, you also have to set the `locationType` value to `"coordinates"`. To see this, you can use the `AlamedaCountyFarmersMarket_TestLatLongField.csv` and publish it using the same method as above. All you need to add is the following:

  ```
  publishParam = {
      "locationType":"coordinates",
      "latitudeFieldName":"LatX",
      "longitudeFieldName":"LongY"
  }
  ```

  to the start of the cell containing this:

  ```
  farmersMarketFL = addCsvFM.publish()
  farmersMarketFL
  ```

  and change the following:

  ```
  farmersMarketFL = addCsvFM.publish()
  ```

  to

  ```
  farmersMarketFL = addCsvFM.publish(publish_parameters =
  publishParam)
  ```

- **CSVs without latitudes and longitudes but with addresses can be geocoded.** To geocode data from a CSV, you would again use the `publish_parameters` dictionary:

  ```
  publishParam = {
      "locationType":"address",
      "addressTemplate":"{address},{city},{state},{zip}"
  }
  ```

  The `locationType` field is set to `"address"`. The `addressTemplate` field is then set to the fields that contain the different address components. In this example, there is a field with the street address, a field with the city, a field with the state, and a field with the zip code. The setup of this will depend on how you have your data in your CSV.

- **Shapefiles and file geodatabases can be added and published using this same method, but they must be zipped.** If you have a large number of shapefiles or file geodatabases that are unzipped, you could automate the process of zipping them up and publishing them.

ArcGIS API for Python can be very useful for quickly adding data to your organization's ArcGIS Online or ArcGIS Enterprise account. You have seen how to add and publish a CSV, and that CSVs can be geocoded on publishing using the ArcGIS Online geocoder. In the next section, you will see how to organize data into folders, create groups, and manage access to groups.

Organizing data and managing groups and users

Organizing your data within your ArcGIS Online account or ArcGIS Enterprise is important; you want to be able to find your data. In addition to folders to hold your data, you can also create groups to share specific data with. In large organizations this is important, as not everyone needs to have access to the same data.

In this section, you will see how to create folders and move data into them, create groups and manage access to them, and create and manage users.

Organizing data into a folder

One of the first things to do after adding data or publishing it should be to find a folder to place it in. It is good practice to use folders to organize your data. This helps you and other members of your organization to find data. You can add folders and move data over using the ArcGIS API for Python. In the exercise below, you will create a new folder and move the farmers' market data from the previous exercise:

1. If you closed ArcGIS Pro, open it up, navigate to where you unzipped the `Chapter5.zip` folder, and open up `Chapter5.aprx`.
2. Right-click on the `Chapter5` folder and select **New > Notebook**. Rename the Notebook to `CreateFolderMoveData`.
3. In the first cell, type in your import statements and create your GIS. You are going to create the GIS logged in to your ArcGIS Online account that you are logged into ArcGIS Pro with:

```
from arcgis.gis import GIS
from IPython.display import display
gis = GIS('home')
```

4. In the next cell, you will create a new folder:

```
gis.content.create_folder(folder="AlamedaFarmersMarkets")
```

5. In the next cell, you will search for the data that you need to move to the folder you are creating. Type in the following:

```
alamedaFM = gis.content.search(query="title:Farmers Markets in
Alameda County")
```

6. Remember that the `search()` method returns a list of items. To confirm what you have in your list, you will run a `for` loop to iterate through the list and display the data. In the same cell as above, type the following:

```
alamedaFM = gis.content.search(query="title:Farmers Markets in
Alameda County")
for item in alamedaFM:
    display(item)
```

7. Click **Cell** > **Run All** to run all your cells up to this point. Your Notebook should look similar to this now:

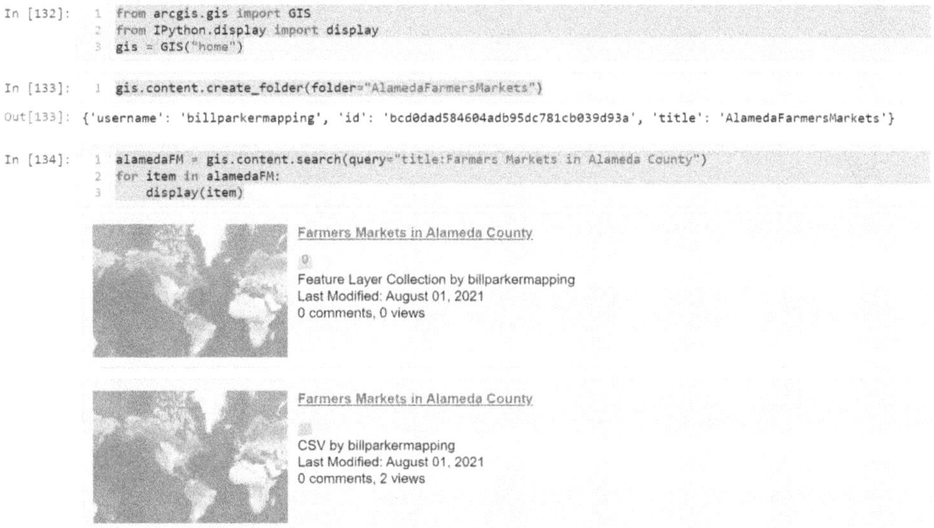

```
In [132]:  1 from arcgis.gis import GIS
           2 from IPython.display import display
           3 gis = GIS("home")

In [133]:  1 gis.content.create_folder(folder="AlamedaFarmersMarkets")

Out[133]: {'username': 'billparkermapping', 'id': 'bcd0dad584604adb95dc781cb039d93a', 'title': 'AlamedaFarmersMarkets'}

In [134]:  1 alamedaFM = gis.content.search(query="title:Farmers Markets in Alameda County")
           2 for item in alamedaFM:
           3     display(item)
```

Farmers Markets in Alameda County

Feature Layer Collection by billparkermapping
Last Modified: August 01, 2021
0 comments, 0 views

Farmers Markets in Alameda County

CSV by billparkermapping
Last Modified: August 01, 2021
0 comments, 2 views

Figure 5.3: Output of creating the folder and finding data to move

8. In the next cell, you will move the feature layer and CSV into the new folder by looping through the search results and using the `move()` method. Type the following:

```
for item in alamedaFM:
    item.move(folder="AlamedaFarmersMarkets")
    print(item)
```

Run the cell. You should see output like so:

```
<Item title:"Farmers Markets in Alameda County" type:Feature Layer
Collection owner:billparkermapping>
<Item title:"Farmers Markets in Alameda County" type:CSV
owner:billparkermapping>
```

This confirms that your data has been moved. If you go to your ArcGIS Online account, you will see that you now have a new folder and that both datasets are inside.

Creating a folder and moving data into that folder is a process that ArcGIS API for Python can be used for. In the exercise, you were able to find all the datasets by their name and move them to a newly created folder. Being able to search your GIS for data and move it to folders using the ArcGIS API for Python is a valuable tool that will save you time.

If you need to move data back to the root directory, simply use the following code: `item.move("\")`

Accessing and managing groups

Groups are spaces where you share data with other users. They are how you can create a collaborative GIS by allowing other users access to your data and maps. Using the ArcGIS API for Python, you can create and manage groups in a programmatic way to save you time, while fostering better collaboration within and outside of your team.

In this section, you will see how to create a new group, manage a group by sharing data to it, and add and remove users from a group.

Creating a group

You can create groups to share data publicly or just with members of the group. In this exercise, you are going to walk through how to create a group for sharing data publicly. You will also see the arguments needed to create private groups and how to change the sharing settings of a group:

1. If you closed ArcGIS Pro, open it up, navigate to where you unzipped the `Chapter5.zip` folder, and open up `Chapter5.aprx`.
2. Right-click on the `Chapter5` folder and select **New** > **Notebook**. Rename the Notebook to `CreateGroupMoveData`.
3. In the first cell, you are going to log into your organization's GIS through the user you are currently logged into ArcGIS Pro with. Type in the following:

   ```
   from arcgis.gis import GIS
   from IPython.display import display
   gis = GIS("home")
   ```

   Run the cell.

4. In the next cell, you will create a group by using the `create()` method of the `groups` module. The `create()` method takes five arguments (explained below). Type in the following:

```
farmerMarketGroup = gis.groups.create(title="Alameda County Farmers
Markets",
tags="Alameda County, Farmers Market",
description = "Group with data for Alameda County Farmers Markets.",
access = "public",
is_invitation_only = "False"
)
```

Run the cell. You should get no output message, but you have created a new public group.

5. To check this, you can go to your ArcGIS Online account and look in **Groups**. You can also type in `farmerMarketGroup` in a new cell and run it to see the group. It should look something like below:

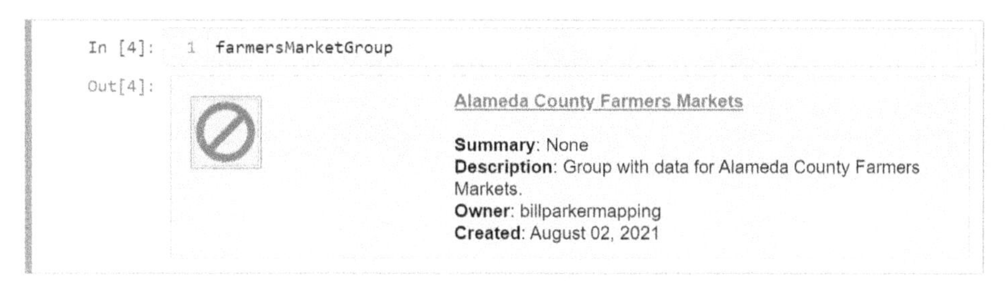

Figure 5.4: Newly created group

To create the group, you used five arguments: `title`, `tags`, `description`, `access`, and `is_invitation_only`. Those are the minimum arguments you should use when setting up a new group, as they give the group a title, tags, description, and set basic access. The list below summarizes these arguments along with the values they can take:

- `title`: A string between single or double quotes that will be the title of your group.
- `tags`: A string between single or double quotes with a comma separating all the tags. When returned, it is a list.
- `description`: A string between single or double quotes that will be the description of your group.

- `access`: A string between single or double quotes that sets the access. The `access` values can be `"org"`, `"private"`, or `"public"`. `"org"` is a group that everyone in your organization can see. `"private"` is a private group that only invited users can see. `"public"` is a public group available to everyone.
- `is_invitation_only`: A string between single or double quotes that is a Boolean value. When set to `"True"`, users can only gain access if invited. When set to `"False"`, users can request access or be invited.

6. You can verify any of the settings by typing the variable for the group, a dot, and the value. For example, to verify the access of the group you just created, type the following in the next cell:

```
farmerMarketGroup.access
```

Run the cell. You should see `'public'` in the output cell.

7. To change any of the values of a group, you can use the `update()` function. The `update()` function takes all the same arguments used to create the group. For example, to update the access, type in the following:

```
farmerMarketGroup.update(access = "private")
```

Run the cell. You should see `"True"` in the output cell.

You have now created a new group and have seen how to change the values of that group. The next step is sharing data with a group.

Sharing content to a group

An empty group is not very useful. The point of creating a group is to share data either publicly or with other users. In this section, you will see how to share data with a group. You will continue in the same Notebook from above in the steps below:

1. You are going to need to access the feature layer containing the Alameda County farmers' markets. In the next cell of the same Notebook from above, you will use the `search()` method to get the **Farmers Markets in Alameda County** feature layer. The `search()` method returns a list; you only have one item with that title, so you can add a `[0]` at the end of the `search()` method to return just the first value from the list to your variable. Type the following:

```
alamedaFM = gis.content.search(query="title:Farmers Markets in
Alameda County")[0]
alamedaFM
```

Run the cell. You should see the display for the **Alameda County Farmers Market** feature layer.

2. Now you can check the access of the feature layer by typing in the following:

```
alamedaFM.access
```

Run the cell. You should see it return `'private'`.

3. Now that you have a feature layer, you can share it with your group using the `share()` method. You are going to use two arguments to set the organization sharing level. The `org` argument can be set to `True` or `False`. When set to `True`, it shares the item with your entire organization; when `False`, it is just shared with the group. The `groups` argument takes the ID from the group. Since you have a variable holding the group, you just have to access its ID by using the `id()` method. Type in the following:

```
alamedaFM.share(org=False,groups=farmerMarketGroup.id)
```

Run the cell. You should see the following result in the **Out** cell, but with a different `itemId`, as it is auto-generated by ArcGIS Online:

```
{'results': [{'itemId': 'df1b9d3df42e4ba8b0634f439ed8dd48',
'success': True, 'notSharedWith': []}]}
```

4. You can check the sharing level of any of your items by calling the `shared_with` property on it. To see that, type in the following:

```
alamedaFM.shared_with
```

Run the cell. You should see the result with your username as the owner:

```
{'everyone': False, 'org': False, 'groups': [<Group title:"Alameda
County Farmers Markets" owner:billparkermapping>]}
```

The result is a dictionary with values for everyone, org, and groups that the item is now shared with.

Now that you have shared some data with the group, you need to add or invite users to join your group.

Adding, inviting, and removing users from a group

Adding, inviting, and removing users from a group all use similar code; they all apply a method to the group. The methods take a list of strings of usernames as their argument. The table below displays the syntax, using our farmers' market group from the example:

`add_users`	`farmerMarketGroup.add_users(["user1","user2", . . .])`
`invite_users`	`farmerMarketGroup.invite_users(["user1","user2", . . .])`
`remove_users`	`farmerMarketGroup.remove_users(["user1","user2", . . .])`

Table 5.1: Code for adding, inviting, and removing users from a group

When any of the above code is run, the output is a dictionary with a list of users that were not added. If all users were added, the output would be `{'notAdded':[]}`. If a user was not added, the output would contain the username and details about why the user was not added. Note that *the group owner cannot be removed from the group.*

Which users can belong to a group is dependent on your organization type. In some cases, organizations do not allow users from outside your organization or public users to be part of a group.

Now you have created a group, shared data with it, and added or invited new users to that group. Next, you will see how to use the `features` module to query and update feature layers.

Using the features module to work with feature layers

ArcGIS Online displays your geographic layers as a web layer. The following are the many types of web layers that can be published in ArcGIS Online: map image layer, imagery layer, tile layer, elevation layer, feature layer, scene layer, and table. **Feature layers** are the primary web layers of vector data that you will use in ArcGIS Online. They are the feature data that you publish to your web GIS and what you display on your web maps. Feature layers can be grouped into a collection called a **feature layer collection**. You have already worked with a feature layer in this chapter, when you published the CSV of farmers' markets.

In this section, you will work with feature layers, querying a feature layer, editing the data in it, appending data to it, downloading attachments, downloading data, and deleting the feature layer.

Querying feature layers

You have seen how to search for data in *Chapter 3, ArcGIS API for Python*. You are going to search for data again to get the farmers' market feature layer you uploaded earlier in this chapter. You will find yourself querying for data a lot with `search()`, as it is a good way to get the data you need. The `get()` method can also get a feature layer, but it takes the item ID as its argument. The problem is that item IDs are long and difficult to remember; most people find it easier to remember a feature layer's name.

In this exercise, you will create a Notebook to query the farmers' market feature layer for the farmers' markets in Oakland:

1. If you closed ArcGIS Pro, open it up, navigate to where you unzipped the `Chapter5.zip` folder, and open up `Chapter5.aprx`.
2. Right-click on the `Chapter5` folder and select **New > Notebook**. Rename the Notebook to `QueryAndEditFeatureLayer`.
3. In the first cell, you are going to log into your organization's GIS through the user you are currently logged into ArcGIS Pro with. Type in the following:

   ```
   from arcgis.gis import GIS
   from IPython.display import display
   gis = GIS('home')
   ```

   Run the cell.
4. In the next cell, you will create a variable to hold the list returned from the search result, and print out the search result to see what your search returned. You are going to search for a feature layer based on a query of the title. Type in the following:

   ```
   fmSearch = gis.content.search(query="title:Farmers Markets in
   Alameda County",item_type="Feature Layer")
   fmSearch
   ```

   Run the cell.
5. In the next cell, to get the feature layer, you will use the list index and assign the index of 0 to a new variable. That variable will contain your feature layer object. Then, you will display the feature layer. Type in the following:

   ```
   farmersMarkets = fmSearch[0]
   farmersMarkets
   ```

Run the cell. You should see output that looks something like this:

```
In [35]:  1 farmersMarkets = fmSearch[0]
          2 display(farmersMarkets)
```

Farmers Markets in Alameda County

Feature Layer Collection by billparkermapping
Last Modified: January 05, 2022
0 comments, 1 views

Figure 5.5: Feature layer collection retrieved from search

6. Now you have your farmers' market feature layer. If you look at the description in the display, you can see that the farmers' market feature layer is actually a feature layer collection. You cannot run a query on a feature layer collection; you have to select an individual layer within it. To query it, you need to access each of the feature layers within the collection. First, you need to know which feature layer you want. In the next cell, create a variable to hold all the layers, and then loop through the layers, printing out the name of each layer:

   ```
   fmLayers = farmersMarkets.layers
   for layer in fmLayers:
       display(layer.properties.name)
   ```

   Run the cell. You should see the following code in the **Out** cell:

   ```
   'Farmers_Markets_in_Alameda_County'
   ```

7. Since only one name was outputted in the above step, you can see that **Farmers Market in Alameda County** only has one layer called `Farmers_Markets_In_Alameda_County`. In the next cell, create a variable to hold that layer object, and print out some of the properties:

   ```
   alamedaFM = fmLayers[0]
   print(alamedaFM.properties.geometryType)
   print(alamedaFM.properties.type)
   print(alamedaFM.properties.fields)
   ```

   Run the cell. You should see the geometry type of `esriGeometryPoint` printed out, followed by `Feature Layer` for the layer type, and a list of data dictionaries for each field.

You can see all the different properties available to you by pressing *Tab* after calling a method. Try this by typing in `alamedaFM.properties`, then press *Tab* and see all the different properties of the feature layer that you can access.

8. Now that you have a single feature layer, you can query it. To query a feature layer, you call the `query` method and pass a query to the `where` argument. In the next cell, type in the following:

```
oaklandFM = alamedaFM.query(where="City='Oakland'")
oaklandFM
```

Run the cell. You should see the following printed in the **Out** cell:

```
<FeatureSet> 8 features
```

9. What is returned to you is not a feature layer but a **feature set**. A feature set allows you to look at the properties of the feature layer you have queried. You can view the different fields and geometry of each attribute you have selected. Take a look at the geometry of each feature in the feature set. Type in the following in the next cell:

```
i = 0
while i < len(oaklandFM):
    print(oaklandFM.features[i].geometry)
    i+=1
```

Run the cell. The results are a data dictionary with `'x'`, `'y'`, and `'spatialReference'` as keys. The value for the spatial reference key is another dictionary with a **well-known ID (WKID)** and latest WKID. You should have eight data dictionaries printed to the **Out** cell, with the first two looking like below:

```
{'x': -13608573.86722754, 'y': 4552721.437081691,
'spatialReference': {'wkid': 102100, 'latestWkid': 3857}}
{'x': -13611486.986982107, 'y': 4551375.895114394,
'spatialReference': {'wkid': 102100, 'latestWkid': 3857}}
```

What is the difference between the WKID and latest WKID?

ArcGIS Online uses WGS 1984 Web Mercator (Auxiliary Sphere) projection for web layers. At the release of ArcGIS 10, the WKID for WGS 1984 Web Mercator (Auxiliary Sphere) was changed from `102100` to `3857`. Because of this, the `latestWKID` property was added to all ArcGIS releases after 10.1 to ensure backward compatibility with older versions using the older WKID.

10. In addition to the geometry, you can access the attributes using the same code as above. Just replace `geometry` with `attributes`:

    ```
    i = 0
    while i < len(oaklandFM):
        print(oaklandFM.features[i].attributes)
        i+=1
    ```

    Run the cell. The results are again a dictionary with the attributes as keys and the attribute values as the pairs. You should have eight data dictionaries in the **Out** cell, with the first two looking like below:

    ```
    {'MarketName': 'Grand Lake ', 'Days': 'Saturday', 'Time': '9 am - 2 pm', 'Location': 'Splash Pad Park', 'City': 'Oakland', 'Latitude': 37.810721, 'Longitude': -122.247899, 'ObjectId': 4}
    {'MarketName': 'Old Oakland', 'Days': 'Friday', 'Time': '8 am - 2 pm', 'Location': '9th Street and Broadway', 'City': 'Oakland', 'Latitude': 37.801171, 'Longitude': -122.274068, 'ObjectId': 5}
    ```

Now that you have seen how to query data, you will see how to edit the data that you have searched for and queried.

Editing features

In looking at the data, you notice that you have a difference in the street name field in the Oakland data. All of the Berkeley locations use the full road type of `Street`, `Way`, or `Avenue`. However, in Oakland, `Street` has been abbreviated to `St`. You are going to update all of those to change `St` to `Street`. You can edit a single feature in the edit mode within ArcGIS Online quickly; but if you had the same typo many times in a file, the method you will use in this exercise would be much more efficient.

To edit data, you create a feature set, isolate the features and attributes that need to be edited, make the edit, and then update the feature layer with your edited feature set.

1. Continue working in the `QueryAndEditFeatureLayer` Notebook from the previous exercise by checking to see if the feature layer is editable. Type in the following:

   ```
   alamedaFM.properties.capabilities
   ```

   Run the cell. The results in the **Out** cell are the capabilities enabled for the feature layer and should look like the following:

   ```
   'Create,Delete,Query,Update,Editing'
   ```

2. Once you know the edit capabilities are enabled, you can query the layer to create a feature set. You already have one created from the previous exercise with all the Oakland farmers' markets, so you can reuse that one here. You will create an empty list to hold the features that need to be edited. You will loop through the features in the feature set looking for the typo and, when found, add those features to an empty list. Do this by typing in the following:

   ```
   fmFeature = []
   for f in oakFM_features:
       print(f.attributes["Location"])
       if "St" in f.attributes["Location"]:
           fmFeature.append(f)
   ```

   Run the cell. The print statement will give you the results of each `"Location"` attribute as you loop through the features.

3. Now that you have that list of features to be updated, you can apply the update. You want to replace the word `"St"` with `"Street"`. To do that, you need to access the attribute of the `Location` field. That is done by passing the field name as the key to the attributes dictionary of each feature as you loop through them, then setting that to the new value. Since you just want to replace one word, you can use the `replace()` function. Do this by typing in the following in a new cell:

   ```
   featEditList = []
   for feat in fmFeature:
       featEdit = feat
       featEdit.attributes["Location"] = featEdit.
   attributes["Location"].replace("St","Street")
       featEditList.append(featEdit)
   ```

Run the cell. The last line of your code will print out the list of features. The features are stored as a dictionary so you see all the values. You should see that the value for `Location` has been updated, and now contains `"Street"` instead of `"St"`, as you wanted:

```
{'MarketName': 'Old Oakland', 'Days': 'Friday', 'Time': '8 am - 2
pm', 'Location': '9th Street and Broadway', 'City': 'Oakland',
'Latitude': 37.801171, 'Longitude': -122.274068, 'ObjectId': 5}
{'MarketName': 'Jack London Square', 'Days': 'Sunday', 'Time': '9
am - 2 pm', 'Location': '44 Webster Street', 'City': 'Oakland',
'Latitude': 37.793834, 'Longitude': -122.274985, 'ObjectId': 6}
{'MarketName': 'Fruitvale Farmers Market', 'Days': 'Tuesday,
Thursday', 'Time': '11 am - 7 pm', 'Location': 'Avenida de la
Fuente and 12th Street', 'City': 'Oakland', 'Latitude': 37.775899,
'Longitude': -122.224058, 'ObjectId': 9}
```

4. To edit the original feature layer, you call the `edit_features` method on it, passing through your list of updated features. By setting that equal to a variable in a new cell, you can call the variable to see the results and verify that it worked:

```
updateFM = alamedaFM.edit_features(updates = featEditList)
updateFM
```

Run the cell. You should see a dictionary returned with keys for `addResults`, `updateResults`, and `deleteResults`. `addResults` and `deleteResults` are empty lists, as you didn't add or delete anything. `updateResults` lists the `objectID`, `uniqueID`, `globalID`, and success status for each update you made. From the **Out** cell above, you see that the only features that were edited have `objectId` values of 5, 6, and 9. The results should look like below, showing an `updateResults` with an `objectID` of 5, 6, and 9 indeed having been updated:

```
{'addResults': [], 'updateResults': [{'objectId': 5, 'uniqueId': 5,
'globalId': None, 'success': True}, {'objectId': 6, 'uniqueId': 6,
'globalId': None, 'success': True}, {'objectId': 9, 'uniqueId': 9,
'globalId': None, 'success': True}], 'deleteResults': []}
```

In this section, you have seen how to edit the attributes of a feature layer. While this was only done for one field, if you had the same typo many times you could loop through all the features containing errors and edit them. This gives you base code to modify for when you have multiple fields that need to be changed. You could also edit the geometry in the same way; you would just need to access the geometry field and make edits to the *x* and *y* values. In the next section, you will see how to append new features to an existing feature layer.

Appending features

The farmers' market data says "Alameda County", but is only for Berkeley and Oakland. You have now collected the rest of the market locations in Alameda County and want to add them to your feature layer. To do that, you will upload the file geodatabase with the new features, publish it, then append them to your existing feature layer.

1. Right-click on the `Chapter5` folder and select **New** > **Notebook**. Rename the Notebook to `AppendDataToFeatureLayer`.
2. In addition to the usual GIS import statement and logging into your organization's ArcGIS Online account, you will also need to import the `zipfile` module to zip the file geodatabase, and `arcpy` and `os` to use the walk functions of each that we used in *Chapter 4*, *The Data Access Module and Cursors*. In the first cell, type in the following:

   ```
   from arcgis.gis import GIS
   import zipfile
   import arcpy
   import os
   gis = GIS('home')
   ```

   Run the cell.
3. The data you are going to append is the `AlamedaCountyAdditionalFarmersMarkets` in `Chapter5.gdb`. It contains the remaining farmers' markets in Alameda County. Since the feature class in the geodatabase has the same schema, you can append the data once you have it on your GIS. To upload a file geodatabase, it needs to be zipped. In the next cell, set the variables for the geodatabase, ZIP filename, ZIP file location, and full path of the ZIP file:

   ```
   gdb = r"C:\PythonBook\Chapter5\Chapter5.gdb"
   zipName = "AdditionalAlamedaFarmersMarket"
   zipLoc = r"C:\PythonBook\Chapter5"
   zipFull = os.path.join(zipLoc,zipName+".zip")
   ```

   Run the cell.
4. **The code for these next two steps will be written into the same cell.** You will zip the geodatabase using the `zipfile` module. This is done by calling the `ZipFile` class of the `zipfile` object and passing in the full path of the ZIP file to be created, along with `'w'` to signify writing the ZIP file. You will also create an `os.walk()` to walk through the folder location where the geodatabase is stored.

In the next cell, type in the following:

```
writeZip = zipfile.ZipFile(zipFull, 'w')
walk = os.walk(zipLoc)
```

5. Geodatabases are not normal files or folders that Python sees well. `os.walk` sees a geodatabase as the `dirpath`, and then the individual files in it as the `filenames` of the walk. To zip up a geodatabase using Python, you will loop through the walk and use a conditional `if` to find any `dirpath` values that are geodatabases. When a geodatabase is found, you will loop through its filenames. You will use a conditional `if` to test if the filename is a lock file, and if not, write it to the ZIP file using the `write` method of the `writeZip` object. You need to set the `arcname` argument of the `write` property for each filename to the full name of the file that includes the geodatabase and the file. This will ensure the ZIP file contains just the geodatabase when zipped, and not all of the folders for the full path. This is done by using the `os.path.basename()` function on the full path of the geodatabase to get just the geodatabase name. In the next cell, type in the following:

```
for dirpath, dirnames, filenames in walk:
    if dirpath == gdb:
        for filename in filenames:
            if filename[-5:] != ".lock":
                writeZip.write(os.path.join(gdb,filename),

arcname=os.path.join(os.path.basename(gdb),filename))
writeZip.close()
```

Run the cell.

A lock file can occur in a geodatabase when you have it open in ArcGIS Pro. They always end in `.lock`, and you cannot zip a lock file. Therefore, by testing for lock files and not writing them, you can zip a geodatabase that you have open in ArcGIS Pro.

6. Next, you need to create the properties of the geodatabase to be loaded into ArcGIS Online and add the item. Recall from the previous exercise that the properties are stored as a dictionary. The properties you will write in this dictionary are the title, type, tags, snippet, and description. You will pass the properties to the `add()` function, along with the path to the zipped geodatabase and the folder to store the data in. In the next cell, type in the following:

```
fmNewProperties = {
    "title":"Additional Farmers Markets In Alameda County",
    "type":"File Geodatabase",
    "tags":"Alameda County, Farmers Market, Additional",
    "snippet":"Alameda Farmers Markets to be added",
    "description":"Farmers Markets outside Oakland and Berkeley to
be added to the full feature layer"
}
fmNewGdb = gis.content.add(item_properties=fmNewProperties,
                                        data=zipFull,

folder="AlamedaFarmersMarkets")
```

Run the cell.

You do not have to publish the geodatabase since you are just using it to append data to an already published feature layer. If you wanted to display the geodatabase data on its own without appending it, you would need to publish it.

7. In the next cell, you are going to get the ID of the geodatabase item you just added. You will need this in the append function, as it takes the ID of the source data to be appended as one of the arguments. Type in the following:

```
newFmGDBId = fmNewGdb.id
```

Run the cell.

8. You need to get the feature layer to append the data to. To do this, you will use the search code that you have used previously in this chapter. You need to get the layer within the feature layer collection that has the Alameda farmers' markets locations. In both cases, you know that there is just one feature layer with the title **Farmers Markets in Alameda County** and just one layer within that. In the next cell, type in the following:

```
fmSearch = gis.content.search(query="title:Farmers Markets in
Alameda County",item_type="Feature Layer")
farmersMarkets = fmSearch[0]
fmLayer = farmersMarkets.layers[0]
fmLayer
```

Run the cell. You should see the following code in the **Out** cell, showing you that you have just a single layer now:

```
<FeatureLayer url:"https://services3.arcgis.com/HReqYJDJNUe3sQwB/
arcgis/rest/services/Farmers_Markets_in_Alameda_County/
FeatureServer/0">
```

9. Now you can append the data from the geodatabase to that feature layer. You will use the `append()` function and will need to supply it with four arguments:

   - The `item_id` of the item to append to the feature layer.
   - `upload_format`, which can take the following values: `sqlite`, `shapefile`, `filegdb`, `featureCollection`, `geojson`, `csv`, or `excel`.
   - `source_table_append` is needed when appending a file geodatabase, as you need to specify which feature class within the file geodatabase to append, even when there is only one.
   - `upsert` is used to determine if the append is also going to update the data within the feature layer. When `upsert` is set to `True` it will update the data; when `False` it will simply append it.

   In the next cell, type in the following:

```
fmLayer.append(item_id=newFmGDBId,
               upload_format='filegdb',
               source_table_name=
                            'AlamedaCountyAdditionalFarmersMarkets',
               upsert = False
              )
```

   Run the cell. If the run is successful, you will get an output of `True`.

Be careful with the `upsert` argument. The default is `True`, and if left that way you may overwrite all of your data in the feature layer with the new data instead of appending it.

10. You can now check the data in ArcGIS Online. You will see that your feature layer now has the additional farmers' markets. Now that you have this data, you can download the full set of farmers' markets. To do that, you create an export and then download the export.

The `export()` function works on a feature layer or feature layer collection, not on the individual layers of a feature layer collection. The `export()` function has two required arguments:

- `title`: A string that will be the name of the zip folder you download
- `export_format`: This can be the following types: `"Shapefile"`, `"CSV"`, `"File Geodatabase"`, `"Feature Collection"`, `"GeoJson"`, `"Scene Package"`, `"KML"`, `"Excel"`, `"geoPackage"`, or `"Vector Tile Package"`.

The `download()` function is called on the export item and takes the location of the export item to be downloaded to. You are going to download this to the same location you zipped up the geodatabase to earlier.

In the next cell, type in the following:

```
fmUpdateExport = farmersMarkets.
export(title="AllFarmersMarketsAlameda",
export_format="File Geodatabase")
fmUpdateExport.download(zipLoc)
```

Run the cell.

11. The zip file `AllFarmersMarketsAlameda` has been downloaded to your `Chapter5` folder. Now you can clean up your ArcGIS Online account by deleting data you don't need. This will help you save storage space and credits. You will delete the export item you just created, and the file geodatabase of the additional farmers' markets. The `delete()` function takes no arguments and will only work on items that do not have delete protection turned on. Type in the following:

```
fmUpdateExport.delete()
fmNewGdb.delete()
```

Run the cell. If run successfully, you will see `True` returned.

In this section, you have seen how to upload a file geodatabase to ArcGIS Online and append that data to an existing feature class. The process involves zipping the file geodatabase, adding the item to ArcGIS Online, and then appending it to an existing feature layer. You then downloaded the new feature layer to a geodatabase and deleted the export and uploaded file geodatabase to save space.

In this exercise, the existing feature layer and the geodatabase to be appended both had the same schema. If they do not, there are arguments that can be used in the `append()` function to allow for appending data with differing schemas. For more information, check the ArcGIS API for Python documentation at `https://developers.arcgis.com/python/api-reference/arcgis.features.toc.html?highlight=append#featurelayer`.

Using the mapping module to visualize your data

So far, you have been managing and updating data, creating folders and moving data there, and creating groups for sharing, all through ArcGIS API for Python. While that has been useful, all of the data is geospatial and it might be helpful to see that data displayed on a map. By working in the Jupyter Notebook environment with ArcGIS API for Python, you can visualize all of the data. In this exercise, you are going to display the farmers' market data and symbolize it by the day it is open within the Notebook environment.

1. Right-click on the `Chapter5` folder and select **New > Notebook**. Rename the Notebook to `CreateMap`.
2. You will start with the standard code to import the `arcgis` module and create a connection to your ArcGIS Online account. You will also import the `pandas` library. This will allow you to create a **Spatially Enabled DataFrame** (**SEDF**). SEDFs are objects that can easily manipulate geometric and attribute data. The `pandas` package and SEDFs will be explored in more detail in *Chapters 8*, *9*, and *10*. Type the following in the first cell:

```
from arcgis.gis import GIS
import pandas as pd
import arcgis
gis = GIS('home')
```

   Run the cell.

3. In the next cell, you will create the variable to hold the map widget and display the map. You can set multiple arguments when you create the map, such as the zoom, extent, and basemap. For this exercise, you will just set the location by typing in the city and state. Type in the following:

```
m = gis.map("Oakland,CA")
m
```

Run the cell. You will see a map displayed that is centered on Oakland, California:

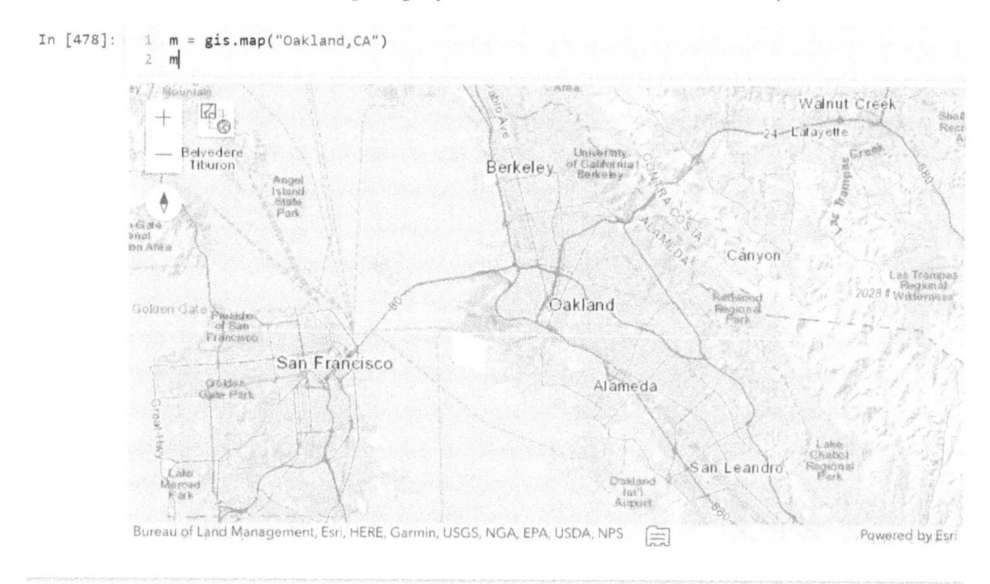

Figure 5.6: Map widget display

4. You can find out a lot of information about this map by calling the properties. Find the zoom level by typing the following in the next cell:

   ```
   m.zoom
   ```

   Run the cell. The **Out** cell will return to you the current zoom level:

   ```
   11.0
   ```

5. You can also set the zoom level by calling the property and setting it equal to an integer. Since the map will display the farmers' market data for all of Alameda County, it would look better if it were one more zoom level out. In the next cell, type in the following:

   ```
   m.zoom = 10
   ```

Run the cell. The map will update to the new zoom level:

Figure 5.7: Map with zoom level set to 10

6. You can also review the center of your map. Find the center by typing the following in the next cell:

```
m.center
```

Run the cell. The **Out** cell will return to you the center of the map as a dictionary with the spatial reference and the *x* and *y* coordinates:

```
{'spatialReference': {'latestWkid': 3857, 'wkid': 102100}, 'x':
-13611375.89013029, 'y': 4551936.947763765}
```

7. This zoom level looks good, but centering the map on Dublin would probably work better for displaying the entire county. You can use the geocoding module to find the *x* and *y* coordinates of Dublin and then set the map's *x* and *y* values to those. To first find the *x* and *y* values for Dublin, you call the `geocode()` function from the `geocoding` module. It can take many different arguments; in this example, you are going to pass it the name of the city and state, and a maximum number of locations to return. You will set the max to 1. Since the geocode function returns a list, you use list indexing to take the first, and only, value. The variable will store a dictionary of values. In the next cell, type in the following:

```
dublinLoc = arcgis.geocoding.geocode('Dublin, CA', max_locations=1)[0]
dublinLoc
```

Run the cell. The **Out** cell will be the dictionary of all the values in the `dublinLoc` variable:

```
{'address': 'Dublin, California', 'location': {'x':
-121.91634999999997, 'y': 37.70423000000005}, 'score': 100,
'attributes': {'Loc_name': 'World', 'Status': 'T', 'Score':
100, 'Match_addr': 'Dublin, California', 'LongLabel': 'Dublin,
CA, USA', 'ShortLabel': 'Dublin', 'Addr_type': 'Locality',
'Type': 'City', 'PlaceName': 'Dublin', 'Place_addr': 'Dublin,
California', 'Phone': '', 'URL': '', 'Rank': 8.67, 'AddBldg':
'', 'AddNum': '', 'AddNumFrom': '', 'AddNumTo': '', 'AddRange':
'', 'Side': '', 'StPreDir': '', 'StPreType': '', 'StName': '',
'StType': '', 'StDir': '', 'BldgType': '', 'BldgName': '',
'LevelType': '', 'LevelName': '', 'UnitType': '', 'UnitName':
'', 'SubAddr': '', 'StAddr': '', 'Block': '', 'Sector': '',
'Nbrhd': '', 'District': '', 'City': 'Dublin', 'MetroArea': 'San
Francisco Bay Area', 'Subregion': 'Alameda County', 'Region':
'California', 'RegionAbbr': 'CA', 'Territory': '', 'Zone': '',
'Postal': '', 'PostalExt': '', 'Country': 'USA', 'LangCode': 'ENG',
'Distance': 0, 'X': -121.91634999999997, 'Y': 37.70423000000005,
'DisplayX': -121.91634999999997, 'DisplayY': 37.70423000000005,
'Xmin': -121.96834999999997, 'Xmax': -121.86434999999996, 'Ymin':
37.65223000000005, 'Ymax': 37.75623000000005, 'ExInfo': ''},
'extent': {'xmin': -121.96834999999997, 'ymin': 37.65223000000005,
'xmax': -121.86434999999996, 'ymax': 37.75623000000005}}
```

8. Within the dictionary, there is a key called `"location"` that has a value of a dictionary containing the *x* and *y* locations. You can use that to set the values of the center of the map. To set the center of the map, you will enter the latitude and longitude values as a list. Remember that latitude is the *y* value and longitude is the *x* value. In the next cell, type in the following:

```
m.center = [dublinLoc["location"]["y"],dublinLoc["location"]["x"]]
```

Run the cell. The map will have updated with its center moved to Dublin:

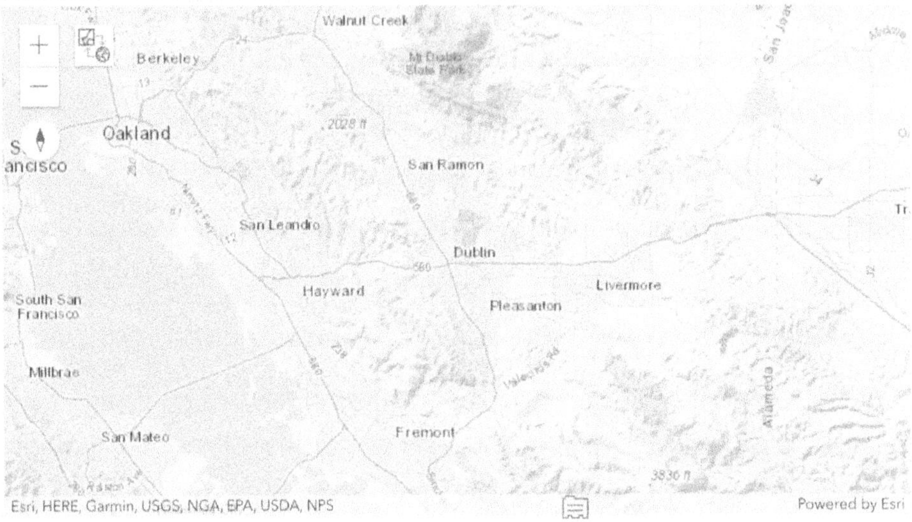

Figure 5.8: Map centered on Dublin

9. So far, you have been using the basemap that comes with the map widget. You can check what that is called by calling the `basemap` property. Type the following in the next cell:

```
m.basemap
```

Run the cell. The **Out** cell will be the basemap currently displayed:

```
'default'
```

10. You can see a list of available basemaps by calling the `basemaps` property. Type the following in the next cell:

```
m.basemaps
```

Run the cell. The **Out** cell will be a list of the basemaps available to be used:

```
['dark-gray', 'dark-gray-vector', 'gray', 'gray-vector', 'hybrid',
'national-geographic', 'oceans', 'osm', 'satellite', 'streets',
'streets-navigation-vector', 'streets-night-vector', 'streets-
relief-vector', 'streets-vector', 'terrain', 'topo', 'topo-vector']
```

11. Since this map shows the farmers' markets in Alameda County, you want to use a street map to help people navigate to them. To set the basemap to the street map, you assign the `basemap` property to `"streets"`. Type the following in the next cell:

```
m.basemap = "streets"
```

Run the cell. The map will be updated to the streets basemap:

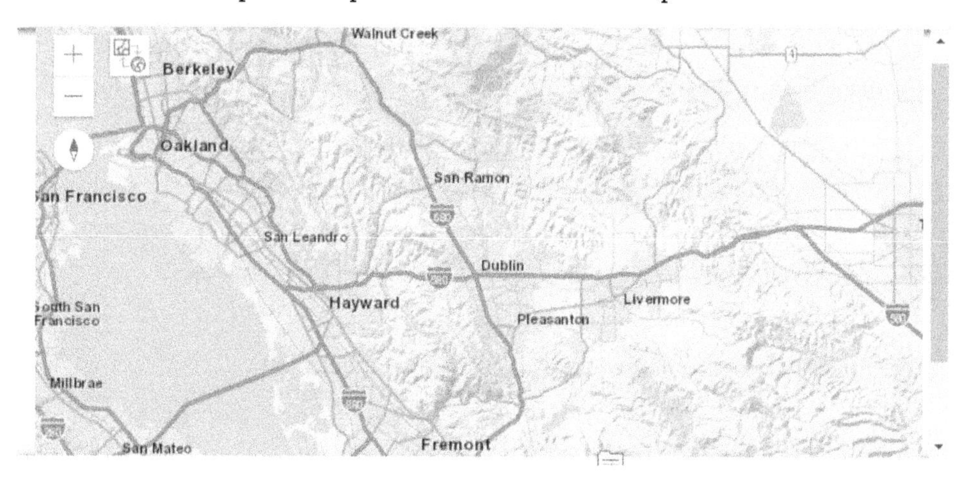

Figure 5.9: Streets basemap

12. Now that you have your basemap set up, you can add in the farmers' market layer. You do this by either getting it if you know the item ID or searching for it using the title. You can reuse the code from above to search for and get the feature layer. Type the following in the next cell:

```
fmSearch = gis.content.search(query="title:Farmers Markets in
Alameda County",item_type="Feature Layer")
farmersMarkets = fmSearch[0]
fmLayer = farmersMarkets.layers[0]
fmLayer
```

Run the cell. The output returned will be the details of the farmers' market feature layer.

At this point, you could just add the data to the map, as you have seen in other examples. But you can also take the time to create a renderer to render the data with a different visual display than the defaults. You have two options for renders. You can use **ArcGIS API for JavaScript** or create an SEDF from your feature layer and use the `plot()` method on the spatial property.

ArcGIS API for JavaScript for visualizing data will be explored in *Chapter 13, Case Study: Predicting Crop Yields*. Here, you will explore the other option:

1. You are first going to create an SEDF. Type the following in a new cell:

```
sdf = pd.DataFrame.spatial.from_layer(fmLayer)
```

2. In the same cell, you will use the `plot()` method of the SEDF's spatial property to plot the data frame on the map. The `plot()` method has a number of arguments you can set depending on the type of data and renderer used. Below is a list of those you will be using:

    - `map_widget`: The map to display the data on.
    - `renderer type`: This can be set to `'s'` for a simple renderer, `'u'` for a unique values renderer, `'c'` for a class breaks renderer, `'h'` for a heatmap renderer, or `'u-a'` for a unique values renderer that will use arcade expressions.
    - `palette`: The color map to use.
    - `col`: The column in your SEDF to be symbolized.
    - `marker_size`: The size of your marker.
    - `line_width`: The outline width of your marker.

For a full list and description, see the Plot argument here: `https://developers.arcgis.com/python/api-reference/arcgis.features.toc.html#spatialdataframe`

3. Below the code you wrote in the previous step, type in the following:

```
sdf.spatial.plot(
    map_widget=m,
    renderer_type="u",
    palette = "nipsy_spectral",
    col= "Days",
    marker_size=10,
    line_width = 0.5,
)
```

Run the cell. Scroll back up to your map and you will see something like the figure below but with different colors, as the color map chooses colors randomly:

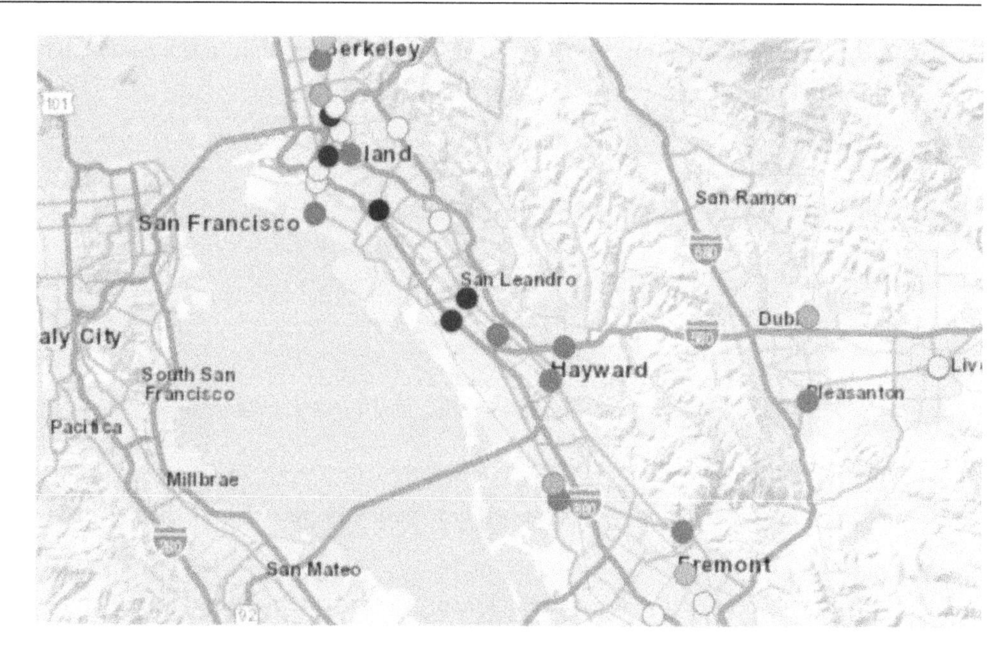

Figure 5.10: Farmers' market data displayed on the map

How do I know what color map options are available?

There are two ways to see the available color maps. Within your Notebook, you can type the following into a cell and run it:

```
from arcgis.mapping import display_colormaps
display_colormaps()
```

The **Out** cell will display all of the available color maps.

You can also find them at this website: `https://matplotlib.org/stable/tutorials/colors/colormaps.html`

4. You can save a web map from a Notebook to your GIS. Like with adding CSVs, file geodatabases, and other items, you need to create a data dictionary with the properties of the item. Then, you call the `save()` function on the map, passing the properties and the folder you want to save it to as arguments.

You will save the map to your `AlamedaFarmersMarkets` folder. In the next cell, type the following code:

```
fmMapProperties = {
    "title":"Alameda County Farmers Market - Map",
    "snippet": "Alameda County Farmers Market Map from Jupyter
Notebook",
    "tags":["Alameda County","Farmers Market","Jupyter Notebook"]
}

fmMapItem = m.save(fmMapProperties, folder="AlamedaFarmersMarkets")
```

Run the cell. You won't get any values returned to the output window, but when you navigate to the `AlamedaFarmersMarket` folder in your GIS, you will see the map saved there.

If you are unhappy with your color map, you need to remove the layer before picking a new color map. To do this, create a new cell and type in the following:

```
m.remove_layers(m.layers[0])
```

This will remove the top layer; if you want to remove a different layer, you will need to change the index value.

In this section, you have explored the map widget in the Notebook. You have seen how to add a map, and how to find and change the center, zoom, and basemap properties. You created an SEDF out of your feature layer and used the arguments of the `plot()` method to display the farmers' markets with different colors denoting the days they are open. Finally, you were able to save the map you created in your Notebook as a web map in your ArcGIS Online organization.

Summary

In this chapter, you have seen the value of managing your organizational GIS with ArcGIS API for Python. You uploaded data to your ArcGIS Online account from a Notebook. You created a group and shared data with that group. You created a folder and moved data to that folder to help organize your content. You have also seen how to find and edit attributes in a feature layer, and how to upload and append data to a feature layer. Finally, you created a map in a Notebook, added a feature layer to that as an SEDF and styled that data, and then saved the map to your ArcGIS Online account.

In the next chapter, you will continue to expand your skillset and learn how to create script tools. These are Python scripts that can be accessed through a standard ArcToolbox interface and can be made available for use by other team members. Script tools can access ArcGIS Online or local resources, and standardize your custom scripts so that non-Python experts can benefit from your code.

6

ArcToolbox Script Tools

This chapter will show you the process of turning a Python script into a **script tool**. You can take standalone scripts or Notebooks that you have written and turn them into script tools. Script tools can be run as standalone tools or integrated into models. They have a **dialog box** that looks like ArcGIS tools and contains the parameters for the tool. The parameters in the dialog box can be set to accept only certain data types, with a dropdown list of acceptable parameters to choose from and other ways to assist the user with the tool. This control over how the user interacts with the tool can reduce errors. Creating script tools is a great way to share your scripts, as it allows non-Python users in your organization to run tools you developed for specific tasks.

This chapter will cover:

- What script tools are and why they are used
- How to create a script tool
- Exercise: Turning scripts into tools

To complete the exercises in this chapter, please download and unzip the `Chapter6.zip` folder from the GitHub repository for this book: `https://github.com/PacktPublishing/Python-for-ArcGIS-Pro/tree/main/Chapter6`

Introduction to script tools

As discussed, a script tool is a tool written in Python with a tool dialog box where the user can input the parameters they want. The tool is added to an **ArcGIS Pro toolbox**, where the parameters and properties for the dialog box are set.

It has a different icon to an ArcGIS system tool or ModelBuilder model in a toolbox; the icon looks like a little scroll, and the title is something you can set:

Figure 6.1: A script tool in a toolbox

You use an interface to manually set the properties and parameters to align with how the script was written:

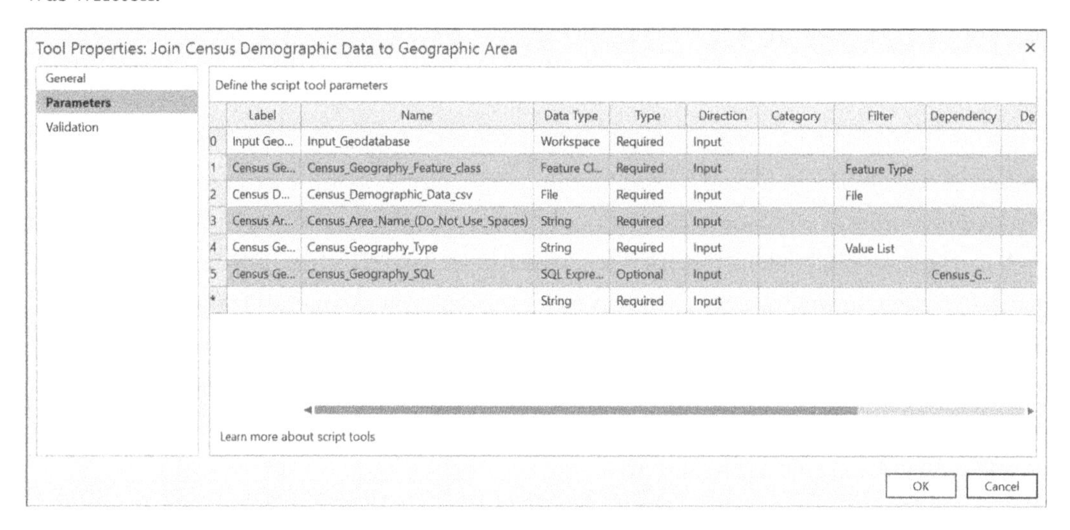

Figure 6.2: Script tool parameters

Once the script tool is created and tested, it can be used over and over by any user with access to the toolbox. The script tool will have the same interface as an ArcGIS Pro tool. This makes it so people who are new to Python within your organization can use the custom tools, as they look and feel familiar to them.

A script tool can be thought of as a way to extend the tools you already use in ArcGIS Pro. You can combine multiple geoprocessing steps into a single script tool that can be run with different inputs, like a ModelBuilder tool. Like models built in ModelBuilder, they can help you to automate tasks and can be run as standalone tools. By using Python, though, you have access to some of the ArcPy modules that are not available in ModelBuilder. You have previously seen how the Data Access module can simplify your workflow by using search, update, and insert cursors. However, these have to be used in a Python script and require knowledge of Python and ArcPy to use.

Creating a script tool offers many benefits over a standalone script. Some of these benefits are:

- A script tool has a user interface in the form of a tool dialog box. This box makes it easy to set the input and output parameters.
- It also allows for error checking, as you can set data validation to ensure the tool will work when run.
- A script tool can be implemented in your geoprocessing workflows. It can be called to run a single instance, or can be part of a model in ModelBuilder.
- You can craft custom messages as part of the tool to output information to the user through the tool dialog.
- Sharing a script tool is an easy way to share a complete geoprocessing task.
- Script tools use a tool dialog box that users unfamiliar with Python are familiar with. This allows those users to use the tool and take advantage of the added functionality Python provides without knowing Python.

If you plan to share your script or incorporate it into an existing geoprocessing workflow, you will want to make it into a script tool.

How to create a script tool

Creating a script tool is a multi-step process. Beyond writing a script to execute a set of geoprocessing tasks, you will need to do the following steps:

1. Write and test that the script will complete the desired analysis and save it as a Python file, with a `.py` extension.
2. Modify the script to take user parameters.
3. Identify or create the toolbox you will store your script tool in.
4. Add a script tool to the toolbox.
5. Associate your script to that script tool.
6. Set the parameters and properties of the script tool.
7. Test the script tool to ensure it works as intended. Make any modifications to the script or the script tool parameters or properties as needed.

A script tool must be created in a toolbox. It is where the script tool will live. Toolboxes are created as part of a project when you create a new project using a template. In this chapter, you will be working in a project that already has a toolbox. If your project does not contain a toolbox, you can create a new toolbox in the following ways:

- As a standalone toolbox in any folder on your computer. This is a useful way to organize script tools that could be used across many projects. For these tools, it can be a good idea to have a folder that contains custom toolboxes with your script tools. These tools can then be easily found and used for different projects. To create the toolbox, right-click on the folder and click **New > Toolbox.**
- Within your project. There are two ways to create a toolbox within your project:
    - Within the **Project** tab of the **Catalog** pane, right-click on the folder or geodatabase for your project and click **New > Toolbox.**
    - Within the toolboxes in the Project tab, right-click on **Toolboxes** and click **New > Toolbox.**

No matter how you decide to create your toolbox or where you store it, the process for creating a script tool within that toolbox is the same. What is important is finding a place to store your script tool where you or others in your organization can find it again to complete the geoprocessing task. This will depend on if the tool is specific to a project or is a universal tool to be used across projects.

Once you have identified the toolbox to store your script tool in, you can create the script tool by right-clicking on the toolbox and selecting **New > Script**. This will bring up the script dialog box, where you can start inputting your script tool information and parameters:

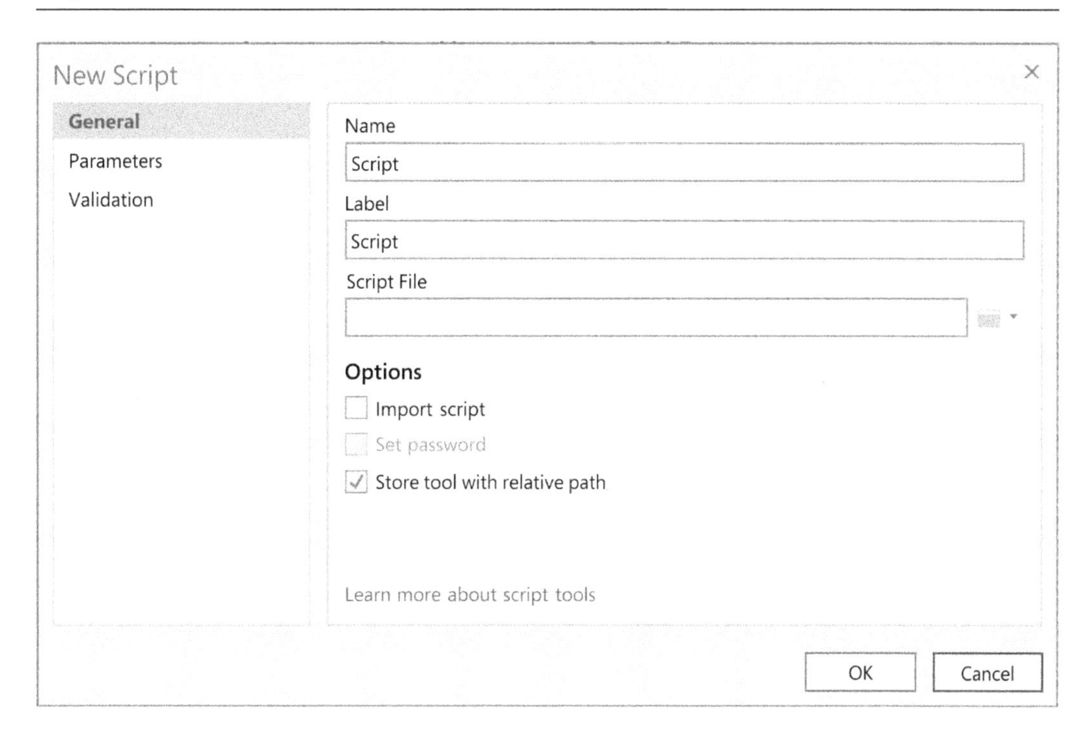

Figure 6.3: New Script dialog box

We will now take a more detailed look at the settings available for your script tool.

Script tool General settings

The **General** tab is where you will input all the information about your tool. You will need to fill in the following fields:

- **Name:** This is the name of your script tool. Just like ModelBuilder names, it cannot contain spaces or special characters.

It is a good idea to only use alphabet characters and CamelCase in the name field; this will ensure you always have a valid name.

- **Label**: This is the label of the script tool that will be displayed in the toolbox. It should be a short and descriptive name of the script tool that other people can easily read. Just like in ModelBuilder, it can contain spaces and special characters.
- **Script File**: This is where you link your Python script file to the script tool. There are two options when you click the folder button. You can either **browse to a script** or **create a new script**:
  - Creating a new script will open up a window in which you navigate to the location to save your new script to. It will not open up an editor for your new script. *It will simply create a blank template script for you to work on that you have to navigate to and open to use.*

```
"""
Tool:               <Tool label>
Source Name:        <File name>
Version:            <ArcGIS Version>
Author:             <Author>
Usage:              <Command syntax>
Required Arguments: <parameter0>
                    <parameter1>
Optional Arguments: <parameter2>
                    <parameter3>
Description:        <Description>
"""
import arcpy


def ScriptTool(parameter0, parameter1, parameter2, parameter3):
    """ScriptTool function docstring"""

    return


if __name__ == '__main__':
    # ScriptTool parameters
    parameter0 = arcpy.GetParameterAsText(0)
    parameter1 = arcpy.GetParameter(1)
    parameter2 = arcpy.GetParameterAsText(2)
    parameter3 = arcpy.GetParameterAsText(3)

    ScriptTool(parameter0, parameter1, parameter2, parameter3)
```

Figure 6.4: Template script created by selecting New Script

    - Selecting **Browse** will open a browse window, where you can browse to your script and select it to add to your script tool. This is the way you will add scripts to script tools most of the time.

- Checking **Import Script** will turn the link to your script in the script file to **embedded**. This will embed the script in the tool and store the script within the toolbox.

Script File

... embedded ...

Options

Import script

Set password

Figure 6.5: Import script option selected

When **Import Script** is checked, it allows you to check **Set Password**. Doing this will prompt you for a password; a password dialog box will pop up and you will need to enter and confirm your password. The password will be anonymized as you enter it:

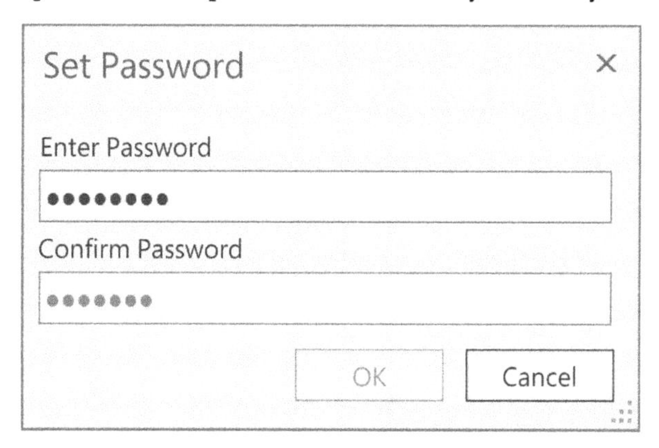

Figure 6.6: Passwords do not match (Confirm Password field is red)

The **Confirm Password** characters are red until they match the password:

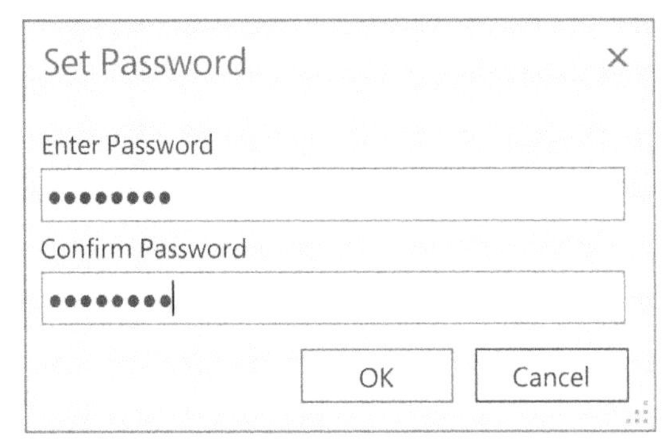

Figure 6.7: Passwords match (Confirm Password field is black)

Setting a password on a script tool allows only users with the password to see and modify the script. This is useful for script tools that are for use among a large number of individuals in your organization, as it protects against accidental changes.

- **Store tool with relative path**: This option is for if you want to store your script as a relative path. This can be beneficial if your script tool and toolbox may move around in their folder locations; it allows the script tool to look for the script by relative path, rather than absolute path.

It is a good idea to check **Store tool with relative path.** It helps keep scripts linked to script tools and is especially useful if you will be sharing script tools by sending them to other users.

Script tool Parameters tab

Once you have set the **General** settings for your script tool, you will set the script tool parameters in the **Parameters** tab.

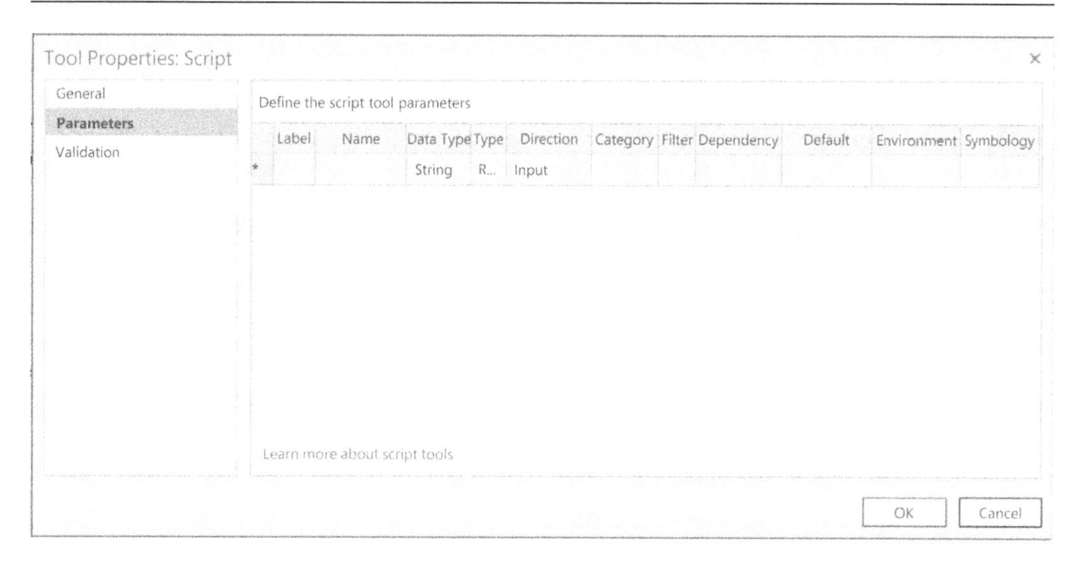

Figure 6.8: Parameter dialog box

Not all of the parameters are mandatory. Many of them have more options than will be discussed in this book. The following are the most common parameter settings:

- **Label:** The label that will appear for the input or output data in the tool dialog box when you open the script tool to run. This is where you can tell the user specific directions for input or output data.
- **Name:** The name of the parameter. It will be created from the **Label** parameter, with spaces replaced by underscores for a default value. You can change the default value if you need to, but most of the time the default value will be enough.
- **Data Type:** The type of data the parameter is. The default is **String**, but there is a list of data types to choose from. Some of the common ones you will use are as follows:

When setting a data type, the script tool expects to see that data type and will not run if provided with a different data type. For example, if you set it to **Feature Dataset** and attempt to input a feature class, the script will reject that input.

    - **String, Long, Double, Shapefile, Feature Class, Feature Dataset,** and **Raster Dataset** ensure that only these types can be input.
    - **Workspace**: This ensures a workspace is input. This can be set to a folder, or is what you would use for a geodatabase.
    - **Table**: This ensures a table is input. It can be a CSV, DBF, or geodatabase table.
    - **Field**: This ensures a field from a table, shapefile, or feature class is input. You can set the field to be chosen from a table, shapefile, or feature class that is being input into the script tool.
    - **SQL Expression**: This ensures a SQL statement is input. This statement can be created by pulling in fields and data from an input table, shapefile, or feature class. The SQL field can be used to verify SQL statements and to access data within fields of a table, shapefile, or feature class.
    - **File**: This ensures a file is input. You need to specify the file extension to be written. This is useful for reading or writing CSVs.

- **Type**: Determines if the parameter is **required**, **optional**, or **derived**:

    - **Required** parameters must be filled in, or the script tool will not run.
    - **Optional** parameters can be filled in or not, as the script will run with or without the value.
    - **Derived** parameters are output parameters that are not created in the script. Often, this is used when the output is the same as the input. This is the case when adding or calculating a field; the input and output parameters are the same.

- **Direction**: Determines if the parameter is an **input** to the script or is an **output** that will be created in the script:

    - **Input** direction are files that already exist and are being fed into the script tool.
    - **Output** direction are files that are being created by the script.

- **Category**: Allows you to group parameters into a dropdown group in the tools dialog box. This can be useful when creating a tool with a lot of parameters, some of which can be grouped together.

- **Filter**: Can be used to set different types of filters and is dependent on the **Data Type** parameter. It is an optional parameter that can be left blank, which lets the user input any values consistent with the data type. The following are some examples of filters on some of the common data types:
  - The String data type has an option for a **Value List Filter**. The Value List Filter opens a dialog box where you can input different choices for the user. When the script tool is run, the user will only be able to choose from the values in the value list:

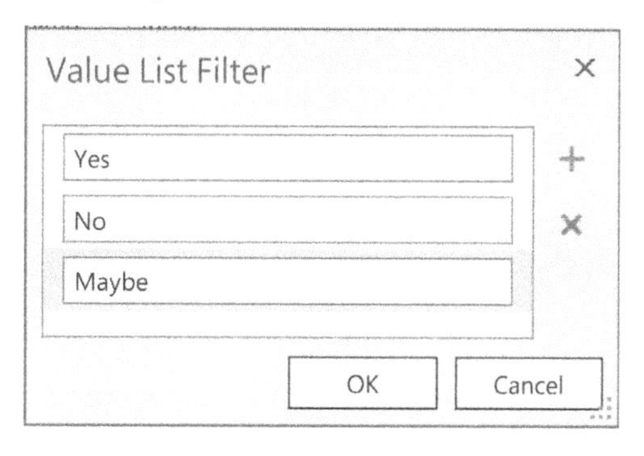

Figure 6.9: Value List Filter

  - The Long data type has an option for a Value List Filter or a **Range Filter**. The Range Filter provides the option for a minimum value and a maximum value. The input value must be within the minimum and maximum values:

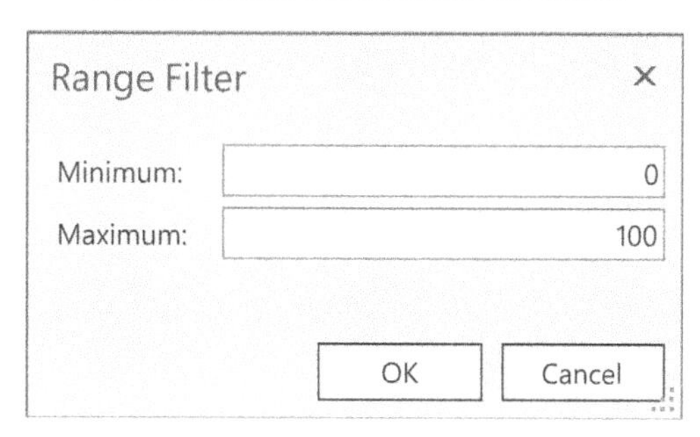

Figure 6.10: Range Filter

  - The Double data type has an option for a Value List Filter or Range Filter.

- The Shapefile data type has an option for a **Feature Type Filter**. This allows you to set a specific type of data that will be allowed as input:

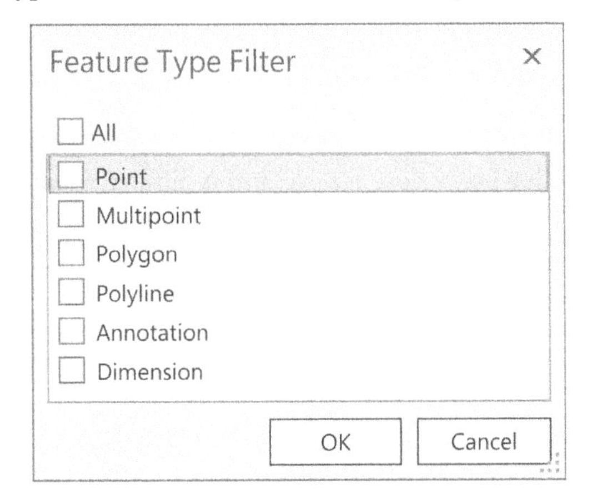

Figure 6.11: Feature Type Filter

- The Feature Class data type also has an option for a Feature Type Filter.
- The Workspace data type has an option for a **Workspace Filter**. This allows you to set a type of workspace that will be allowed. It can limit the user to a **File System**, **Local Database**, or **Remote Database**.

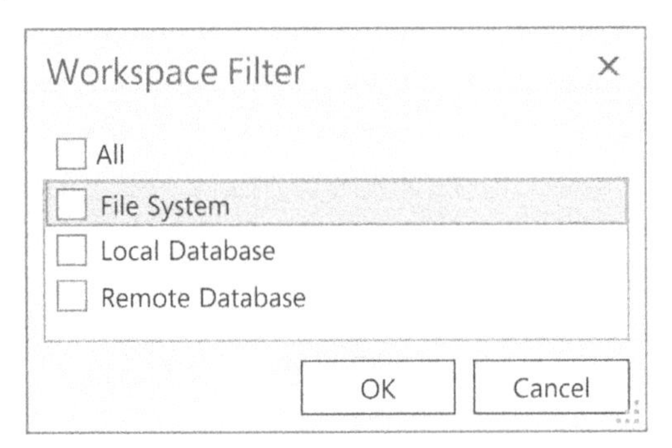

Figure 6.12: Workspace Filter

- The Field type has an option for a **Field Filter**. This allows you to set a specific field type as the only type of field that will be allowed:

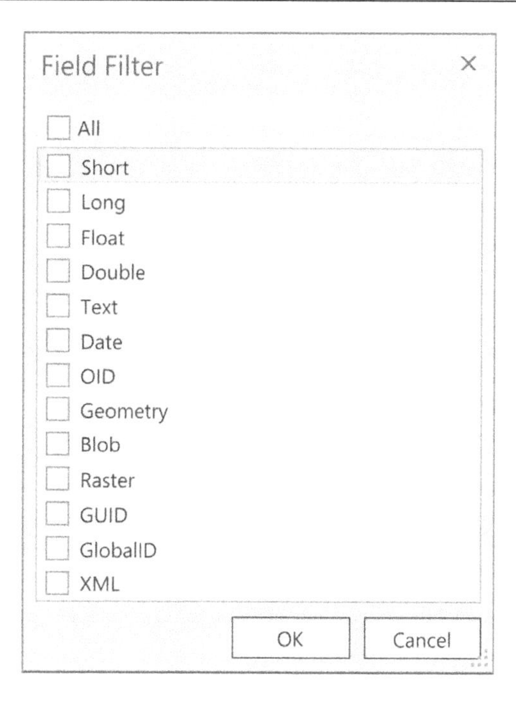

Figure 6.13: Field Filter

- The File type has an option for a **File Filter**. This allows you to specify different file types that will be allowed. The file types should be written without a period and should be separated by a semicolon if you want to allow more than one file type.

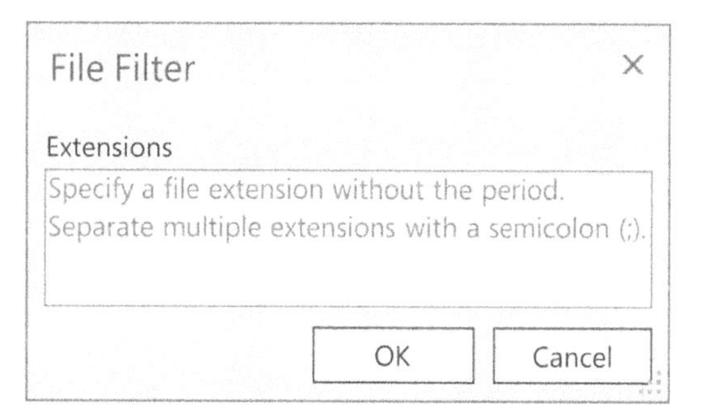

Figure 6.14: File Filter

- The Feature Dataset, Raster Dataset, Table, and SQL Expression data types have no filters.

- **Dependency**: Used to allow access from other input data. You set this to the name of a parameter above it and it will allow you to access the data in that parameter. This can be used to extract a specific field from a shapefile, feature class, or table. It can also be used to allow a SQL Expression access to the shapefile, feature class, or table, and build and validate expressions using the fields and data.

When using the Field or SQL Expression data types, it is a good idea to use the dependency to link them an input parameter. This allows your user to access the fields or to build and validate a query from the dataset they will be working on.

- **Default**: Allows you to set a default value when the user opens the script tool. It can take any type of value, but it needs to be consistent with the data type and any filters you have set.
- **Environment**: Allows you to set a parameter as a geoprocessing environment. All of the geoprocessing environment options available to you are available in this as a dropdown.

The **Environment** parameter is very useful when doing raster analysis, as you can use it to set things like the snap raster, cell size, extent, and other raster environment settings.

- **Symbology**: Allows you to set the symbology of an output dataset to the same as that of a layer file. The input to the symbology is the location of a layer file with the symbology you wish to apply.

You have now seen all of the parameter options available to you in the script tool dialog box for creating a script tool. You won't need to use each setting for every parameter, but **Label**, **Name**, **Data Type**, **Type**, and **Direction** will need to be set for each parameter.

Script tool Validation

The **Validation** panel allows you to set custom tool behavior. These custom behaviors act on parameters, allowing you to customize the parameter values even more. For instance, you could set parameters to be enabled or disabled based on the input from other parameters. You could set default values for a parameter based on the input value from other parameters. In addition to customizing parameters, you can also set custom error and warning messages.

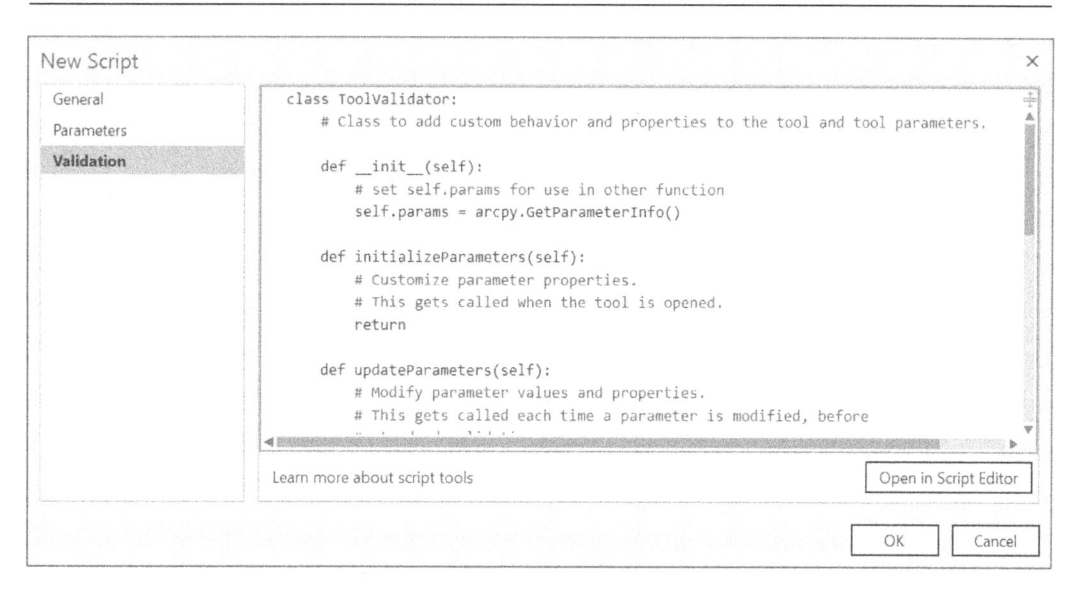

Figure 6.15: Validation panel

Tool validation is written using a Python class called `ToolValidator`. This class is what controls the look of the dialog box and how it changes based on user input. The `ToolValidator` class can only be accessed in the **Validation** panel. The code is written in Python and can be written directly into the panel or by clicking **Open in Script Editor**. *Even though you can write the code in a Python editor, the code is stored in the toolbox and not in a separate script file.*

This book will not discuss how to use the `ToolValidator` class. For more information on `ToolValidator`, explore the details in the ArcGIS Pro help.

Writing messages

As you have seen, when running standalone scripts and Notebooks, messages are printed out just to the interpreter or the Notebook. Script tools work in the same way as other geoprocessing tools and print out basic start and stop messages. In addition to these basic messages, when you create a script tool you can add **custom messages** to be printed to the dialog box. As for geoprocessing tools, these messages are stored in the geoprocessing history. The following are the ArcPy functions you can use to write custom messages:

- `arcpy.AddMessage()`: This will output general information. You can write a message as a string between quotes, you can output variables as messages, or do a combination of both.

- `arcpy.AddWarning()`: This will output a warning message. You can write a message as a string between quotes, output variables as messages, or a combination of both.
- `arcpy.AddError()`: This will output an error message. You can write a message as a string between quotes, an output variable, or a combination of both.

When using `arcpy.AddError()`, you will usually be running it after an `if` statement test. In this case, you will typically want your script set up to end after the error message. This will allow your user to see the error, fix it, and rerun the script tool.

- `arcpy.AddIDMessage()`: This allows you to output specific Esri system messages as error, informative, or warning messages. It has two required and two optional parameters. Required are the type of message (`"ERROR"`, `"INFORMATIVE"`, or `"WARNING"`) and the message ID. The message ID can be a number from `0` to `999999`. The two optional parameters are arguments that may be required depending on the message ID. This is not a message type you will explore in this book.
- `arcpy.AddReturnMessage()`: This will return all the messages from a previously run geoprocessing tool. Within scripts, the standard output messages from geoprocessing tools are not written to the dialog box. This function allows you to output them after the tool has run.

You will most often find yourself using `arcpy.AddMessage()` to write out informative messages to the user. These messages can be used to check the progress of the script or to check that it is working as you expected. It is good practice to insert messages throughout the scripts as you convert them to script tools, and you will be doing so in the following exercise.

Exercise: Turning scripts into tools

Now that you are familiar with the steps to turn a script into a script tool, and with how the dialog box is used to set up a script tool in ArcGIS Pro, it is time to create a script tool. In this section, you will work through an exercise in which you will convert a script from a previous chapter into a script tool. You will work through all the steps from the previous section to create a script tool you can share within your organization.

The script tool you will create will be from `CreateCensusTableInsertRows.ipynb` from *Chapter 4, The Data Access Module and Cursors*. In this Notebook, you took a census CSV, extracted the data you needed from it, inserted that data into a table, and then joined that table to the corresponding census geography.

You will use it to create a script tool that can be run by anyone in your organization, the result being a feature class of the simplified Hispanic/race data from the census. This tool will work with any census polygon geography, meaning you could create a census geography polygon with the Hispanic/race data of a block group, tract, place, county, or state.

The process will start with getting the script out of the Notebook and into a Python interpreter. You will modify the script to work with user parameters defined in the script dialog box. Next, you will create a new script tool in ArcGIS Pro and associate it with the script. You will add output messages, and finally, you will test it.

Exporting a Notebook to a script in ArcGIS Pro 2.8

Starting in ArcGIS Pro 2.8, an option to export your Notebook to a Python file or HTML file was added. To export your Notebook to a Python file, do the following:

If you are using ArcGIS Pro 2.7, skip down to the next section, *Copying and pasting cells to script in ArcGIS Pro 2.7*.

1. Open up ArcGIS Pro, navigate to where you unzipped the `Chapter6.zip` folder, and open up `Chapter6.aprx`.
2. Open the `CreateCensusTableInsertRows.ipynb` Notebook. This is the same Notebook you created in *Chapter 4*.
3. Click on the **Notebook** tab in the ribbon.
4. Click on the **Export** button.

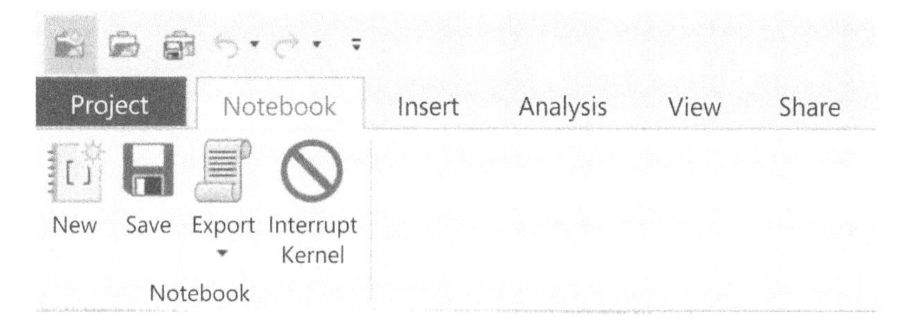

Figure 6.16: Export button

5. Click on **Export To Python File**.

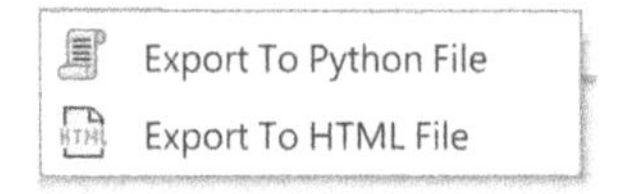

Figure 6.17: Export to Python File option

6. Navigate to the `Chapter6` project folder and add a new folder, naming it `PythonScripts`.
7. Save the Python script as `CreateCensusTableInsertRows.py`.

Once you have saved the Python file, open it up in the following way:

1. Open IDLE from your Desktop icon.
2. Click **File > Open**.
3. Navigate to where you saved the file, select it, and click **Open**.

Your file should look like this:

```
CreateCensusTableInsertRows.py - C:\PythonBook\Chapter6\PythonScripts\CreateCensusTableInsertRows.py (3.7.10)
File Edit Format Run Options Window Help
import csv, os
arcpy.env.overwriteOutput = True

gdb = r"C:\PythonBook\Chapter4\Chapter4\California.gdb"
tract = r"C:\PythonBook\Chapter4\Chapter4\California.gdb\Tracts_AlamedaCounty"
csvFile = r"C:\PythonBook\Chapter4\CensusCSV\ACSDT5Y2019.B03002_2021-07-22T010004\ACSDT5Y2019.B03002_data_with_overlays_2021-07-22T010002.csv"
table = "AlamedaCounty_RaceHispanic"
tablePath = os.path.join(gdb,table)
censusPoly = os.path.join(gdb,table+"_Tract")

fields = {"geoid_census":["GeoID_Join","STRING"],
          "total_pop":["Total Population","LONG"],
          "white":["White","LONG"],
          "prct_white":["Percent White","FLOAT"],
          "black":["Black","LONG"],
          "prct_black":["Percent Black","FLOAT"],
          "am_indian_nat_alaska":["American Indian/Native Alaskan","LONG"],
          "prct_am_indian_nat_alaska":["Percent American Indian/Native Alaskan","FLOAT"],
          "asian":["Asian","LONG"],
          "prct_asian":["Percent Asian","FLOAT"],
          "nat_hawaiian_pac_island":["Native Hawiian/Pacific Islander","LONG"],
          "prct_nat_hawiian_pac_island":["Percent Native Hawiian/Pacific Islander","FLOAT"],
          "some_other":["Some Other Race","LONG"],
          "prct_some_other":["Percent Some Other Race","FLOAT"],
          "two_or_more":["Two Or More Races","LONG"],
          "prct_two_or_more":["Percent Two Or More Races","FLOAT"],
          "hispanic_latino":["Hispanic/Latino","LONG"],
          "prct_hispanic_latino":["Percent Hispanic/Latino","FLOAT"],
          "total_minority":["Total Minority","LONG"],
          "percent_minority":["Percent Minority","FLOAT"],
          }


arcpy.management.CreateTable(gdb,table)

tableFields = []
for field in fields:
    name = field
    alias = fields[field][0]
    dataType = fields[field][1]
    print(name)
    print(alias)
    tableFields.append(field)
    arcpy.management.AddField(tablePath,name,dataType,field_alias = alias)

fileRef = open(csvFile)
csvRef = csv.reader(fileRef)

for row in csvRef:
    if csvRef.line_num <= 2:
Ln: 1 Col: 0
```

Figure 6.18: Census script exported from Notebook

This looks just like the Notebook but with all of the cells placed into a single Python file. This will be very useful for modifying the script to become a script tool.

The script exported directly from the *Chapter 4* Notebook is also provided in the `PythonScripts` folder in the `Chapter4.zip` as `CreateCensusTableInsertRows_ExportNotebook.py`.

Copying and pasting cells to a script in ArcGIS Pro 2.7

If you do not have ArcGIS Pro 2.8 and are instead working with ArcGIS Pro 2.7, you will need to use a more manual process to convert your Notebook to a script:

1. Open IDLE from your Desktop icon.
2. Click **File** > **New**.
3. Within the new script, click **File** > **Save**.
4. Navigate to the `Chapter6` project folder and add a new folder, naming it `PythonScripts`.
5. Save the Python script as `CreateCensusTableInsertRows.py`.
6. Open up ArcGIS Pro, navigate to where you unzipped the `Chapter6.zip` folder, and open up `Chapter6.aprx`.
7. Open the `CreateCensusTableInsertRows.ipynb` Notebook. This is the same Notebook you created in *Chapter 4*.
8. Click in the first cell, highlight all the text, and copy it.
9. Click in your Python script and paste the text.
10. Repeat *steps 8-9* with each cell, making sure the indentation after any `for` loops is correct.

When you are finished, your file should look like this:

```
CreateCensusTableInsertRows_Copy.py - C:/PythonBook/Chapter6/PythonScripts/CreateCensu...
File Edit Format Run Options Window Help
import csv, os
arcpy.env.overwriteOutput = True

gdb = r"C:\PythonBook\Chapter4\Chapter4\California.gdb"
tract = r"C:\PythonBook\Chapter4\Chapter4\California.gdb\Tracts_AlamedaCounty"
csvFile = r"C:\PythonBook\Chapter4\CensusCSV\ACSDT5Y2019.B03002_2021-07-22T010004\ACSDT5Y2019.B03002_data_with_overlays_20
table = "AlamedaCounty_RaceHispanic"
tablePath = os.path.join(gdb,table)
censusPoly = os.path.join(gdb,table+"_Tract")

fields = {"geoid_census":["GeoID_Join","STRING"],
          "total_pop":["Total Population","LONG"],
          "white":["White","LONG"],
          "prct_white":["Percent White","FLOAT"],
          "black":["Black","LONG"],
          "prct_black":["Percent Black","FLOAT"],
          "am_indian_nat_alaska":["American Indian/Native Alaskan","LONG"],
          "prct_am_indian_nat_alaska":["Percent American Indian/Native Alaskan","FLOAT"],
          "asian":["Asian","LONG"],
          "prct_asian":["Percent Asian","FLOAT"],
          "nat_hawaiian_pac_island":["Native Hawiian/Pacific Islander","LONG"],
          "prct_nat_hawiian_pac_island":["Percent Native Hawiian/Pacific Islander","FLOAT"],
          "some_other":["Some Other Race","LONG"],
          "prct_some_other":["Percent Some Other Race","FLOAT"],
          "two_or_more":["Two Or More Races","LONG"],
          "prct_two_or_more":["Percent Two Or More Races","FLOAT"],
          "hispanic_latino":["Hispanic/Latino","LONG"],
          "prct_hispanic_latino":["Percent Hispanic/Latino","FLOAT"],
          "total_minority":["Total Minority","LONG"],
          "percent_minority":["Percent Minority","FLOAT"],
          }

arcpy.management.CreateTable(gdb,table)

tableFields = []
for field in fields:
    name = field
    alias = fields[field][0]
    dataType = fields[field][1]
    print(name)
                                                                  Ln: 83  Col: 43
```

Figure 6.19: Census script copied from the Notebook

Both are acceptable ways to get a script from a Notebook. If you have ArcGIS Pro 2.8, it is better to export, as it removes the possibility of you missing a cell when copying and pasting. Both ways give you the same result, as you can see. The rest of the process will be the same, no matter which version of ArcGIS Pro you have.

Modifying a script to accept user input in the script tool

Now that you have a script open in a Python editor, you will need to modify it to allow user input. To do this, you need to decide what you will ask the user for and what needs to be hardcoded. We provide some guidelines for this here.

The following should be accepted as **user input**:

- Path for any input data
- Path for a workspace
- Path for any working location for intermediate data
- Path for output data when using an iterative process to create multiple outputs with names derived from variables
- SQL expressions that may change based on the data being analyzed
- Variables that will change how the script runs, especially when the variables can be value list parameters

The following should be **hardcoded**:

- Intermediate data names
- Data names that are created from other input values
- Output data names when using an iterative process to create multiple outputs with names derived from variables

Based on this, when you look at the script we have, it appears that some of the variables declared at the top are good candidates to be user input:

```
gdb = r"C:\PythonBook\Chapter4\Chapter4\California.gdb"
tract = r"C:\PythonBook\Chapter4\Chapter4\California.gdb\Tracts_
AlamedaCounty"
csvFile = r"C:\PythonBook\Chapter4\CensusCSV\ACSDT5Y2019.B03002_2021-07-
22T010004\ACSDT5Y2019.B03002_data_with_overlays_2021-07-22T010002.csv"
table = "AlamedaCounty_RaceHispanic"
tablePath = os.path.join(gdb,table)
censusPoly = os.path.join(gdb,table+"_Tract")
```

Let's go through them one by one:

- `gdb`: This is the workspace defined in the Notebook as the path for all the data. You will set this as a user-defined parameter in the script.
- `tract`: This is the input tracts geometry feature class that will be copied before having the tabular data joined to it. You will set this as a user-defined parameter in the script.
- `csvFile`: This is the input CSV file from the census bureau that will be used to create a table. You will set this as a user-defined parameter in the script.

- `table`: This is the name of the table created. If this script is always going to create the same table, you could hardcode it. If this script is going to use the name of other data to name the table, you could also hardcode it. In this case, the script will be able to take any type of census polygon geometry, and this will vary depending on your study site. Because of this, you will set *part* of the name as a user-defined parameter in the script. This will allow the user to name the table based on the area they are analyzing.
- `tablePath`: This is the full path to where the table will be written. It is currently created from the workspace and table name. This can be left as a hardcoded value, as the path will change based on user input values.
- `censusPoly`: This is the full path of the new census polygon created with the joined demographic data. It does have a value at the end for the geometry type: tract. Because of this, you will set part of the name as a user-defined parameter in the script. This will allow the user to name the file based on the geometry they are using.

Now that you have identified the variables that need to be changed to user input, you can update the code:

1. Starting with `gdb`, find the following line:

   ```
   gdb = r"C:\PythonBook\Chapter4\Chapter4\California.gdb"
   ```

   Replace it with:

   ```
   gdb = arcpy.GetParameterAsText(0)
   ```

   The `(0)` parameter is the index value. Since Python is 0-based, the first parameter is at position 0.

2. For `tract`, replace the following line:

   ```
   tract = r"C:\PythonBook\Chapter4\Chapter4\California.gdb"
   ```

   with:

   ```
   tract = arcpy.GetParameterAsText(1)
   ```

3. For `csvFile`, replace the following line:

```
csvFile = r"C:\PythonBook\Chapter4\CensusCSV\ACSDT5Y2019.
B03002_2021-07-22T010004\ACSDT5Y2019.B03002_data_with_overlays_2021-
07-22T010002.csv"
```

with:

```
csvFile = arcpy.GetParameterAsText(2)
```

4. For `table`, you are going to create a new variable to split this variable into two, so part of it can be user input and the other part hardcoded:

    a. Place your cursor at the beginning of the line with the `table` variable and press *Enter* to create a new line.

    b. On the new line, create a new variable that will hold the user-defined census geography area name. Type in the following:

    ```
    areaName = arcpy.GetParameterAsText(3)
    ```

    c. Change the value for the `table` attribute to contain the variable from the user and the name of the demographic data. Type in the following:

    ```
    table = "{0}_RaceHispanic".format(areaName)
    ```

5. For the `censusPoly` variable, you are also going to create a new variable to split this variable into two, so one part can be user input and one part hardcoded:

    a. Place your cursor at the beginning of the line with the `censusPoly` variable and press *Enter* to create a new line.

    b. On the new line, create a new variable that will hold the user-defined census geography area type. Type in the following:

    ```
    censusType = arcpy.GetParameterAsText(4)
    ```

    c. Change the value for the `censusPoly` attribute to contain the variable from the user:

    ```
    censusPoly = os.path.join(gdb,table+"_{0}".format(censusType))
    ```

You have now identified the variables to be set as user input and set them to `arcpy.GetParameterAsText()`. The next step is to create the new tool in a toolbox and define all the settings.

Creating your script tool in ArcGIS Pro

Now that you have finished updating your script to get it ready to be used in a script tool, it is time to switch to ArcGIS Pro. You have already seen the different settings available in the script tool dialog box, and now you will set them up based on the data in your script:

1. Open ArcGIS Pro, open up the `Chapter6` Project you downloaded from the GitHub repository, and find the `Chapter6.tbx` file:

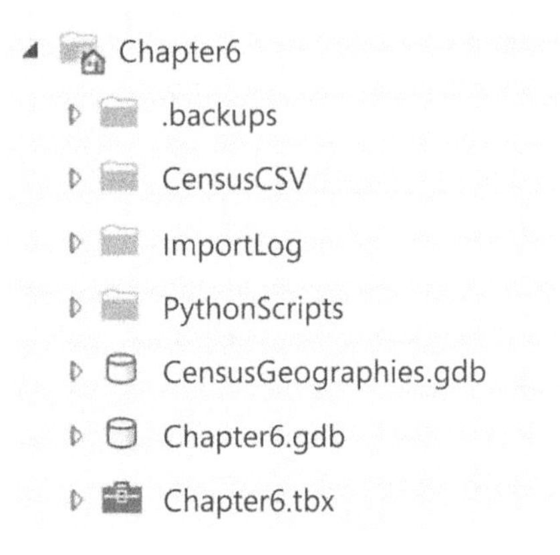

Figure 6.20: Chapter 6 Project and toolbox

2. Right-click on `Chapter6.tbx` and select **New** > **Script**:

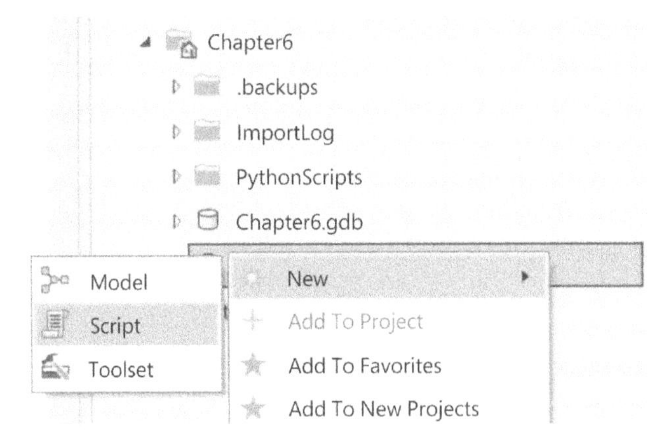

Figure 6.21: Creating a new script tool

3. You will see the **New Script** tool dialog box. Start with the **General** tab and fill in the following:

    a. For **Name**, give your script a simple name without special characters. Type in `JoinCensusDemographicData`.

    b. For **Label**, give your script tool a short label that describes what the script tool does. This label is what will be displayed in the toolbox for the script tool. Type in `Join Census Demographic Data to Geographic Area`.

    c. For **Script File**, click the folder icon, navigate to where you stored your `CreateCensusTableInsertRows.py` script, and select it.

    d. Leave **Import Script** unchecked, as you don't want to import the script until you have tested it and verified that it works properly.

    e. Leave **Store tool with relative path** checked to keep the script stored as a relative path within the script tool.

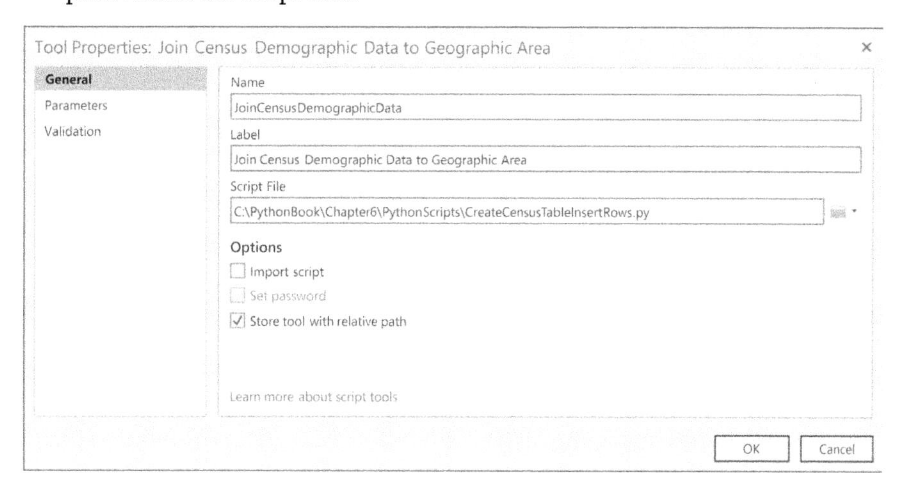

Figure 6.22: Script tool General tab setup

4. Now click on the **Parameters** tab to set all of the tool parameters.

5. The first parameter to set is the `gdb` parameter from the script. Define the script tool parameter for it by entering the following:

    a. **Label**: `Input Geodatabase`

    b. **Name**: Take the default created.

    c. **Data Type**: Workspace

d. **Type**: Required
e. **Direction**: Input
f. **Category**, **Filter**, **Dependency**, **Default**, **Environment**, and **Symbology** will all be left blank.

6. The second parameter is the `tract` feature class from the script. Define the script tool parameter for it by entering the following:

   a. **Label**: `Census Geography Feature class`

   "Tract" is the census geography, but since this tool will work with any geography you will call the label `Census Geography Feature Class`.

   b. **Name**: Take the default created.
   c. **Data Type**: Feature Class
   d. **Type**: Required
   e. **Direction**: Input
   f. **Filter**: Select the **Feature Type** filter and select **Polygon** to ensure only polygon data is allowed as input.
   g. **Category**, **Dependency**, **Default**, **Environment**, and **Symbology** can all be left blank.

7. The third parameter is the `csvFile` from the script. Define the script tool parameter for it by entering the following:

   a. **Label**: `Census Demographic Data CSV`
   b. **Name**: Take the default created.
   c. **Data Type**: File
   d. **Type**: Required
   e. **Direction**: Input
   f. **Filter**: Select the **File** filter type and type in `CSV` to ensure only a CSV file is allowed as input.
   g. **Category**, **Dependency**, **Default**, **Environment**, and **Symbology** will all be left blank.

8. The fourth parameter is the `areaName` from the script. Define the script tool parameter for it by entering the following:

    a. **Label:** `Census Area Name (Do Not Use Spaces)`

    You can put hints and directions in the label to the tool user. In this case, you are telling them not to use spaces in the name because that would cause the tool not to work. Later you will see how to check for spaces, throw a warning, and fix the issue within the script.

    b. **Name:** Take the default created.
    c. **Data Type:** String
    d. **Type:** Required
    e. **Direction:** Input
    f. **Category**, **Filter**, **Dependency**, **Default**, **Environment**, and **Symbology** can all be left blank.

9. The fifth parameter is the `censusType` from the script. Define the script tool parameter for it by entering the following:

    a. **Label:** `Census Geography Type`
    b. **Name:** Take the default created.
    c. **Data Type:** String
    d. **Type:** Required
    e. **Direction:** Input
    f. **Filter:** Select the **Value List Filter** type and type in the following values:

        - `Block`
        - `BlockGroup`
        - `Tract`
        - `Place`
        - `County`
        - `State`

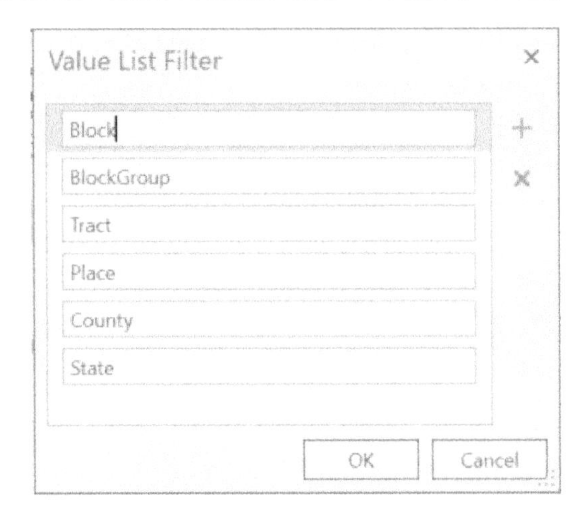

Figure 6.23: Value List Filter

g. **Category**, **Dependency**, **Default**, **Environment**, and **Symbology** can all be left blank.

10. Check that your tool parameters look like the following figure and click **OK**:

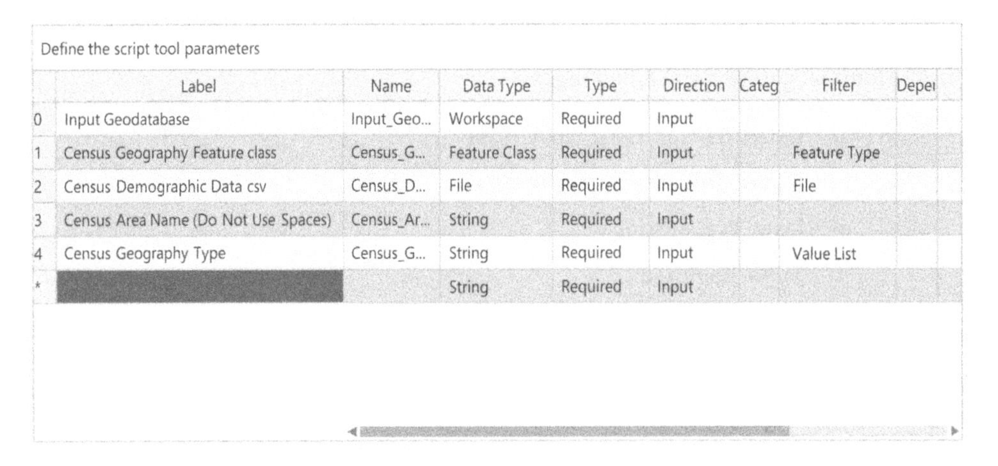

Define the script tool parameters

	Label	Name	Data Type	Type	Direction	Categ	Filter	Depe
0	Input Geodatabase	Input_Geo...	Workspace	Required	Input			
1	Census Geography Feature class	Census_G...	Feature Class	Required	Input		Feature Type	
2	Census Demographic Data csv	Census_D...	File	Required	Input		File	
3	Census Area Name (Do Not Use Spaces)	Census_Ar...	String	Required	Input			
4	Census Geography Type	Census_G...	String	Required	Input		Value List	
*			String	Required	Input			

Figure 6.24: Tool properties - Parameters

Your script tool is now ready to run and test.

Running and testing the script tool

The script tool should be displayed in your `Chapter6` toolbox, ready to run for testing. It will run just like a geoprocessing tool. To run a test, you will run the same data through it as you ran in *Chapter 4*.

You will be using the data in `CensusGeographies.gdb` and the `CensusCSV` folder that you downloaded from the GitHub site for this chapter:

1. Double-click on the script tool to open up its geoprocessing window:

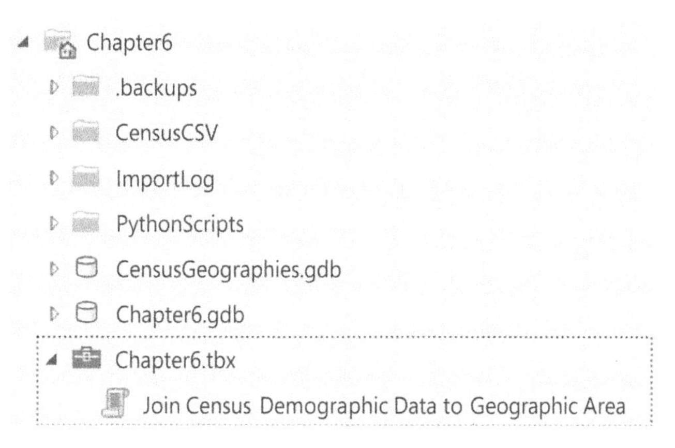

Figure 6.25: Script tool location

2. A geoprocessing window will open, with **Join Census Demographic Data to Geographic Area** (your script tool label) as the header, and all the tool parameters labeled:

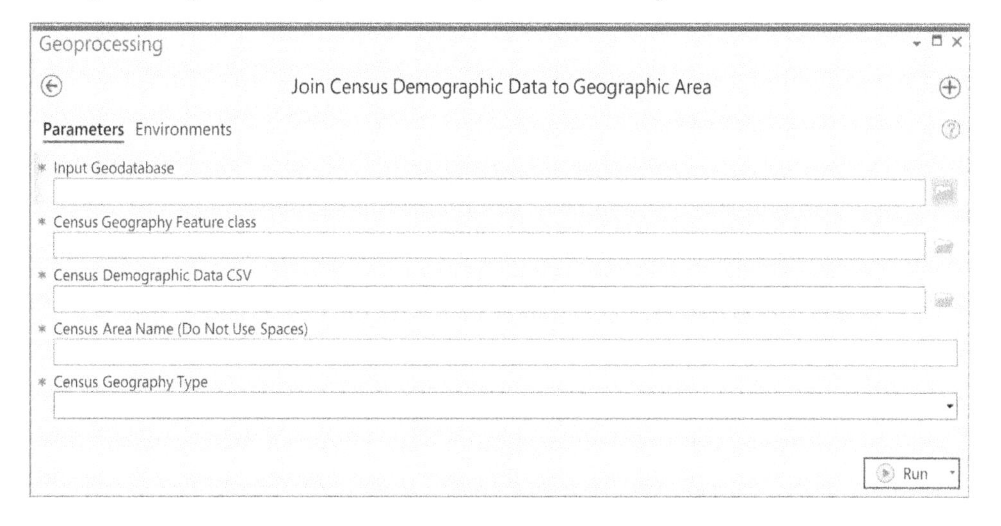

Figure 6.26: Script tool Geoprocessing window

Fill in the tool parameters with the following:

a. **Input Geodatabase:** `C:\PythonBook\Chapter6\Chapter6.gdb`
b. **Census Geography Feature Class:** `C:\PythonBook\Chapter6\CensusGeographies.gdb\AlamedaCounty`
c. **Census Demographic Data CSV:** `C:\PythonBook\Chapter6\CensusCSV\ACSDT5Y2019.B03002_2021-07-22T010004\ACSDT5Y2019.B03002_data_with_overlays_2021-07-22T010002.csv`
d. **Census Area Name:** `AlamedaCounty`
e. **Census Geography Type:** `Tract`

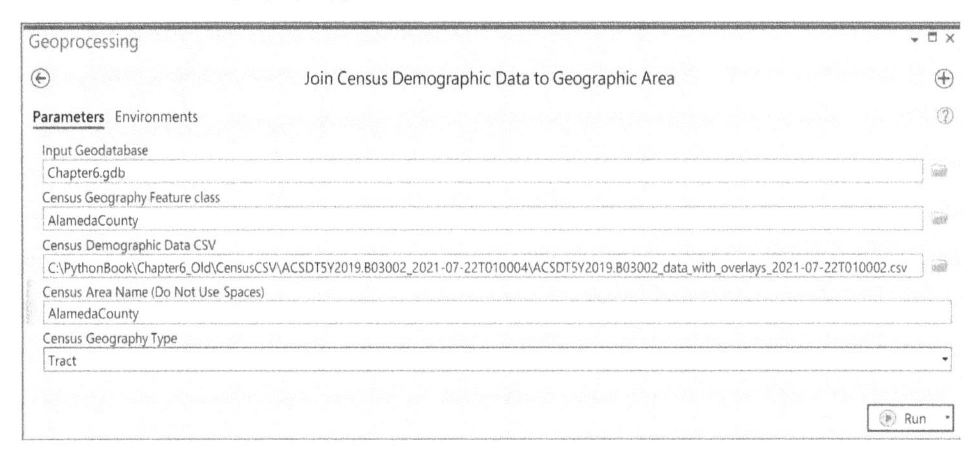

Figure 6.27: Script tool Geoprocessing window with parameters filled in

3. Click **Run** and the tool will take 1-3 minutes to run. When it is finished running, you will have a new table and feature class in the `Chapter6.gdb` file. Add them to a map and view the data to see that the demographic data from the CSV has now been joined to a new feature class.

If you do not see the new feature class and table after running, right-click on the `Chapter6.gdb` file and select **Refresh**. When a script tool is run, it does not always refresh the geodatabase the data was written to like a geoprocessing tool, so sometimes you need to do that manually after running to see the datasets.

You now have a script tool that a user with no knowledge of Python can run to create a new feature class of a census geography joined with Hispanic/race demographic data for that geography.

You may notice, though, that the input census feature class was one that you had already done some geoprocessing on. In addition, there is no test to ensure the user didn't use spaces in the `areaName` field. You also haven't added any custom messages, so you didn't have any output written to the geoprocessing window as the script tool ran. In the next section, you will work on all of these.

Updating the script tool to take census geography files

In the preceding exercise, you used a census geography file for the tracts in Alameda County. That file was one you created in *Chapter 4*. However, the census geography for tracts that you download from the census contains all the tracts in the state. If you were to run that file through the script tool, you would still get the CSV data for Alameda County joined. However, the output feature class would contain all the tracts in California, and those not in Alameda County would have null values. That is not what you want; it is a larger area than you need and the null values are not ideal.

In this exercise, you will update the script tool to allow the user to create an optional SQL query from the input census geography when creating the new feature class, and join the table to that new feature class:

1. Open up the `CreateCensusTableInsertRows.py` file that is linked to your script tool from ArcGIS Pro by right-clicking on the **Join Census Demographic Data to Geographic Area** script tool in the **Project** pane and selecting **Edit**.
2. Add a new line after:

   ```
   censusPoly = os.path.join(gdb,table+"_{0}".format(censusType))
   ```

   by clicking at the end of the line and pressing *Enter*.
3. On this new line, you will declare a variable for the SQL statement and set it to `arcpy.GetParameterAsText()`. Type in the following:

   ```
   sql = arcpy.GetParameterAsText(5)
   ```

4. Scroll to the bottom of the script. You will replace `CopyFeatures()` with `Select()`. Delete the following line:

   ```
   arcpy.management.CopyFeatures(tract,censusPoly)
   ```

5. On the same line you just deleted, you will write a `Select()` function. As we've seen before, the `Select()` function takes three arguments:

- **Input feature**: This can be a feature class or layer and is a mandatory field.
- **Output feature**: This can be a feature class or layer and is a mandatory field.
- **SQL statement**: This is an SQL statement that will be applied to the input feature class, and is an optional field.

Type in the following:

```
arcpy.analysis.Select(tract,censusPoly,sql)
```

6. Save the script file and close it.
7. In ArcGIS Pro, you will update the script tool to take the new `sql` parameter. Right-click on the **Join Census Demographic Data to Geographic Area** script tool and select **Properties**. Select the **Parameters** tab in the **Tool Properties** dialog box.
8. In the **Parameters** tab, add the following to create a new parameter at the bottom:
    a. **Label**: `Census Geography SQL`
    b. **Name**: Take the default created.
    c. **Data Type**: SQL Expression
    d. **Type**: Optional
    e. **Direction**: Input
    f. **Dependency**: Select **Census_Geography_Feature_Class**. This will allow you to access the attributes for that input feature class to build your SQL Expression.
    g. **Category**, **Default**, **Environment**, and **Symbology** can all be left blank.
9. Click **OK** to close the dialog box.

The script tool has been updated. You can now create a query of the input feature class to select only the areas you have downloaded a CSV for. Take a look at the new script tool **Geoprocessing** dialog box by double-clicking on the script tool:

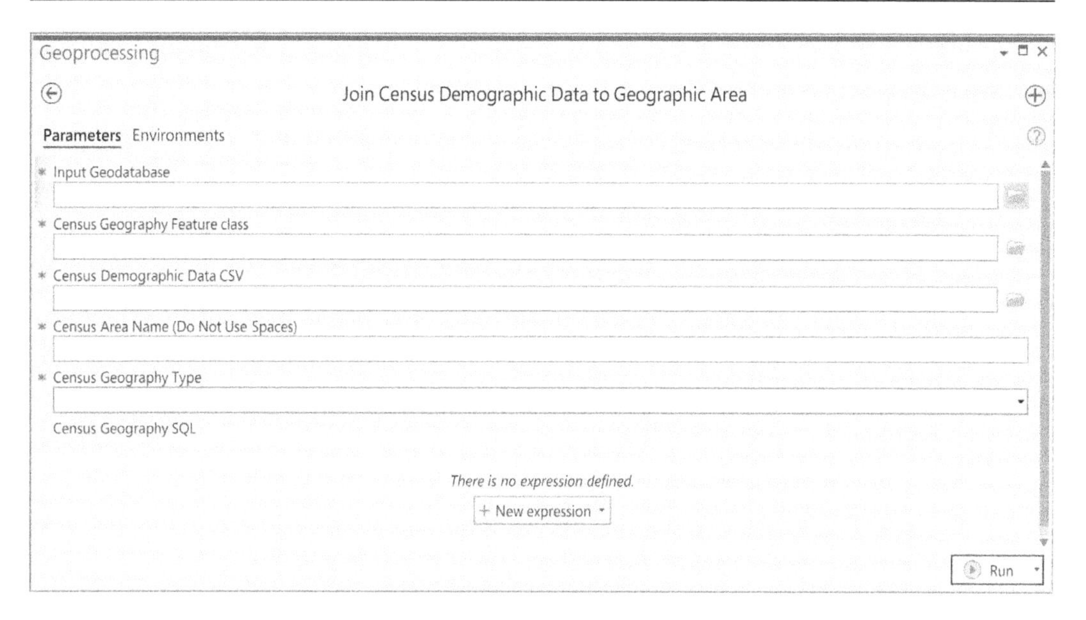

Figure 6.28: Join Census Demographic Data to Geographic Area, with a SQL statement

If your geographic area already matches the data in the CSV, you can leave the SQL statement blank. The `Select()` tool will run with a blank SQL expression and just select everything, making it work just like the `CopyFeatures()` tool.

Before running a test of this, you will add a test and warning message to make sure that the **Census Area Name** parameter does not contain spaces.

Testing input parameters

The input parameter for the **Census Area Name** is currently going directly into the code to be part of the name for the table and feature class being created. If there are spaces in that value, it will cause an error. You have already placed a reminder to the user in the label about this, but it might not always be followed. Therefore, in this exercise, you will create a test in Python to see if a space is present.

If it is, you will send a warning message and fix the space. This will ensure that the script will still run even if the user makes a mistake.

1. Open up the `CreateCensusTableInsterRows.py` file that is linked to your script tool from ArcGIS Pro by right-clicking on the **Join Census Demographic Data to Geographic Area** script tool and selecting **Edit**.
2. Add a line after `areaName = arcpy.GetParameterAsText(3)`.
3. In the new line, you will test for the presence of a space in the `areaName` variable. To do this, you will use the `.find()` method. The `.find()` method returns the position of a character in a string. If a character is present more than once, `.find()` returns just the first position. If a character is not present in the string, `.find()` returns `-1`. You will use this to write a conditional to test if there are any spaces. If there are, then you will use the `.replace()` method to remove the spaces, replacing them with nothing. You will also use the `arcpy.AddWarning()` message to print out a warning message when the script is running. Type in the following:

```
if areaName.find(" ") != -1:
    areaName = areaName.replace(" ","")
    arcpy.AddWarning("areaName input had spaces and has been updated
to the following {0}".format(areaName))
```

You don't need to put an `else` statement after the `if`; if there is no space in the `areaName` attribute, the script tool can just run with it as it is.

4. Save the script, but do not close it. You will continue working on it in the next section.

The script tool has now been updated to fix an input value if it is entered incorrectly and print a warning message. The next step to finishing the script tool is adding custom messages to print out its progress.

Adding custom messages

Adding custom messages to a script tool is valuable. It helps the user to see what has been run. It can also help you to troubleshoot the script tool as you build it by showing you the output at areas that are not working. Ultimately, adding messages makes your script tool appear much more like a geoprocessing tool, allowing you to track your progress. Let's see how you can do this for your script tool:

1. If you closed the script tool after the last section, open it up by right-clicking on the **Join Census Demographic Data to Geographic Area** script tool and selecting **Edit**.
2. You are going to add some custom messages, but first you will convert any `print` statements from when the code was in a Notebook to `arcpy.AddMessage` statements. In the script header, click **Edit** > **Replace** to get the **Replace Dialog** box.
3. In the **Find** box, type in `print`.
4. In the **Replace with** box, type in `arcpy.AddMessage`.

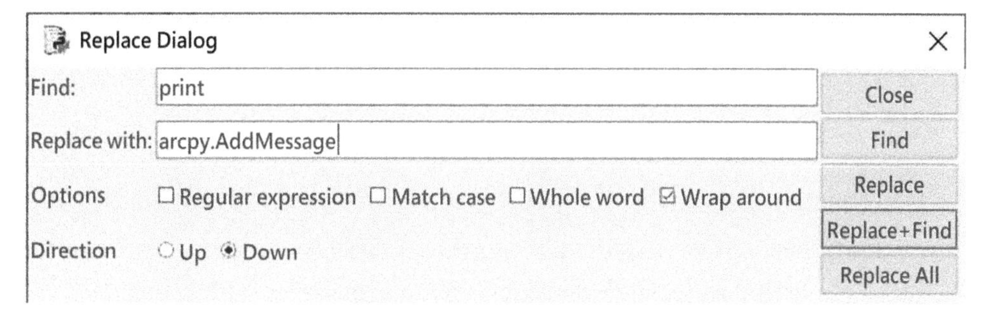

Figure 6.29: Replace Dialog box

5. Click the **Replace** button to replace each instance of `print`, or click the **Replace All** button to replace them all.

If you have the word "print" anywhere other than print statements, it will replace them as well when you click **Replace All**, so be careful when doing that. By clicking **Replace** to replace each instance, you can see what you are replacing.

6. Now you will add some more custom messages so the user can view what is being done in the tool:

   a. Below `arcpy.env.overwriteOutput = True`, type in the following to let the user know the script is starting:

   ```
   arcpy.AddMessage("Starting . . .")
   ```

   b. Above `arcpy.management.CreateTable(gdb,table)`, type in the following to let the user know the script is creating a new table:

   ```
   arcpy.AddMessage("Creating a new table for the csv data")
   ```

c. Within the `for field in fields:` loop, find the following line:

```
arcpy.management.AddField(tablePath,name,dataType,field_alias
= alias)
```

Below it, type in the following to let the user know the script is adding fields to the table:

```
arcpy.AddMessage("Adding field {0} to the table".
format(field))
```

d. Above `for row in csvRef:`, type in the following to let the user know the script is inserting the CSV data into the table:

```
arcpy.AddMessage("Inserting csv values into table,
{0}".format(tablePath))
```

e. Above `arcpy.analysis.Select(tract,censusPoly,sql)`, type in the following to let the user know the script is running the `Select()` function:

```
arcpy.AddMessage("Selecting out geographies to join table to")
```

f. Above `arcpy.management.JoinField(censusPoly,"GEOID",tablePath,"geoid_census",tableFields)`, type in the following to let the user know the select is done and the join is being run:

```
arcpy.AddMessage("Select has finished, joining the table to
the new feature class")
```

g. Below `arcpy.management.JoinField(censusPoly,"GEOID",tablePath,"geoid_census",tableFields)`, type in the following to let the user know the script has finished:

```
arcpy.AddMessage("Finished")
```

7. Save the script and close it.

Why add arcpy.AddMessage() statements?

Just like how print statements can be useful in debugging a standalone script, you can use `arcpy.AddMessage()` to debug your script tool. This is useful when you become comfortable creating script tools and start writing your scripts from the start for use in a script tool. It can allow you to skip the step of testing them as standalone scripts.

Testing the finished script tool

An important part of creating a script tool is testing it. In this section, you will test your script tool on tract, place, and county CSVs. In addition, for one of these test runs you will add a space to the `areaName` parameter to check that the script tool runs properly and fixes the input.

Testing SQL with Contra Costa Tract data

In many situations, you will have downloaded the census geography files for a larger area than you have downloaded your demographic data for. Instead of joining the demographic CSV to the full geography, you can run an SQL query on the geography file to create a geography file that only has the same area as your demographic data. In this exercise, you will do this by using the Contra Costa Tract data in the `CensusCSV` folder along with the full State of California tract geography file in `CensusGeographies.gdb`.

1. Double-click on the **Join Census Demographic Data to Geographic Area** script tool to open up its geoprocessing window.
2. Fill in the tool parameters with the following:

   - **Input Geodatabase**: `C:\PythonBook\Chapter6\Chapter6.gdb`
   - **Census Geography Feature Class**: `C:\PythonBook\Chapter6\CensusGeographies.gdb\tl_2019_06_tract`
   - **Census Demographic Data CSV**: `C:\PythonBook\Chapter6\CensusCSV\ContraCostaTract_HispanicRace\ACSDT5Y2019.B03002_data_with_overlays_2021-08-21T201533.csv`
   - **Census Area Name**: `ContraCostaCounty`
   - **Census Geography Type**: Tract

- **Census Geography SQL**: Click **+ New Expression** and build the following expression: **Where COUNTYFP is equal to 013** (the FIPS County code for Contra Costa County is 013)

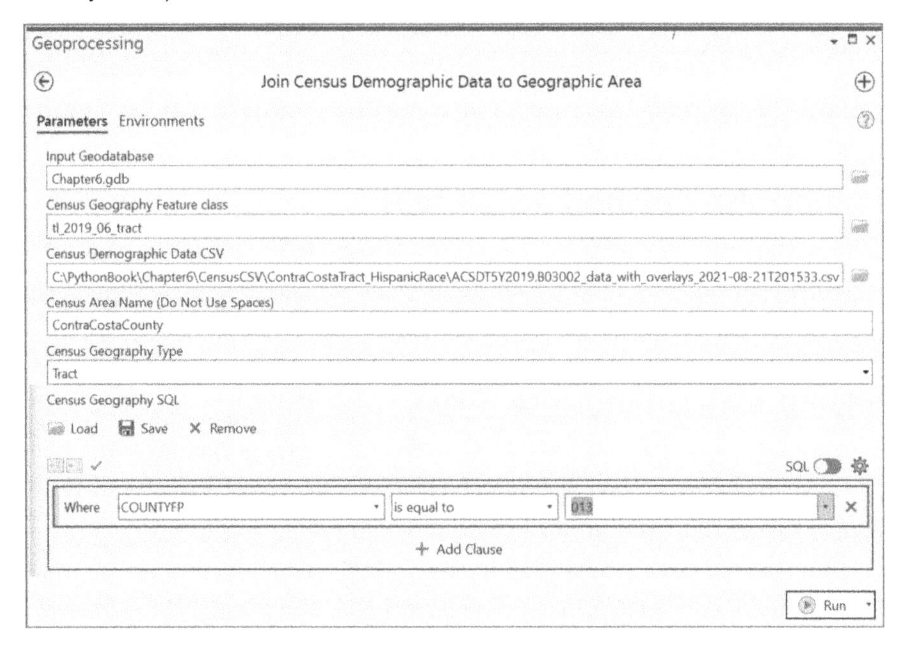

Figure 6.30: Script tool Geoprocessing window with parameters for our Contra Costa County test

3. Click **Run** to run. To view the messages when running, you need to click **View Details** in the **Geoprocessing** window followed by **Messages** in the **Details** window:

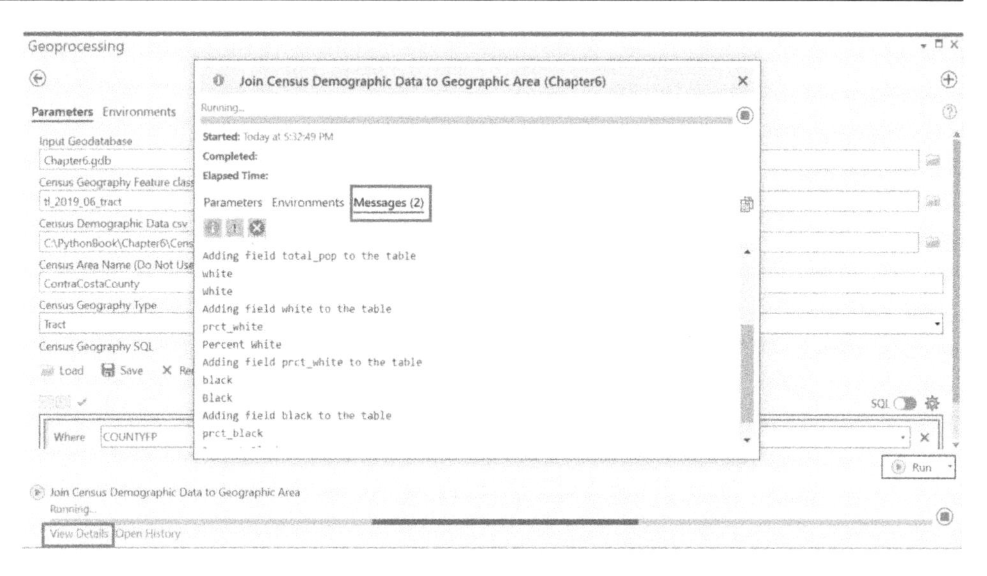

Figure 6.31 Custom messages printed out

4. When finished, you can load the `ContraCostaCounty_RaceHispanic_Tract` feature class into a map to view the data. You will see that the script tool has created a new feature class of just Contra Costa County tracts with Hispanic/race data.

This test shows that you can use tract-level census geography and a SQL statement to create a new feature class that contains just the demographic data for a specific area. In the next section, you will test the script on a different census geography.

Testing the script with California county geography

The census geography file contains all the counties within the United States. In this example, you only have demographic data for California. Like the previous example, you will use the SQL statement so that your output feature class is just the counties in California.

You will use the US county data in the `CensusCSV` folder along with the full United States county geography file in `CensusGeographies.gdb`.

1. Double-click on the **Join Census Demographic Data to Geographic Area** script tool to open up its geoprocessing window.
2. Fill in the tool parameters with the following:

    a. **Input Geodatabase:** `C:\PythonBook\Chapter6\Chapter6.gdb`

    b. **Census Geography Feature Class:** `C:\PythonBook\Chapter6\CensusGeographies.gdb\tl_2019_us_county`

    c. **Census Demographic Data CSV:** `C:\PythonBook\Chapter6\CensusCSV\CaliforniaCounty_HispanicRace\ACSDT5Y2019.B03002_data_with_overlays_2021-08-21T201639.csv`

    d. **Census Area Name:** `California`

    e. **Census Geography Type:** County

    f. **Census Geography SQL:** Click **+ New Expression** and build the following expression: **Where STATEFP is equal to 06** (the FIPS State code for California is 06)

Figure 6.32: Script tool Geoprocessing window with parameters for the county test

3. Click **Run** to run. To view the messages when running, you need to click **View Details** in the **Geoprocessing** window and then **Messages** in the **Details** window.
4. When finished, you can load the `California_RaceHispanic_Tract` feature class into a map to view the data. You will see that the script tool has created a new feature class of just the California counties with Hispanic/race data.

This test shows that the script tool will work with census county geography files too. You now know that your script tool will work on tract and county geographies. In the next section, you will test it on place geography and make an error in the input to see how it is handled.

Testing the script with a space in the area name

You will use the `OaklandBerkeley` Place data in the `CensusCSV` folder along with the full California Place geography file in `CensusGeographies.gdb`. In this test, you will leave a space in the **Census Area Name** parameter to check if the script is dealing with it correctly.

1. Double-click on the **Join Census Demographic Data to Geographic Area** script tool to open up its geoprocessing window.
2. Fill in the tool parameters with the following:

    a. **Input Geodatabase**: `C:\PythonBook\Chapter6\Chapter6.gdb`

    b. **Census Geography Feature Class**: `C:\PythonBook\Chapter6\CensusGeographies.gdb\tl_2020_06_place`

    c. **Census Demographic Data CSV**: `C:\PythonBook\Chapter6\CensusCSV\OaklandBerkeleyPlace_HispanicRace\ACSDT5Y2019.B03002_data_with_overlays_2021-08-21T201855.csv`

    d. **Census Area Name**: `Oakland Berkeley`

    e. **Census Geography Type**: Place

f. **Census Geography SQL**: Click **+ New Expression** and build the following expression: **Where NAMELSAD includes the value(s) Berkeley city,Oakland city**

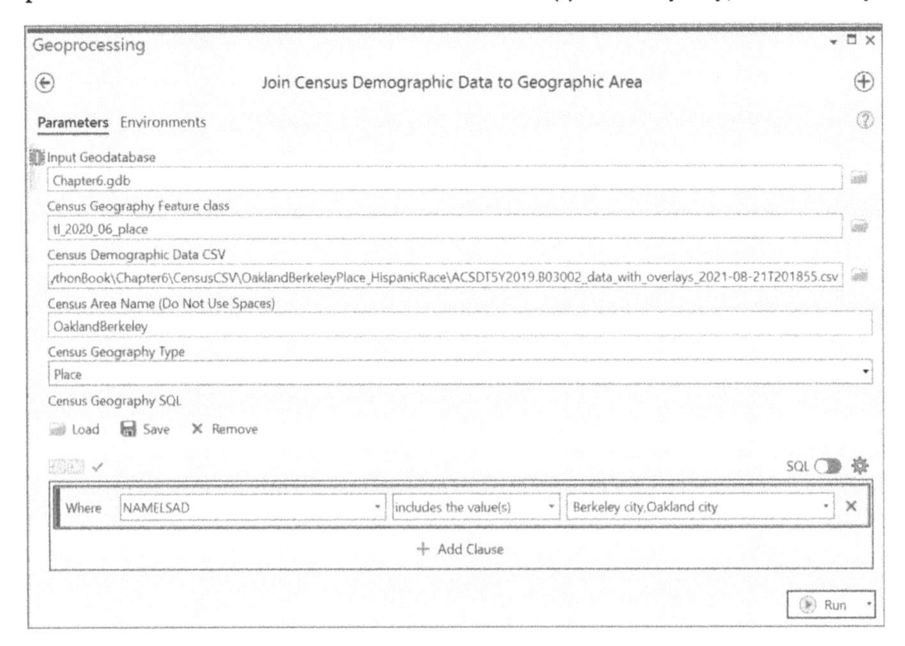

Figure 6.33: Script tool Geoprocessing window with parameters for our area name test

3. Click **Run** to run. To view the messages when running, you need to click **View Details** in the **Geoprocessing** window and then **Messages** in the **Details** window. You will see that a warning message has been output, telling you that the input for `areaName` had spaces and has been changed to `OaklandBerkeley`:

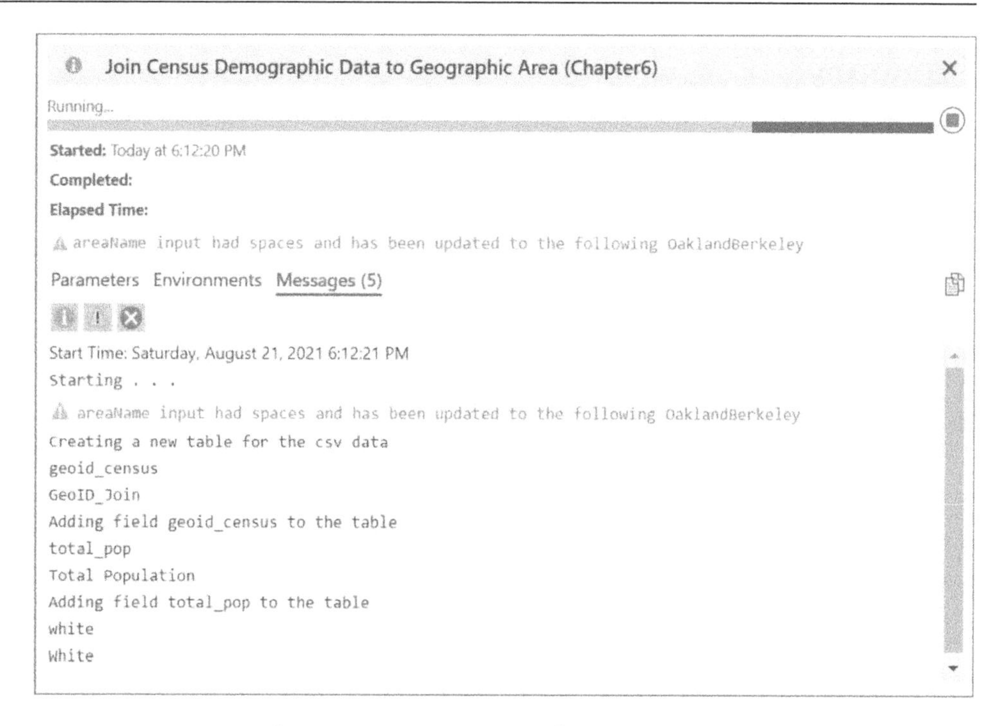

Figure 6.34: Warning message for Oakland Berkeley

4. When the test is finished, you can load the `OaklandBerkeley_RaceHispanic_Tract` feature class into a map to view the data. You will see that the script tool has created a new feature class of just the Oakland and Berkeley places with Hispanic/race data.

You have now tested the script tool with three different geographies and checked the `if` statement for the `areaName`. You can continue to check it with additional geographies to see how robust it is and what it can handle. When you are satisfied that the script tool will work with all the specified geographies in the **Census Geography Type** parameter, you can deploy this script tool to your team. This script tool will now allow anyone on your team to download a census CSV of Hispanic/race data and join it up to the census geography. It can be passed around your organization, and members with no knowledge of Python can complete the task quickly for any area they need.

Summary

In this chapter, you have learned why it is useful to convert a script or Notebook into a script tool, and how to do it. You learned about the different parameters available to you in the script tool dialog box in ArcGIS Pro, and then you took a Notebook you had already written and turned it into a script tool. You saw how to add custom messages to your script tool, which include warning and error messages that can give the user the information they need to change their inputs. Finally, you tested the finished script tool in a variety of different scenarios to check it was working correctly.

In the next chapter, you will learn how to use the `arcpy.mp` module to automate many tasks involved in creating maps.

7

Automated Map Production

The `arcpy.mp` module is used to work with maps, layers, and layouts within an ArcGIS Pro project. It was introduced with ArcGIS Pro to replace the `arcpy.mapping` module from ArcMap. The new features of ArcGIS Pro allow for additional functionality in the `arcpy.mp` module. While the goal is still to assist with map automation, the added functionality allows you more control over symbology settings in map automation. Creating maps that display geospatial analysis is a vital task for the GIS Professional, and the `arcpy.mp` module allows you to automate and update maps en masse. In this chapter, you will learn about the following:

- Referencing projects and maps within projects
- Updating data sources for a layer in a map
- Adding, removing, and moving layers from a map
- Adjusting the symbology of layers in a map
- Working with the different layout elements: legend, north arrow, scale bar, and text
- Exporting maps

All of these tasks will be done in Notebooks to give you sample code to apply to your own projects.

To complete the exercises in this chapter, please download and unzip the `Chapter7.zip` folder in the GitHub repository for this book: `https://github.com/PacktPublishing/Python-for-ArcGIS-Pro/tree/main/Chapter7`

Referencing projects and maps within projects

The `arcpy.mp` module can assist you in automating mapping tasks, but it does not remove the need for creating a map in ArcGIS Pro. You will still want to create your maps in ArcGIS Pro, but once created, `arcpy.mp` can be used to automate tasks such as adding, removing, and styling layers, and exporting maps across maps and projects.

ArcGIS Pro projects are stored as `.aprx` files. The `.aprx` file contains any maps with their associated layers, and any layouts with their associated layout elements. In this section, you will be starting with a project that contains two maps and a layout.

1. Open up ArcGIS Pro, navigate to where you unzipped the `Chapter7.zip` folder, and open up `Chapter7.aprx`. You will see two maps in the project, `Map` and `Map1`.
2. The first map, `Map`, contains the Oakland Vegetation from CalFire (`OaklandFVeg`), AC Transit Routes (`Summer21RouteShape`), and AC Transit Stops (`UniqueStops_Summer21`) from *Chapter 2*, and Alameda County Race/Hispanic Data (`AlamedaCounty_RaceHispanicTract`) from *Chapters 4* and *6*. In addition, it has two basemaps: **World Topographic Map** and **World Hillshade**.

Notice that the AC Transit Routes and Stops data do not display and have a red exclamation point next to them, signifying that the links to those layers are broken. You will see how to fix a broken link using the `arcpy.mp` module in the next section.

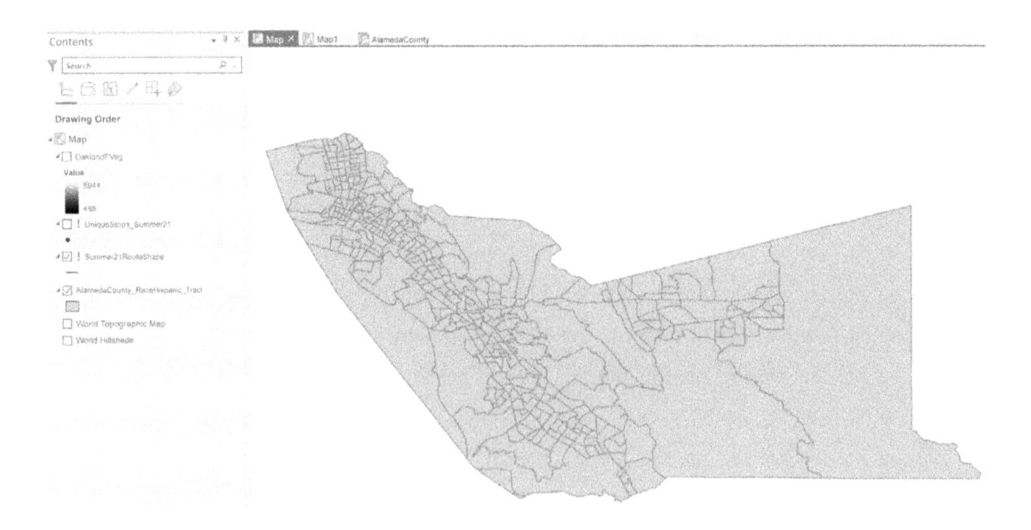

Figure 7.1: Starting map

3. Click over to the layout titled **AlamedaCounty**. It is a basic layout with a title, north arrow, scale bar, and legend to go along with the map. You will make changes to all of these in this chapter.

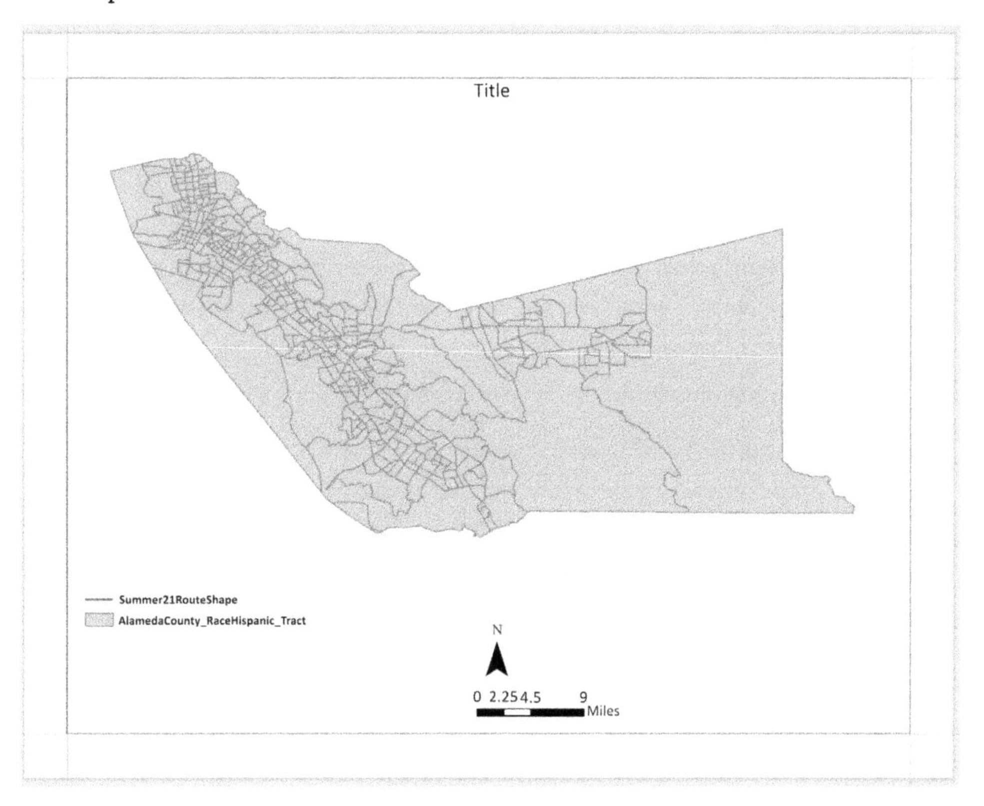

Figure 7.2: AlamedaCounty layout

To make any edits to the map and layout, you will need to **reference** the project and then the map and layout within the project. Referencing a project is done using `arcpy.mp.ArcGISProject()` and can be achieved in two ways:

- Reference a project by using the full path to the project where it is stored. For example, on this project, the code would look like the following:

```
 project = arcpy.mp.ArcGISProject(r "C:\PythonBook\Chapter7\
Chapter7.aprx")
```

- Reference the current project within ArcGIS Pro. That will always look like the following:

```
project = arcpy.mp.ArcGISProject("CURRENT")
```

When referencing a project using "CURRENT", you must be working within ArcGIS Pro. "CURRENT" will not work when running a standalone script.

Which way you reference your project depends on what the goal of your script is:

- If you are writing a standalone script to be run outside of ArcGIS Pro, you must reference the project with the full path. The script will not run the other way, as it does not recognize a "CURRENT" project to be running.
- If you are writing script tools or Notebooks to be run inside an instance of ArcGIS Pro, "CURRENT" will work better as it will always reference the project you have open.

Why not always use the full path when referencing a project?

The full path will always make sure that your script will work. But "CURRENT" does have some advantages when working directly in ArcGIS Pro. For one, the map view will automatically refresh with any changes you made when using "CURRENT". Another reason is that, when designing a script tool to work within ArcGIS Pro on an open project, there is no reason to ask the user for the path of the project, as they are working in it.

Throughout this chapter, you will be working in Notebooks within ArcGIS Pro, so the samples will reference the project using the "CURRENT" method.

An **ArcGISProject object** is created when you use the `arcpy.mp.ArcGISProject()` function. This object allows you to access the different properties, classes, and methods within a project. To work with maps, layers, and layouts, you will access the properties, classes, and methods available to them through the project.

When you use these to modify the project, you will want to save the changes or save a new project. The `save()` method applies to the project object that is open and will save the changes to that project. The `saveACopy()` method works as the **Save As** option in ArcGIS Pro and takes a full path, including the name and `.aprx` extension, to save a new project.

A **lock** is put on the project when it is referenced in a script. This lock will prevent anyone else from modifying the project while the script is running. If the script is run to completion, the lock is removed.

The lock can also be deleted within the script by using the `del` statement to delete the `ArcGISProject` object when the work is completed. It is considered best practice to delete the `ArcGISProject` object when you are finished with it, to ensure the lock is removed.

In the next section, you will reference your project and the map you opened at the start to fix the broken links.

Updating and fixing data sources

Updating data within a map can be a time-consuming process. It can involve many clicks to get into a dataset's properties to change the data source to a new dataset, which becomes frustrating when multiple links are broken due to geodatabases being moved.

Luckily, the `arcpy.mp` module has a class that allows you access to the layers in your map. You will be exploring many of the properties available in the `layers` class throughout this chapter. First, you will look at how to use the `updateConnectionProperties()` method on the `layers` class to fix broken links.

Fixing broken links

It can be frustrating to see the red exclamation point of a broken link on multiple layers when you open a map. The data is not displaying and you have to click through the properties of multiple layers to fix the broken link. The `updateConnectionProperties()` method in the `layers` class of the `arcpy.mp` module can simplify the process and be used to automate the updating of links of a layer.

In the `Map` of the `Chapter7` project, there are two layers with broken links. They are the AC Transit Stops and Routes layers, as we saw at the start of the chapter:

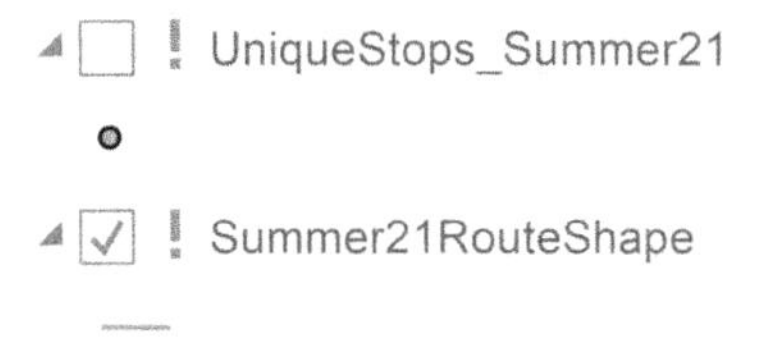

Figure 7.3: Broken links in the map

Walk through the following process to fix those links:

1. If you closed down ArcGIS Pro from the above section, open up ArcGIS Pro, navigate to where you unzipped the `Chapter7.zip` folder, and open up `Chapter7.aprx`.

2. You will see two maps in the project, `Map` and `Map1`. Make sure `Map` is the active map, as it has the broken links from above.
3. Within the **Projects** tab of the **Catalog** tab, right-click on **Chapter7** and select **New > Notebook** to create a new Notebook.
4. Rename the Notebook to `FixBrokenLinks`.
5. The first cell will hold the variable `mapName`. This will be the name of the map that you will search through to find broken links. Type in the following:

```
mapName = "Map"
```

   Run the cell.

6. The second cell will hold the variable `newLinkPath`. This is the path to the geodatabase that contains the layers. In this case, it is just a single geodatabase that contains both layers. Type in the following:

```
newLinkPath =  r"C:\PythonBook\Chapter7\TransitData.gdb"
```

   Run the cell.

7. In the next cell, you are going to reference the `CURRENT` project to create an `ArcGISProject` object, a map object using the `listMaps` object, and a list of all the layers using the `listLayers` object. Then, you will iterate through all the layer objects and use the layer `name` and `isBroken` properties to list the layer names and whether the link is broken or not. Type in the following:

```
project = arcpy.mp.ArcGISProject("CURRENT")
m = project.listMaps(mapName)[0]
layers = m.listLayers()
print(layers)
for layer in layers:
    print(layer.name)
    print(layer.isBroken)
    print("---------")
```

   Run the cell. The `print` statements will help you see the results of the code. Below is the output of the first few lines of code:

```
[<arcpy._mp.Layer object at 0x0000026E243AB588>, <arcpy._mp.Layer
object at 0x0000026E243AB908>, <arcpy._mp.Layer object at
0x0000026E243ABB88>, <arcpy._mp.Layer object at 0x0000026E243AB8C8>,
<arcpy._mp.Layer object at 0x0000026E243ABD48>, <arcpy._mp.Layer
object at 0x0000026E243AB2C8>, <arcpy._mp.Layer object at
0x0000026E25F7A088>]
DimondBridgeViewTrail
False
---------
OaklandFVeg
False
---------
UniqueStops_Summer21
True
---------
Summer21RouteShape
True
```

The layers are stored as objects that are not very useful when printed out. Accessing the properties of a layer object is the best way to get useful information out of the layer object. There are many more properties of a layer that you can access and you will explore more of them in this chapter.

8. In the next cell, you will loop through the layers and use a conditional to test whether the layer `isBroken`. If a layer `isBroken`, you will print out the name and `connectionProperties` property for that layer.

`connectionProperties` is a read-only property, so you cannot update it by writing new values to it; you have to use the `updateConnectionProperties` method for this.

Type in the following:

```
for layer in layers:
    if layer.isBroken is True:
        print(layer.name)
        print(layer.connectionProperties)
```

Run the cell. The results of the `print` statement are the name of the layer as a string and the connection properties as a dictionary:

```
UniqueStops_Summer21
{'dataset': 'UniqueStops_Summer21', 'workspace_factory': 'File
Geodatabase', 'connection_info': {'database': 'C:\\PythonBook\\
Chapter7_old\\Chapter7\\TransitData.gdb'}}
Summer21RouteShape
{'dataset': 'Summer21RouteShape', 'workspace_factory': 'File
Geodatabase', 'connection_info': {'database': 'C:\\PythonBook\\
Chapter7_old\\Chapter7\\TransitData.gdb'}}
```

The `connectionProperties` dictionary has the following key/value pairs:

- **Key**: `'dataset'`, **Value**: A string of the name of the layer.
- **Key**: `'workspace_factory'`, **Value**: A string of the type of workspace the layer is stored in. This can be many things, including a '`Shape File`', '`File Geodatabase`', '`SDE`'.
- **Key**: `connection_info`, **Value**: A dictionary that can contain a number of key/value pairs depending on `'workspace_factory'`. For shapefiles and file geodatabases, it contains just a database key with the value being either the path to the folder for the shapefile or the full path of the geodatabase.

When the workspace factory is an SDE for enterprise geodatabases, the `connection_info` dictionary has many more key/value pairs. Refer to the documentation for updating and fixing data sources for more details, located here: `https://pro.arcgis.com/en/pro-app/latest/arcpy/mapping/updatingandfixingdatasources.htm`

9. In the same cell you just ran, you will add to your loop to create a new connection property for each of the layers. You will first create a copy of the current connection property so you have the dictionary schema correct. Then, you will update just the database value within the `connection_info` key. Below the last line from above, with the same indentation, type in the following:

```
        newConnProp = layer.connectionProperties
        newConnProp["connection_info"]["database"] = newLinkPath
```

```
        print(newConnProp)
        print("---------")
```

Run the cell. Check the results from the `print` statement to make sure that the new connection dictionary has the correct path for the data:

```
UniqueStops_Summer21
{'dataset': 'UniqueStops_Summer21', 'workspace_factory': 'File
Geodatabase', 'connection_info': {'database': 'C:\\PythonBook\\
Chapter7_old\\Chapter7\\TransitData.gdb'}}
{'dataset': 'UniqueStops_Summer21', 'workspace_factory': 'File
Geodatabase', 'connection_info': {'database': 'C:\\PythonBook\\
Chapter7\\TransitData.gdb'}}
---------
Summer21RouteShape
{'dataset': 'Summer21RouteShape', 'workspace_factory': 'File
Geodatabase', 'connection_info': {'database': 'C:\\PythonBook\\
Chapter7_old\\Chapter7\\TransitData.gdb'}}
{'dataset': 'Summer21RouteShape', 'workspace_factory': 'File
Geodatabase', 'connection_info': {'database': 'C:\\PythonBook\\
Chapter7\\TransitData.gdb'}}
---------
```

10. Now you can use the `updateConnectionProperties()` method to update the connection properties. The `updateConnectionProperties()` method has two mandatory parameters:

    - `current_connection_info`: The current connection properties for the layer.
    - `new_connection_info`: The connection properties to be updated for the layer.

    It also has three optional parameters:

    - `auto_updating_joins_and_relates`, which is set to `True` by default. When set to `False`, it will not update the source joined or related to the layer.
    - `validate`, which is set to `False` by default. When set to `True`, the `updateConnectionProperties()` method will not validate that `new_connection_info` exists. This can force it to update to `new_connection_info` even if the connection does not exist yet.
    - `ignore_case`, which is set to `False` by default. When set to `True`, it will make the searches for layers case-insensitive. This can help you find connections if you are unsure of the case of the current connection layer.

To update the connection, in the same cell as above, with the same indentation and below the last line, type in the following:

```
        layer.updateConnectionProperties(
            layer.connectionProperties, newConnProp
        )
```

Run the cell. The result in the **Out** cell will look the same as above, but the red exclamation points next to the layer in the **Contents** pane are gone, and the data is now displayed on the map.

Figure 7.4: Layers with links fixed

11. The last steps are saving the project and deleting the ArcGIS Pro project object from Python to remove the lock. In a new cell below the above cell, type in the following:

    ```
    project.save()
    del project
    ```

    Run the cell.

You have now fixed the broken links in your map. In addition, you have saved the code as a Notebook and can now open this up in any project, change the variables for `mapName` and `newLinkPath`, and run it to update any broken links in that project.

You have just begun to work with all the layer properties. In the next section, you will continue to get comfortable with using layers.

Working with layers

You have already done some work with layers by fixing the broken links. In this section, you will learn more about the layer object and its classes and functions, along with how it interacts with the map object. First, you will learn how to add, move, and remove layers from a map.

Adding, moving, and removing layers

You can add layers to a map using different methods on the map object:

- `addBasemap(basemap_name)` adds a basemap layer to a map.
- `addDataFromPath(data_path)` adds a layer to a map from the local path or a URL.
- `addLayer(add_layer, {position})` adds a layer from another map, or a layer file (`.lyrx`) to a map at a defined position of `'AUTO_ARRANGE'` (default), `'TOP'`, or `'BOTTOM'`.
- `insertLayer(reference_layer, insert_layer, {insert_position})` adds a layer either `'BEFORE'` (default) or `'AFTER'` a reference layer that is in the map.

The `addLayer` and `insertLayer` methods require a layer from another map, referenced in from any map in any project, or a layer file. They will not work with a shapefile, feature class, or URL. Those must use the `addDataFromPath` method instead.

In addition to adding layers to a map, you can move a layer using `moveLayer(reference_layer, move_layer, {insert_position})`. This will move a layer up or down in the table of contents. Like the `insertLayer()` method, the `insert_position` parameter is either `'BEFORE'` or `'AFTER'` the `reference_layer`. Layers can be removed from a map by using the `removeLayer(remove_layer)` method.

To explore this, you will create a Notebook that will store the sample code for these different methods.

1. If you closed down the ArcGIS Pro session from the previous section, open it again, navigate to where you unzipped the `Chapter7.zip` folder, and open up `Chapter7.aprx`.
2. Within the **Projects** tab of the **Catalog** tab, right-click on **Chapter7** and select **New > Notebook** to create a new Notebook.
3. Rename the Notebook to `AddRemoveData`.
4. The first cell will hold the variable `mapName`, the name of the map that you will add, move, and remove layers to and from. Type in the following:

   ```
   mapName = "Map"
   ```

   Run the cell.

5. The next cell will contain the paths to the different layers you will be adding to the map. Type in the following:

   ```
   cpadUnits = r"C:\PythonBook\Chapter7\CPAD_2020b_Units.shp"
   oaklandBerkeley = r"C:\PythonBook\Chapter7\OaklandBerkeley_
   RaceHispanic_Place.lyrx"
   ```

```
cpadOakland = r"C:\PythonBook\Chapter7\Chapter7.gdb\CPAD_2020b_
Units_Oakland"
```

Run the cell.

6. In the next cell, you will create the `ArcGISProject` object and the map object. Type in the following:

```
project = arcpy.mp.ArcGISProject("CURRENT")
m = project.listMaps(mapName)[0]
```

Run the cell.

7. In the next cell, you will add a new basemap. When you use the `addBasemap()` method, you are actually replacing the basemap already in your map. The `basemap_name` parameter is the same name you see when adding a basemap from the basemap gallery in ArcGIS Pro:

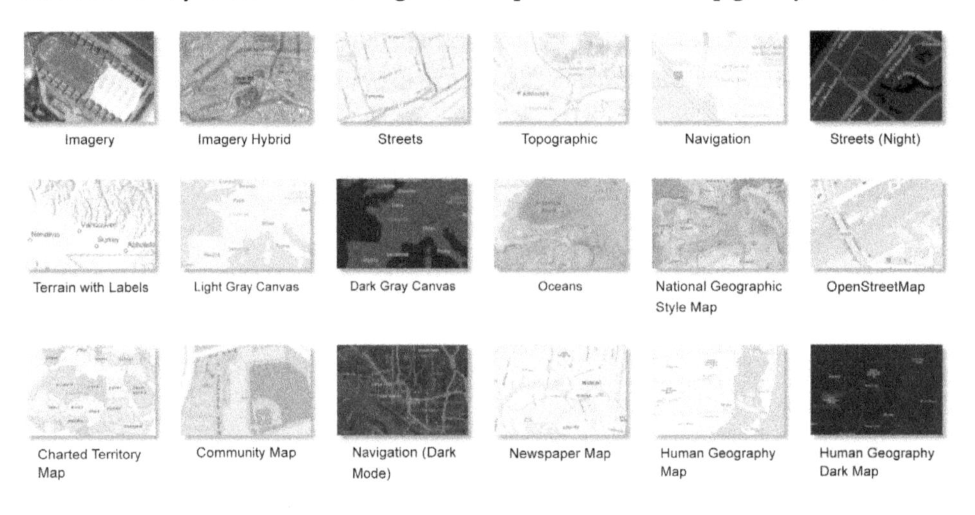

Figure 7.5: Basemap gallery

You will add the **Streets** basemap. Type in the following:

```
m.addBasemap("Streets")
```

Run the cell. You will see that the basemap has been replaced with the **World Street Map** basemap:

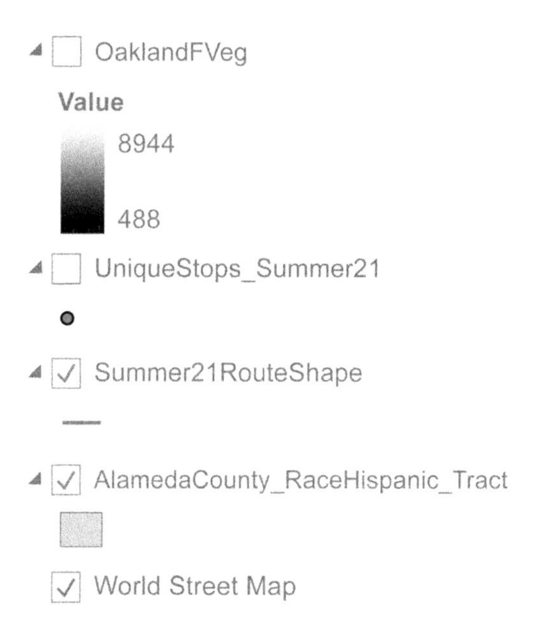

Figure 7.6: Updated basemap

8. In the next cell, you will add a feature class from its full path. Type in the following:

   ```
   m.addDataFromPath(cpadUnits)
   ```

   Run the cell. The **Out** cell will output a layer object and the CPAD layer will be added to your map in the top position:

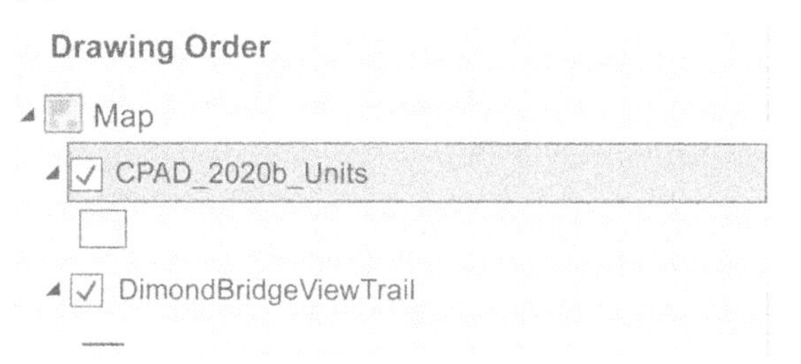

Figure 7.7: Added layer

9. In the next cell, you will add a layer file using `addLayer()`. To add a layer, you need to create a layer from the `.lyrx` file. You can also specify the `"TOP"` (default) or `"BOTTOM"` position to add the layer to.

To create the layer from the layer file and add it to the bottom, type in the following:

```
addLyr = arcpy.mp.LayerFile(oaklandBerkeley)
m.addLayer(addLyr,"BOTTOM")
```

Run the cell. Note that when you add to the "BOTTOM", the layer is added *below the basemap*:

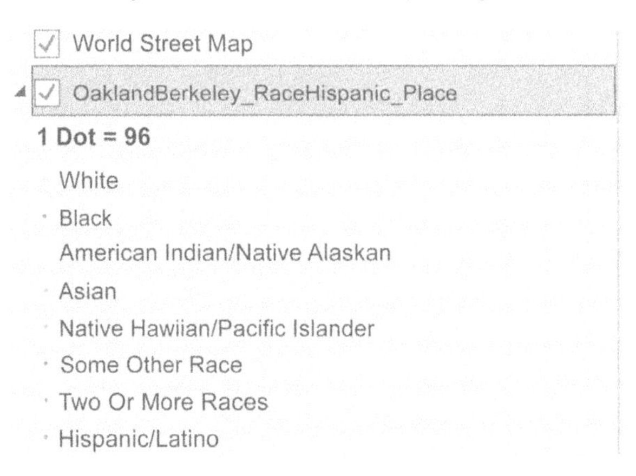

Figure 7.8: Layer added to the bottom of a map

This isn't the best position for the layer; it is below the basemap and won't be seen.

10. To move the layer, you need to reference the layer in the map you want it to be above or below. You will move this layer above the `AlamedaCounty_RaceHispanic_Tract` layer, using the `listLayers()` method with the layer name as the wildcard. This will create a list with just one layer; you will extract that layer from the list using the list index value of 0. You must then do the same to create a layer object for the `OaklandBerkeley_RaceHispanic_Place` layer that was just added.

    In the next cell, type in the following:

    ```
    refLayer = m.listLayers("AlamedaCounty_RaceHispanic_Tract")[0]
    oakBerkLyr = m.listLayers("OaklandBerkeley_RaceHispanic_Place")[0]
    m.moveLayer(refLayer, oakBerkLyr,"BEFORE")
    ```

    Run the cell. Now the `OaklandBerkeley_RaceHispanic_Place` layer has been moved above the `AlamedaCounty_RaceHispanic_Tract` layer.

The keyword "BEFORE" will move a layer above the reference layer, and "AFTER" will move it below the reference layer.

11. In the next cell, you will insert a layer. The `insertLayer()` method takes a layer and not a path. Both `addLayer` and `insertLayer` can use layers already within other maps, either in your project or other projects. Click over to **Map1** and you will see that you have a styled layer, `CPAD_2020b_Units_Oakland`. You will insert this layer into **Map**, positioning it above the `OaklandBerkeley_RaceHispanic_Place` layer that was just added. This will be done by creating another map object from **Map1**, creating a layer object from the `CPAD_2020b_Units_Oakland` layer that is in **Map1**, and using the `insertLayer()` method. Type in the following:

```
m2 = project.listMaps("Map1")[0]
insertLyr = m2.listLayers("CPAD_2020b_Units_Oakland")[0]
m.insertLayer(oakBerkLyr,insertLyr,"AFTER")
```

Run the cell, and observe that `CPAD_2020b_Units_Oakland` is inserted below the `OaklandBerkeley_RaceHispanic_Place` layer:

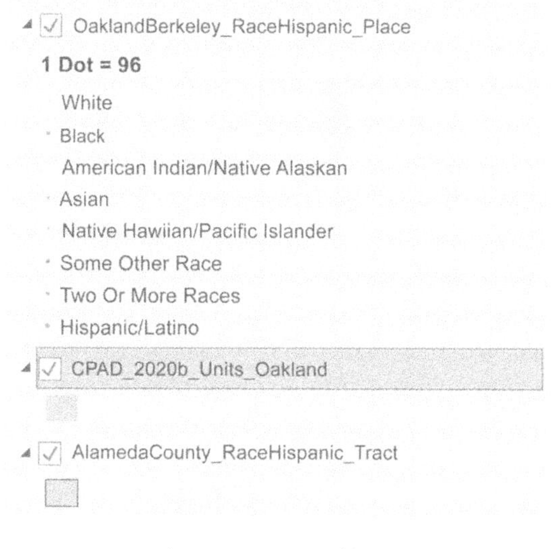

Figure 7.9: Inserted layer

12. Now that you have inserted and moved around layers, it is time to remove those you won't need. Instead of removing them one at a time, you will create a list of all the layers to remove and then iterate through the list to remove each layer. You will remove all the following datasets: `CPAD_2020b_Units`, `CPAD_2020b_Units_Oakland`, and `OaklandBerkeley_RaceHispanic_Place`. You find each layer using the `listLayers` method and add it to a list. Type in the following:

```
cpadUnits = m.listLayers("CPAD_2020b_Units")[0]
cpadOakland = m.listLayers("CPAD_2020b_Units_Oakland")[0]
oakBerRaceHis = m.listLayers("OaklandBerkeley_RaceHispanic_Place")
[0]
removeList = [cpadUnits, cpadOakland, oakBerRaceHis]
for layer in removeList:
    print(layer.name)
    m.removeLayer(layer)
```

Run the cell. All the layers in the list will be removed. The layers in your map should now look like the figure below:

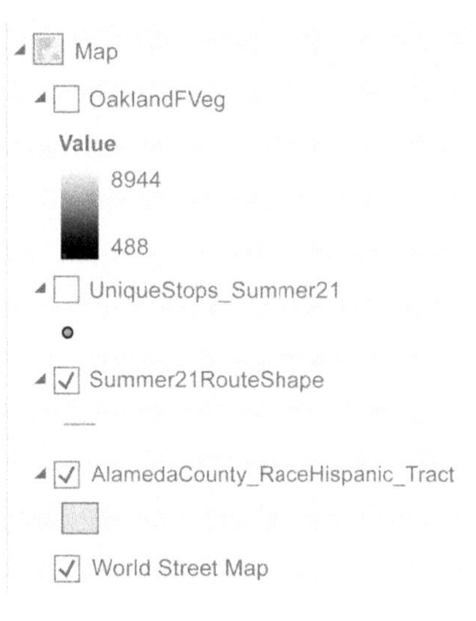

Figure 7.10: Layers remaining

The process of creating a list of layers to remove can also be used to create a list of layers to add. You have to create a list of either layer files or full paths of data, as layer files are added using the `insertLayer` method and full paths for data are added using the `addDataFromPath` method.

13. Now that you are done adding, moving, and removing layers, you can save your project and delete the `ArcGISProject` object to remove any schema locks. Type in the following:

    ```
    project.save()
    del project
    ```

    Run the cell.

You now have sample code in a Notebook that can be used to add, move, and remove layers from a map in your project. This can make map updates quick; when you need to move a layer, you can just open your Notebook, make the modifications, and then save your updated map. It is even more valuable when you need to make the same change to multiple maps as you can loop through the maps, adding, moving, or removing the same layer on those maps. You can also add layers created through an analysis process and export the map; you will see how to do this in *Chapter 12, Case Study: Advanced Map Automation.*

In the next section, you will see how to use Python to change the symbology of layers.

Layer symbology

The way you symbolize the layers in your map is how you can create beautiful cartographic designs. In ArcMap, you were limited in the ways you could do this using Python. The only way to update the **symbology** of a layer was to apply the symbology of a previously created layer file. This meant you had to create a layer file with the symbology you wanted, and then use the `arcpy.mapping` module to update your layer's symbology to that of your layer file. You did not have access to most of the symbology settings in ArcMap through the `arcpy.mapping` module.

In the new `arcpy.mp` module available in ArcGIS Pro, you have many more options for creating symbology. In this section, you will explore how to change the symbology of feature layers and raster layers using the `renderer` and `colorizer` properties of the symbology class. A **renderer** is used for symbolizing the vector data on your maps. Renders are applied to the symbology property of feature layers on your map to create different symbologies. A **colorizer** is used for symbolizing the raster data on your maps. Colorizers are applied to the `symbology` property of raster layers on your maps to create different symbologies.

Feature layers have the following renderers:

- `SimpleRenderer`: Symbolizes a single value using one symbol
- `UniqueValuesRenderer`: Symbolizes unique values based on a single attribute

- `GraduatedColorsRenderer`: Symbolizes graduated colors based on a single attribute
- `GraduatedSymbolsRenderer`: Symbolizes graduated symbols based on a single attribute

Raster layers have the following colorizers:

- `RasterUniqueValueColorizer`: Colorizes by unique values based on a single raster attribute
- `RasterClassifyColorizer`: Colorizes by groups for values based on a single raster attribute
- `RasterStretchColorizer`: Colorizes, creating a stretch of a color scheme across a single raster value

Compare these feature layer renderers and raster colorizers to those available in ArcGIS Pro (shown in the figures below) and you will see that not all are available:

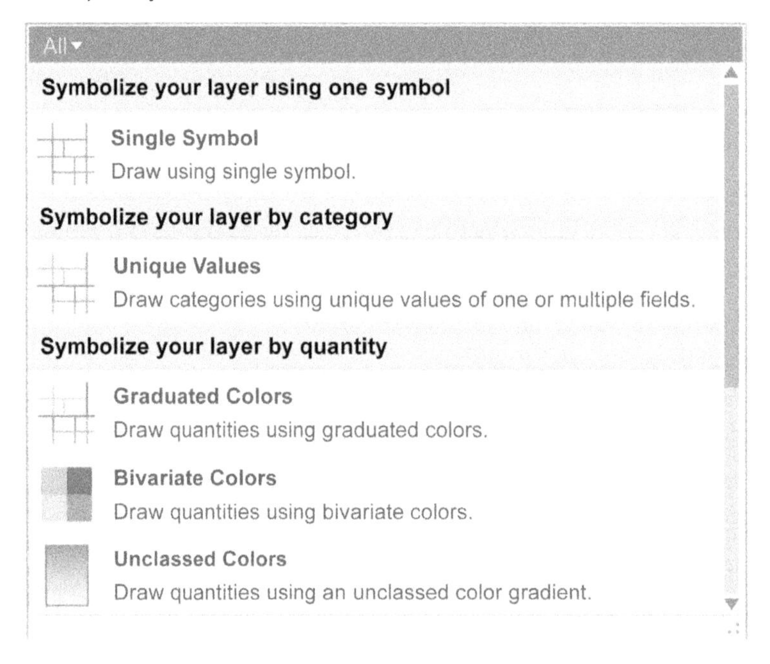

Figure 7.11: Feature class symbology options

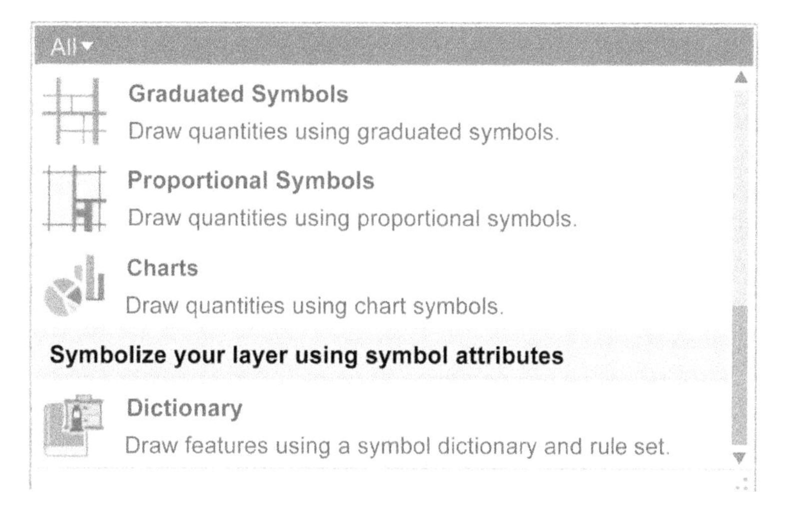

Figure 7.12: Feature class symbology options continued

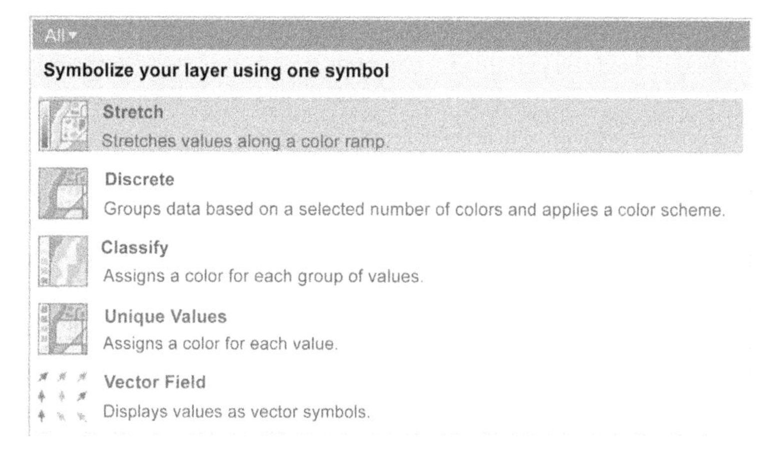

Figure 7.13: Raster symbology options

For feature classes, the following are not available through Python: **Unclassed Colors**, **Proportional Symbols**, **Dot Density**, **Charts**, and **Dictionary**.

For rasters, the following are not available through Python: **Discrete** and **Vector Field**.

In order to use the renderers or colorizers, you need to check to see if the layer or raster supports it, as not all do. You use the **has attribute** method that is built into Python, `hasattr()`, to check if an object has a property.

In this exercise, you will be looking to see whether the `symbology` property of a layer has the `render` property if the layer is a feature layer, or the `colorizer` property if the layer is a raster layer. You will work with `UniqueValuesRenderer`, `GraduatedColorsRenderer`, and `RasterUniqueValuesColorizer`, creating a Notebook that contains sample code for using each of those renderers/colorizers.

1. If you closed down the ArcGIS Pro session from the previous section, open it again, navigate to where you unzipped the `Chapter7.zip` folder, and open up `Chapter7.aprx`.
2. Within the **Projects** tab of the **Catalog** tab, right-click on **Chapter7** and select **New > Notebook** to create a new Notebook.
3. Rename the Notebook to `Symbolize`.
4. The first cell will hold the variable for the project and map. Type in the following:

   ```
   project = arcpy.mp.ArcGISProject("CURRENT")
   mapName = "Map"
   m = project.listMaps(mapName)[0]
   ```

   Run the cell.

5. In the next cell, you will create variables for each of the layer names you will symbolize. Type in the following:

   ```
   census = "AlamedaCounty_RaceHispanic_Tract"
   busRoute = "Summer21RouteShape"
   vegRaster = "OaklandFVeg"
   ```

   Run the cell.

6. In the next cell, you will create variables to hold each of the layers and use the `listLayers` method to access each of those layers. You will add `print` statements at the end to verify you have the correct layers. Type in the following:

   ```
   censusLyr = m.listLayers(census)[0]
   busRouteLyr = m.listLayers(busRoute)[0]
   vegLyr = m.listLayers(vegRaster)[0]
   print(busRouteLyr)
   ```

```
print(censusLyr)
print(vegLyr)
```

Run the cell. The following results should be printed in the **Out** cell:

```
Summer21RouteShape
AlamedaCounty_RaceHispanic_Tract
OaklandFVeg
```

7. Over the next three cells, you will be creating a variable that contains the `symbology` property of each layer, using the `hasattr()` method to test if it has renderer or colorizer properties, and then printing out the existing renderer or colorizer.

   In the next cell, type in the following:

```
censusLyrSym = censusLyr.symbology
print(hasattr(censusLyrSym,"renderer"))
print(censusLyrSym.renderer.type)
```

   Run the cell. The following results should be printed in the **Out** cell:

```
True
SimpleRenderer
```

8. In the next cell, type the following:

```
busRouteLyrSym = busRouteLyr.symbology
print(hasattr(busRouteLyrSym,"renderer"))
print(busRouteLyrSym.renderer.type)
```

   Run the cell. The following results should be printed in the **Out** cell:

```
True
SimpleRenderer
```

9. In the next cell, type the following:

```
vegLyrSym = vegLyr.symbology
print(hasattr(vegLyrSym,"colorizer"))
print(vegLyrSym.colorizer.type)
```

   Run the cell. The following results should be printed in the **Out** cell:

```
True
RasterStretchColorizer
```

10. `UniqueValueRender` needs to have at least two properties assigned for it to work: the **fields**, and **colors** for the values. The `fields` property is a list value as you can use more than one field to symbolize. To select the colors, you can either create `if` statements to apply a specific **RGB** value to each unique field, or use a **color ramp**. The `colorRamp` property is set by selecting a color ramp from a list of color ramps available in the project. This property is called by using the `listColorRamps()` method and passing in the name of a color ramp. To view all the available color ramps, type in the following in the next cell:

    ```
    for ramp in project.listColorRamps():
        print(ramp.name)
    ```

    Run the cell. You will see a long list of all the different color ramps available. Below are the first ten that you should see printed in the **Out** cell:

    ```
    Accent (3 Classes)
    Accent (4 Classes)
    Accent (5 Classes)
    Accent (6 Classes)
    Accent (7 Classes)
    Accent (8 Classes)
    Aspect
    Basic Random
    Bathymetric Scale
    Bathymetry #1
    ```

    These names correspond to names you see when checking the **Show names** box in the **Color Scheme** dropdown in the **Symbology** window:

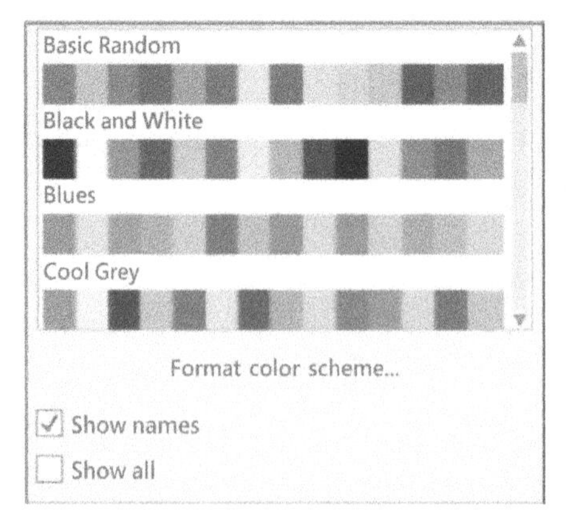

Figure 7.14: Color ramp names

11. In the next cell, you will symbolize the different bus routes with different colors. This will be done using `UniqueValueRenderer` and the `PUB_RTE` field and setting the colors to the `"Basic Random"` color ramp. The `listColorRamps()` method takes a color ramp as an argument and returns a list that contains just that color ramp. You will then use the list index of 0 to extract the color ramp.

    To change the symbology of a layer object, you access the variable that contains the symbology and make the updates to that variable. What you are changing, though, is the symbology properties *of that variable*, not of the layer. Once you have finished updating the variable that contains the symbology properties, you will set the `symbology` property of the layer equal to that newly created symbology variable. In the next cell, type the following:

    ```
    busRouteLyrSym.updateRenderer('UniqueValueRenderer')
    busRouteLyrSym.renderer.fields = ["PUB_RTE"]
    print(busRouteLyrSym.renderer.fields)
    busRouteLyrSym.renderer.colorRamp =
    project.listColorRamps("Basic Random")[0]
    busRouteLyr.symbology = busRouteLyrSym
    ```

    Run the cell. Since you added a print statement to print out the field names being used to symbolize, you should see `['PUB_RTE']` printed out. In addition to this, the `Summer21RouteShape` layer should now have each route symbolized with a different color:

Figure 7.15: Unique value renderer table of contents

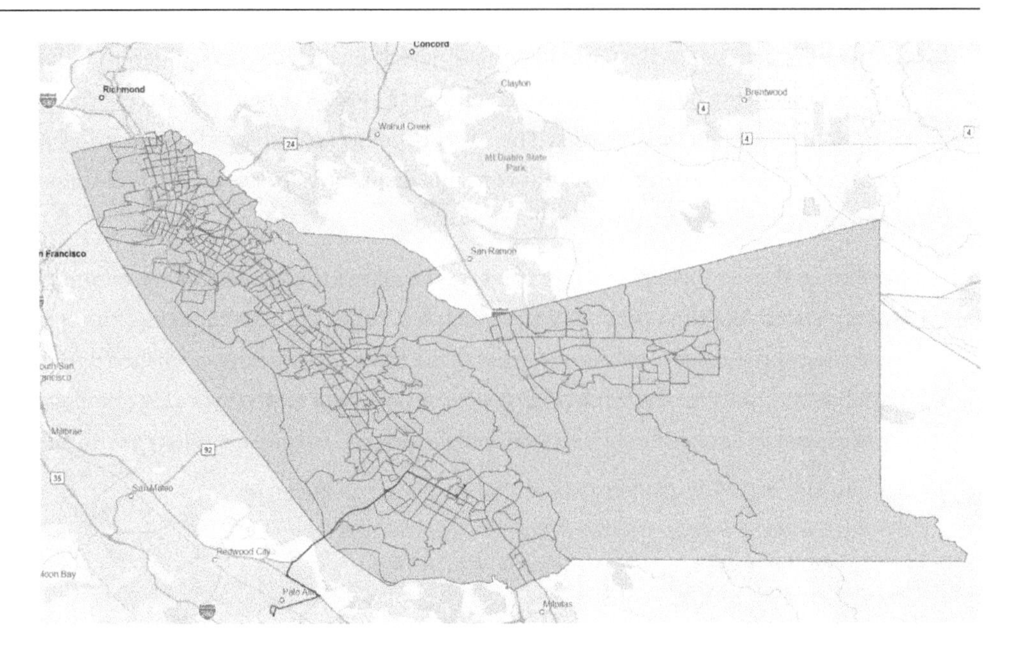

Figure 7.16: Unique value renderer map

12. In the next cell, you will change the symbology of the `AlamedaCounty_RaceHispanic_Tract` layer. You will be updating the renderer in the `symbology` variable to `GraduatedColorsRenderer`, and updating the properties of the renderer that are associated with `GraduateColorsRenderer`.

    To do this, you will set the `classificationField` property to the `"percent_minority"` field. You will then set the number of breaks to 4 using the `breakCount` property. Finally, you will set the `colorRamp` property to the `"Condition Number"` color ramp by using the `listColorRamps` object. You will add in `print` statements to track your code as it runs. Type in the following:

    ```
    censusLyrSym.updateRenderer('GraduatedColorsRenderer')
    print(censusLyrSym.renderer.type)
    censusLyrSym.renderer.classificationField = "percent_minority"
    print(censusLyrSym.renderer.classificationField)
    censusLyrSym.renderer.breakCount = 4
    censusLyrSym.renderer.colorRamp =
    project.listColorRamps('Condition Number')[0]
    censusLyr.symbology = censusLyrSym
    ```

Run the cell. Since you added a print statement to print out the renderer type of `GraduatedColorsRenderer` and the classification field of `percent_minority`, you will see the following in the **Out** cell:

```
GraduatedColorsRenderer
percent_minority
```

The `AlamedaCounty_RaceHispanic_Tract` layer has now been updated and is symbolized in four categories, with the green-to-red color ramp. Your contents and map will look like the figures below:

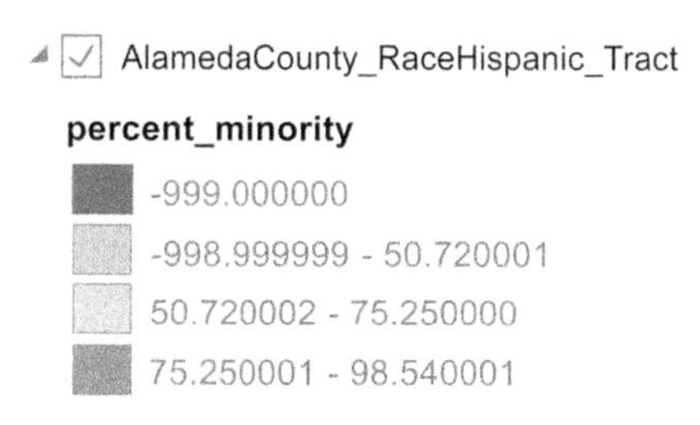

Figure 7.17: GraduatedColorsRenderer table of contents (low values in green, high values in red). Be mindful that this color ramp would not work from an accessibility standpoint for those with red-green color blindness!

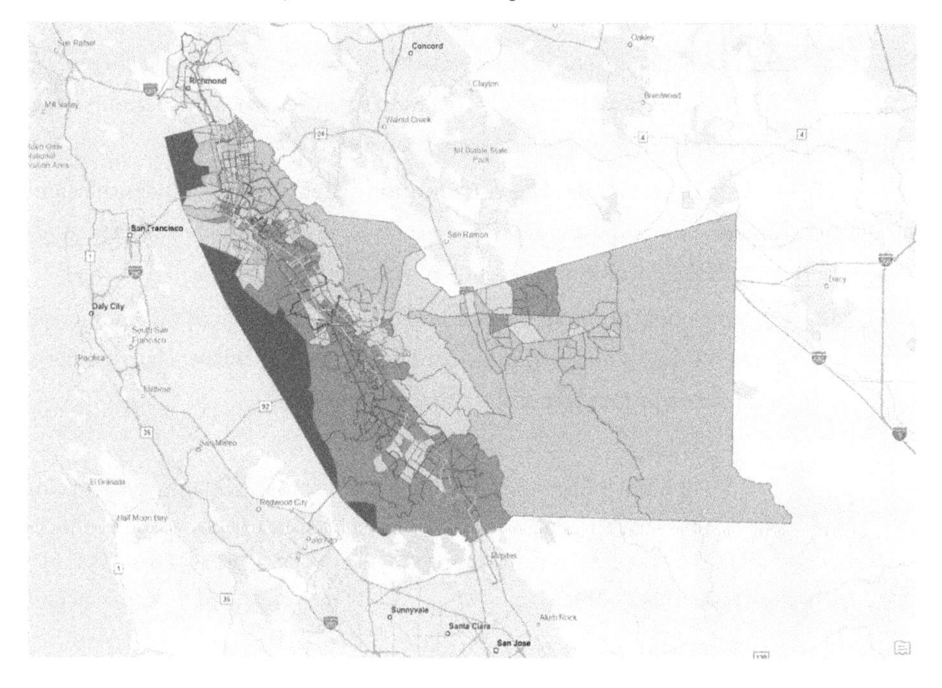

Figure 7.18: GraduatedColorsRenderer map

13. When setting a graduated colors symbology in ArcGIS Pro, you also have the option of setting the **classification method**. You can do that using Python as well. The `classificationMethod` property can be set to any of the following:

    - `DefinedInterval` sets a **defined interval** classification scheme, with an interval size where each class has the same number of units. For example, if the defined interval is 10, each class will have 10 units, and the number of classes is determined by the sample size.
    - `EqualInterval` sets an **equal interval** classification scheme, with an interval size where each class range is the same. It is set by defining the number of classes, and the range is created based on the values. For example, if you have values from 0 to 50 and set it for 5 classes, your class ranges will be 0-10, 11-20, 21-30, 31-40, and 41-50.
    - `GeometricInterval` sets a **geometric interval** classification scheme, with an interval size based on an algorithm that ensures classes have approximately the same number of values in them, and that the change between intervals is somewhat consistent.
    - `ManualInterval` sets a **user-defined** classification scheme, with the intervals specified by user-defined values.
    - `NaturalBreaks` sets a **Jenks Natural Breaks** classification scheme, which uses the Jenks Natural Breaks algorithm to find the best way to group similar values together and show the difference between classes.
    - `Quantile` sets a **quantile** classification scheme, which places the same number of data points in each class. For example, if your sample size is 50 and you set it to 5 classes, each class contains 10 values.
    - `StandardDeviation` sets a **standard deviation** classification scheme, which calculates the data's mean and standard deviation and creates class breaks with equal values as a proportion of the standard deviation.

For more information on all the classification schemes available in ArcGIS Pro, visit the documentation here: `https://pro.arcgis.com/en/pro-app/latest/help/mapping/layer-properties/data-classification-methods.htm`

You will verify what classification method was set as the default, since you did not choose one when setting the renderer to `GraduatedColorsRenderer`. Type in the following:

```
print(censusLyrSym.renderer.classificationMethod)
```

Run the cell. The result will be the following in the **Out** cell:

```
NaturalBreaks
```

The Jenks Natural Breaks classification is the default when one is not specified.

14. The Jenks Natural Breaks classification method does a good job of symbolizing the data, except that its lowest value is -999. Recall that you have set the areas with 0 population to a -999 percent minority to show that no one lives in those tracts.

    You will use the `classBreaks` property to set new break values. This will allow you to display the lowest value as 0. You will define the first break value and the interval to be used, and then iterate through the different class breaks, increasing the first value and break value. When the loop has finished running, you will set the `symbology` property to the new variable containing the updated class breaks. Finally, you will set the transparency of the layer using the `transparency` property of the layer. Type in the following:

```
breakValue = 25
firstVal = 0
for brk in censusLyrSym.renderer.classBreaks:
    brk.upperBound = breakValue
    brk.label = "{0} - {1}".format(str(firstVal),str(breakValue))
    breakValue += 25
    firstVal += 25
censusLyr.symbology = censusLyrSym
censusLyr.transparency = 40
```

    Run the cell. Since you added no print statements, there will be no output, but the symbology will have changed in the table of contents and the map:

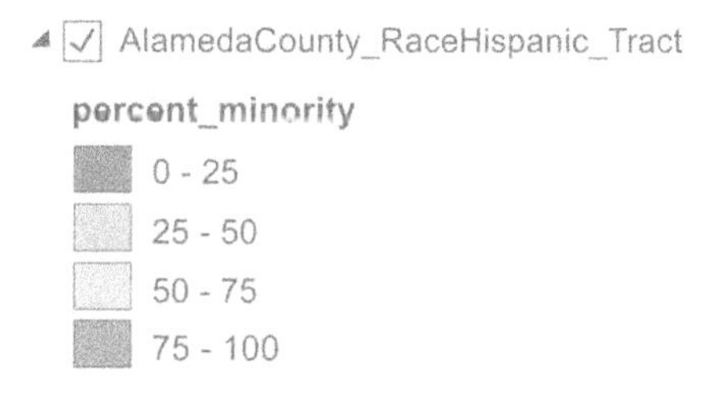

Figure 7.19: Manual breaks in the table of contents

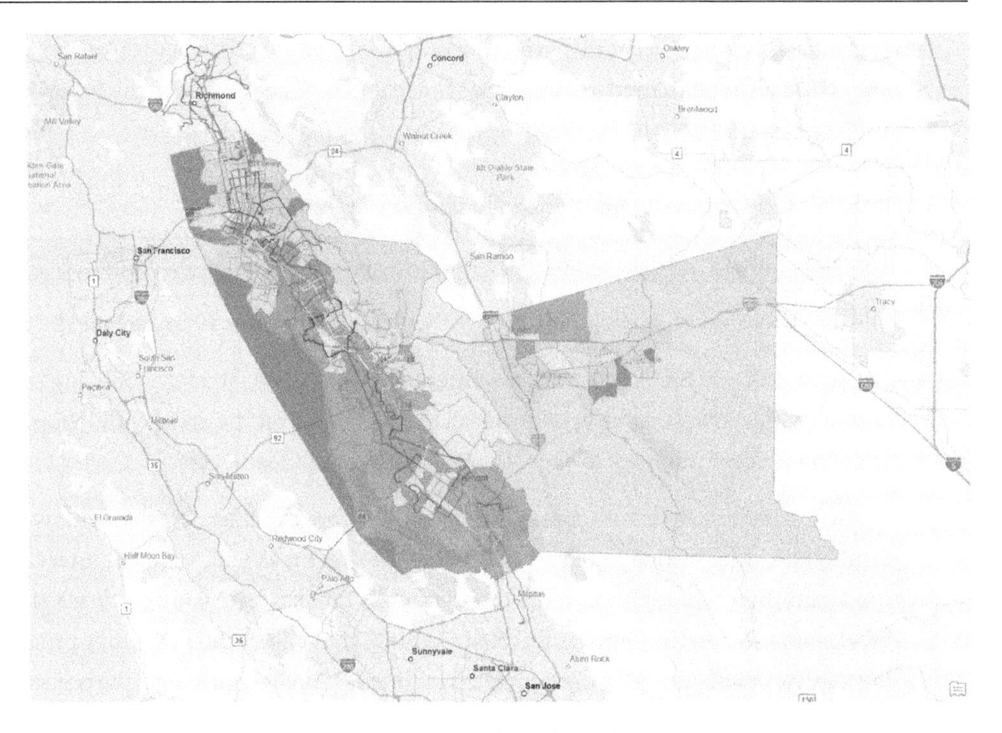

Figure 7.20: Manual breaks on the map

The above map now splits the 0-50% minority groups into 0-25% and 25-50%. This allows you to see those differences better than the previous version, which had just a single class for 0-50%.

15. You have changed the break values to manual values, which will also change the `classificationMethod` property to `ManualInterval`. To check this, type in the following:

```
print(censusLyrSym.renderer.classificationMethod)
```

Run the cell. The value returned will be `ManualInterval`.

16. The next layer to symbolize is the `OaklandFVeg` raster layer, which shows the different land covers in Oakland. You will update the colorizer on the `vegLyrSym` symbology variable to `RasterUniqueValueColorizer`. You will then use the `field` property of the colorizer to select the field to be used for colorizing. Next, you will use the `listColorRamps()` method of the `ArcGISProject` object to extract the `"Basic Random"` color ramp and set it to the `colorRamp` property of the symbology variable. Finally, you will update the symbology of the layer by setting it equal to the new symbology variable you created. You will add in a `print` statement to track the process and ensure the values are what you expect. Type in the following:

```
vegLyrSym.updateColorizer("RasterUniqueValueColorizer")
print(vegLyrSym.colorizer.type)
vegLyrSym.colorizer.field = "WHR10NAME"
vegLyrSym.colorizer.colorRamp =
project.listColorRamps("Basic Random")[0]
vegLyr.symbology = vegLyrSym
```

Run the cell. The **OaklandFVeg** layer in the table of contents and map will have updated symbology. Since a random color ramp was used, yours will not look the same as the figures below:

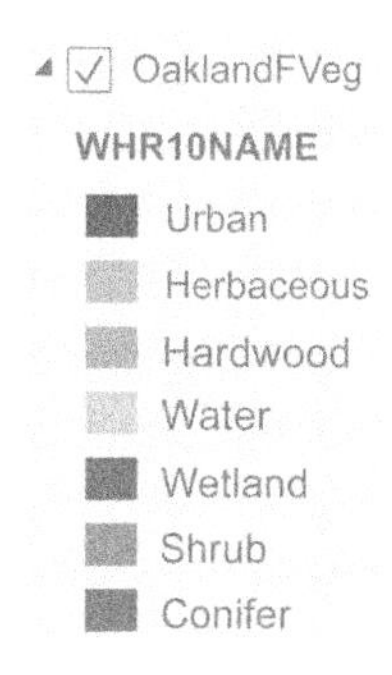

Figure 7.21: Oakland vegetation table of contents

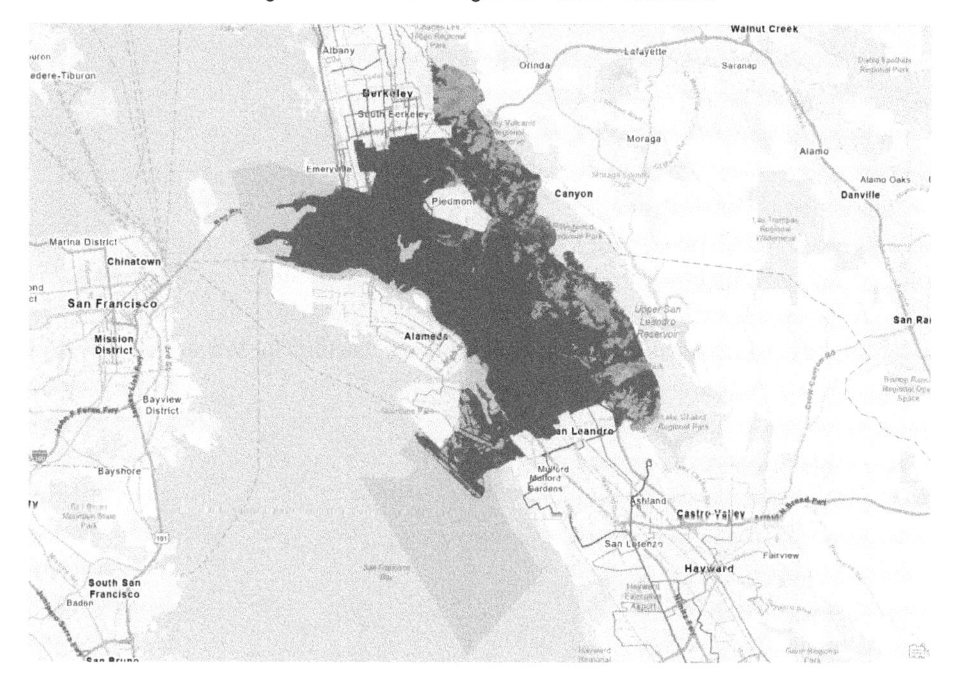

Figure 7.22: Oakland vegetation symbology map

17. Using the Basic Random color ramp can be good when you have a lot of categories and the color choice doesn't matter much. For land cover, though, it can look a little odd to not have water in blue and urban areas in gray, so you will change just those colors by accessing each **item**. To access the items, you have to access the **group** they are in first. You always have at least one group, as your items are displayed in the default group.

Groups are ways to group similar types of data in your symbology and give them all a header to show this similarity. For example, you could create a **Forest** group and place all the forest land cover types in that group.

You will check each item's **label** for Water or Urban and, when found, change the item's color using the RGB values.

You are actually using RGBA, as you are also setting the alpha value for transparency. The alpha value runs from 0 (for fully transparent) to 100 (for fully opaque).

The last step is to set the layer symbology equal to the new symbology variable with the updated colors. Type in the following:

```
for group in vegLyrSym.colorizer.groups:
    for item in group.items:
        if item.label == "Water":
            item.color = {'RGB' : [0, 0, 255,100]}
        elif item.label == "Urban":
            item.color = {'RGB' : [153, 153, 153,100]}
vegLyr.symbology = vegLyrSym
```

Run the cell. The **OaklandFVeg** layer will now use blue for **Water** and gray for **Urban**:

Figure 7.23: Updated colors for Water and Urban

How do you determine what the RGB values are?

There are many different websites that will help you determine the RGB values of colors. The HTML Color Picker from w3.schools at `https://www.w3schools.com/colors/colors_picker.asp` is a good option. You can pick a color and see the RGB value of that color, as well as having the ability to select lighter and darker options.

18. Lastly, to save your map and release the schema lock on the project, type in the following:

```
project.save()
del project
```

Run the cell.

The method used to update individual colors using the colorizer can be modified to use a renderer. You would iterate through the groups in the layer, replacing `colorizer` with `renderer`. Then, when iterating through the items, you would use `item.symbol.color` instead of the `item.color` you used in the colorizer exercise.

You now have seen many different examples of how to update layer symbology within a map. You can use default class breaks in a graduated color renderer or set your own manual breaks. You also have examples of how to access and change specific colors in a raster layer. While these methods might not seem more efficient than doing the same tasks in ArcGIS Pro, the real power in these Notebooks emerges when they are applied to making the same symbology changes across multiple maps and projects.

Layouts

Now that you know how to update the symbology on your layers, you can start working with **layouts**. Layouts are what you create to export your maps to files. They include layers and elements like the legend, north arrow, scale bar, title, and text. You can modify all of these layout elements using Python. In addition to modifying layout elements, you can also use Python to export your layouts to different file types, like PDF, JPG, PNG, and more. Exporting your map to a file is the final step in map production, as you can then print your map or insert it into a document.

In this section, you will see how to make additional modifications to the layers, like creating definition queries and changing layer names to make your maps more informative and easier to read. You will learn how to turn layers on and off in a map. Then, you will see how you can modify all the different layout elements and, finally, how to export your map. You will continue working in your `Chapter7` project using the same **Map** and **AlamedaCounty** layout you have been using so far.

Layers

In the previous sections, you have worked with layers to update their data sources, add and remove them, move them, and change their symbology. In this section, you will see how to change their name, turn them on or off, and apply definition queries. All of these will allow you to continue to manipulate how a layer looks on the map and the legend.

1. If you closed down the ArcGIS Pro session from the previous section, open it again, navigate to where you unzipped the `Chapter7.zip` folder, and open up `Chapter7.aprx`.
2. Within the **Projects** tab of the **Catalog** tab, right-click on **Chapter7** and select **New > Notebook** to create a new Notebook.
3. Rename the Notebook to `LayoutElements`.
4. The first cell will hold the variables for the project, map, and layout. Type in the following:

```
project = arcpy.mp.ArcGISProject("CURRENT")
mapName = "Map"
layoutName = "AlamedaCounty"
```

```
m = project.listMaps(mapName)[0]
layout = project.listLayouts(layoutName)[0]
```

Run the cell.

5. In the next cell, you will create the same variables as in the `SymbolizeNotebook` exercises, for the `AlamedaCounty_RaceHispanic_Tract`, `Summer21RouteShape`, and `OaklandFVeg` layers. Type in the following:

```
census = "AlamedaCounty_RaceHispanic_Tract"
busRoute = "Summer21RouteShape"
vegRaster = "OaklandFVeg"
censusLyr = m.listLayers(census)[0]
busRouteLyr = m.listLayers(busRoute)[0]
vegLyr = m.listLayers(vegRaster)[0]
```

Run the cell.

6. You can turn a layer on and off by changing its `visible` property. The `visible` property is a **Boolean** type that is `True` when the layer is visible and `False` when it is not. You will check the visibility of all three layers by typing the following:

```
print(censusLyr.visible)
print(busRouteLyr.visible)
print(vegLyr.visible)
```

Run the cell. Depending on which layers are turned on or off in your view, you should see a combination of `True` and `False` printed out. If you still have all the layers turned on from the last exercise, you will see the following in the **Out** cell:

```
True
True
True
```

7. You can change the visibility of a layer by setting the `visible` property to `True` or `False`, depending on whether you want the layer to be drawn or not. For the layout you are working on, you do not want the `vegLyr` to be drawn, so you will set that to `False` by typing in the following:

```
vegLyr.visible = False
print(vegLyr.visible)
```

Run the cell. The **OaklandFVeg** layer in the contents will now be unchecked:

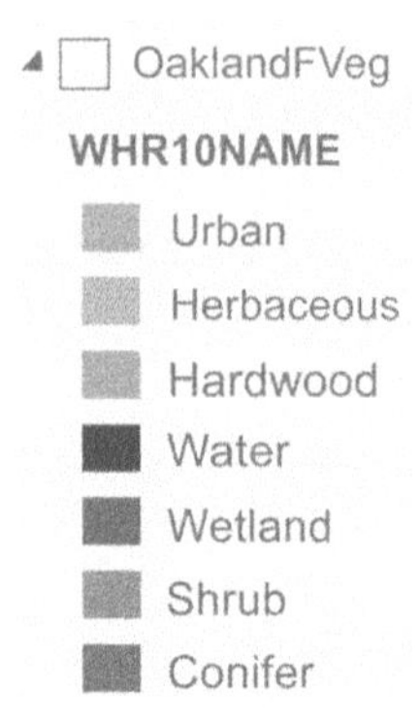

Figure 7.24: Layer with the visible property set to False

8. When viewing the **AlamedaCounty** layout, it is difficult to see all the bus routes at this scale, and there are too many for the legend:

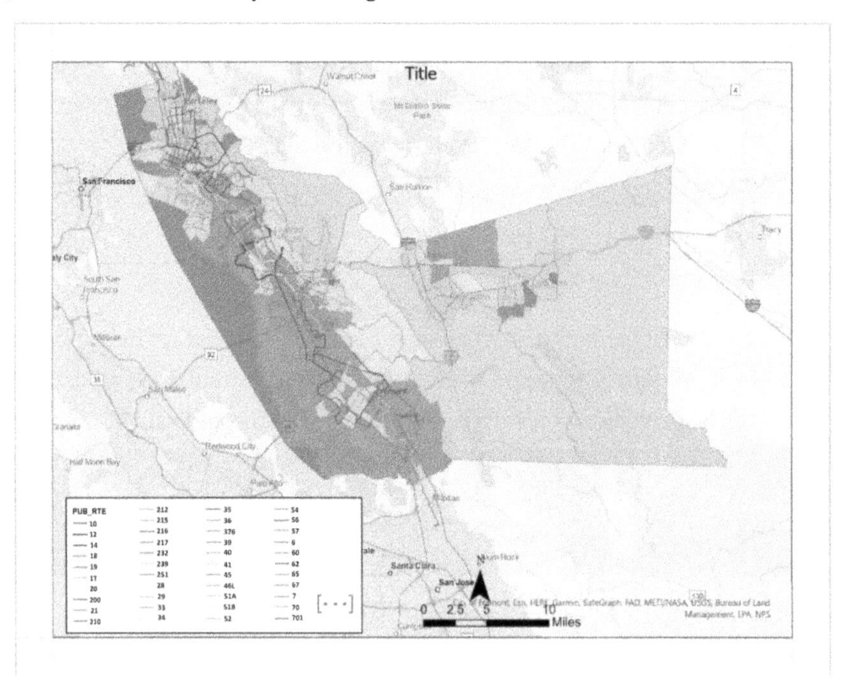

Figure 7.25: AlamedaCounty layout

You will change the bus routes layer so that it only displays the transbay bus routes. To do that, you can write a **definition query** for the `busRouteLyr` layer that queries the `PUB_RTE` field for just the transbay routes.

This is done using the `definitionQuery` property of a layer. Recall from earlier chapters that all of the AC Transit transbay routes start with a letter. Type in the following:

```
busRouteLyr.definitionQuery = "PUB_RTE In
('F','G','J','L','LA','NL','NX','O','P','U','V','W')"
```

Run the cell. You will not see any change in the **Contents** tab as you are still symbolizing with colors. However, in the **AlamedaCounty** layout, you will see only the transbay bus lines, while the legend still shows all of the bus routes. This will be changed later using the `LegendItem` properties of `LegendElement`.

Now that you have reduced the number of bus lines shown, it is time to learn about the different layout elements that are available in ArcPy.

Layout elements

The `layout` object has many properties and methods available to you that allow you to modify it. For instance, you can view and write the name and page dimensions through the `name`, `pageHeight`, `pageUnits`, and `pageWidth` properties.

`pageUnits` must be set to `CENTIMETER`, `INCH`, `MILLIMETER`, or `POINT`; the `pageHeight` and `pageWidth` values are then based on those units.

In addition to those properties, there are methods available for exporting a layout to different file formats that we will look at later in this section.

You can also access all of the elements in the layout by using the `listElements` method. The `listElements` method returns a list of all the layout elements in the `layout` object. It takes two optional arguments to filter the list of returned elements: `element_type` and `wildcard`. The different `element_type` values are:

- `GRAPHIC_ELEMENT`: Filters the list for just graphic elements on the page layout. Graphic elements include any line or polygon elements drawn on your map.
- `LEGEND_ELEMENT`: Filters the list for just legend elements on the page layout.
- `MAPFRAME_ELEMENT`: Filters the list for just map frames on the page layout.
- `MAPSURROUND_ELEMENT`: Filters the list for the different map surround elements on the page layout. Map surround elements include the north arrow, scale bar, and neatline.

- `PICTURE_ELEMENT`: Filters the list for the different picture elements on the page layout. Picture elements include JPG or PNG files added to the map layout.
- `TEXT_ELEMENT`: Filters the list for the different text elements on a page. Text element includes any title or subtitles on your map.

You will continue working in the `LayoutElements` Notebook to explore these different properties and methods.

1. In the next cell of the `LayoutElements` Notebook, you will print out the `name`, `pageUnits`, `pageHeight`, and `pageWidth`. Type in the following:

    ```
    print(layout.name)
    print(layout.pageUnits)
    print(layout.pageHeight)
    print(layout.pageWidth)
    ```

    Run the cell. You will see the following in the **Out** cell:

    ```
    AlamedaCounty
    INCH
    8.5
    11.0
    ```

2. The layout is going to show the transbay bus routes in Alameda County, so you should change the layout name to reflect that. Once the layout `name` property has been updated, you will also want to update the `layoutName` variable to match the name of the property. You will add a `print` statement to check that your code works properly. Type in the following:

    ```
    layout.name = "AlamedaCountyTransbayBus"
    layoutName = layout.name
    print(layoutName)
    ```

    Run the cell. The `print` statement shows the new layout name, and the layout name has changed in the project:

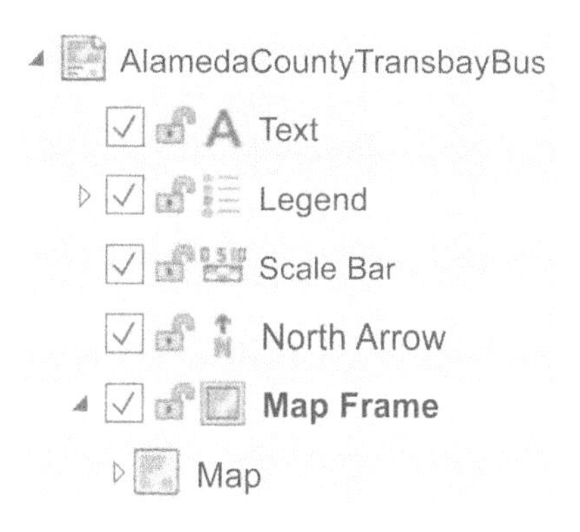

Figure 7.26: Changed layout name

3. In the next cell, you will create a list of all the different layout elements, printing out their name and element_type. Type in the following:

```
lytElems = layout.listElements()
for elem in lytElems:
    print("{0} is a {1} element".format(elem.name,elem.type))
```

Run the cell. The print statement will show that the names printed out match those names in the **Contents** tab for the layout. You will see the following in the **Out** cell:

```
Text is a TEXT_ELEMENT element
Legend is a LEGEND_ELEMENT element
Scale Bar is a MAPSURROUND_ELEMENT element
North Arrow is a MAPSURROUND_ELEMENT element
Map Frame is a MAPFRAME_ELEMENT element
```

4. It is important to have useful names for the elements in your layout. This will help you identify the correct element in your code. The text element named "Text" is not very descriptive. You will change that name to "Title" by accessing the text element and changing its name property. Type in the following:

```
textElem = layout.listElements("TEXT_ELEMENT","Text")[0]
textElem.name = "Title"
print(textElem.name)
```

Run the cell. You will see the following in the **Out** cell:

```
Title
```

The layout element name in the **Contents** tab will also have changed to **Title**:

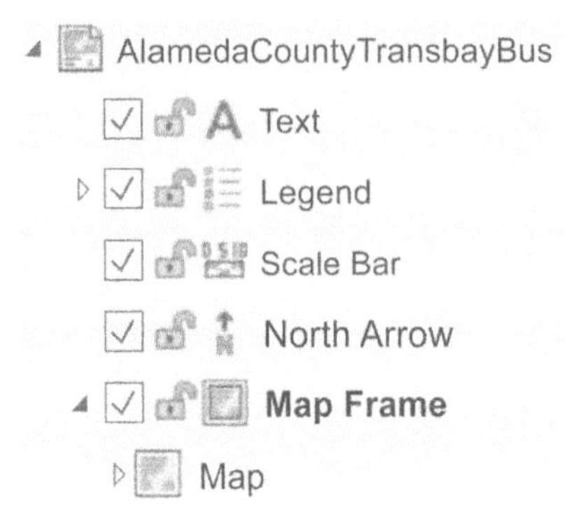

Figure 7.27: Updated text element name

Now you have an understanding of the different layout elements and how to list them, you will look at each one in more detail.

Legend

The first layout element you will work with is `LegendElement`, which has a number of properties and methods associated with it. These properties allow you to make adjustments to the legend. Some of the properties you are likely to use most often are listed below:

- `mapFrame`: The map that the legend is referenced to. This must be set to a map frame data type.
- `name`: The name of the legend.
- `elementPositionX`: The *x* location of the anchor's position. The units are the same as those set in the `pageUnits` property of the layout object.
- `elementPositionY`: The *y* location of the anchor's position. The units are the same as those set in `pageUnits`.

`elementPostionX` and `elementPositionY` are set based on the anchor point. The anchor point can only be set in ArcGIS Pro.

- `elementHeight`: The height of the element. The units are the same as those set in `pageUnits`.
- `elementWidth`: The width of the element. The units are the same as those set in `pageUnits`.
- `fitingStrategy`: The fitting strategy method to be applied to the legend. The accepted values are `AdjustFontSize`, `AdjustColumns`, `AdjustColumnsAndFont`, `AdjustFrame`, and `ManualColumns`.

The `fittingStrategy` methods are the same as are found within the **Fitting Strategy** dropdown in the **Legend Arrangement Options** tab:

Adjust font size
Adjust columns
Adjust columns and font size
Adjust frame
Manual columns

Figure 7.28: Fitting strategy options

- `columnCount`: The number of columns in a legend. Only applies when `fittingStrategy` is set to `AdjustFontSize`, `AdjustFrame`, or `ManualColumns`.
- `title`: The title of the legend.
- `showTitle`: A Boolean that displays the title when set to `True` and removes the title when set to `False`.
- `items`: A list of `LegendItem` classes that can be accessed by using the `LegendItem` class to modify the properties of the legend items.

There are additional properties available for `LegendElement`. For more information about them, refer to the `LegendElement` documentation at `https://pro.arcgis.com/en/pro-app/latest/arcpy/mapping/legendelement-class.htm`.

As well as the properties associated with `LegendElement`, there are also some methods you can use:

- `addItem(layer, {add_position})`: This will add a layer to the legend. The optional `add_position` argument can be set to the `'TOP'` (default) or `'BOTTOM'` of the legend layer stack.

- `moveItem(reference_item, move_item, {move_position})`: This will move `move_item` to a `move_position`, based on the `reference_item`. The `move_position` is either `'AFTER'` or `'BEFORE'` (default). `'AFTER'` places `move_item` below `reference_item`, and `'BEFORE'` places `move_item` above `reference_item`.
- `removeItem(remove_item)`: This will remove `remove_item` from the legend.

The `items` property of the `LegendElements` object returns a list of `LegendItem` objects. There is an object for each item in the legend. Each `LegendItem` has properties that can be modified using Python. The `LegendItem` properties are many of the different legend options you have access to for each legend item within ArcGIS Pro. The properties available are:

- `arrangement`: The arrangement of legend items.
- `column`: The column number position for a legend item. It is only available when the `fittingStrategy` is set to `ManualColumns`.
- `name`: A read-only value of the legend item name. To change the legend item name, you must change the layer name.
- `patchHeight`: The height of the legend item patch. The units are in points.
- `patchWidth`: The width of the legend item patch. The units are in points.
- `showFeatureCount`: A Boolean that will show the count of the feature items next to the legend item. A `True` value will show the feature count, and `False` will remove it.
- `showVisibleFeatures`: A Boolean value; when set to `True`, the legend item will only display the visible features. When set to `False`, all the features in the layer will be displayed in the legend.
- `visible`: A Boolean value; when set to `True`, the legend item will be displayed, and when set to `False` it will be removed.
- `type`: Returns the type, which is `LEGEND_ITEM`.

Take a look at the legend in the **AlamedaCountyTransbayBus** layout and you'll see that it needs to have some work done. It is currently displaying all of the bus lines in the layer even though only the transbay routes are displayed in the figure. It is also not able to show the items for the `AlamedaCounty_RaceHispanic_Tract` data. In this exercise, you will explore some of the properties above and create a better-looking legend.

1. Continuing to work in the `LayoutElements` Notebook, in the next cell, you will access the legend using the `listElements()` method and display the legend title, height, width, and fitting strategy properties. Type in the following:

```
legend = layout.listElements("LEGEND_ELEMENT","Legend")[0]
print(legend.title)
print(legend.showTitle)
print(legend.elementHeight)
print(legend.elementWidth)
print(legend.fittingStrategy)
```

Run the cell. You will see the following in the **Out** cell:

```
Legend
False
1.66
3.86
AdjustColumnsAndFont
```

2. In the next cell, you will change `fittingStrategy` to `AdjustFontSize`. Type in the following:

```
legend.fittingStrategy = "AdjustFontSize"
print(legend.fittingStrategy)
```

Run the cell. You will see the following in the **Out** cell:

```
AdjustFontSize.
```

The legend should have changed in the **AlamedaCountyTransbayBus** layout to look like the following figure:

Figure 7.29: Legend set to AdjustFontSize

This doesn't look any better, and the old fitting strategy, `AdjustColumnsAndFont`, will be more desirable as ArcGIS Pro will automatically change column sizes and font sizes as you remove data from the legend in the next steps.

3. Return the fitting strategy to the old settings by typing in the following:

```
legend.fittingStrategy = "AdjustColumnsAndFont"
print(legend.fittingStrategy)
```

Run the cell. You will see the following in the **Out** cell:

```
AdjustColumnsAndFont
```

The legend should be returned to its original appearance.

4. To adjust the legend so that it only shows the bus routes displayed on the map, you will iterate through all of the `items` objects in the `LegendItem` list. Within the loop, you will test for the name of an item being `"Summer21RouteShape"`. When that name is found, you will print the item name to confirm it has been found, set the `showVisibleFeatures` property to `True`, and then print the `showVisibleFeature` property to verify it was changed. Type the following:

```
for item in legend.items:
    if item.name == "Summer21RouteShape":
        print(item.name)
        item.showVisibleFeatures = True
        print(item.showVisibleFeatures)
```

Run the cell. You should see the following in the **Out** cell:

```
Summer21RouteShape
True
```

The legend should now appear like the figure below:

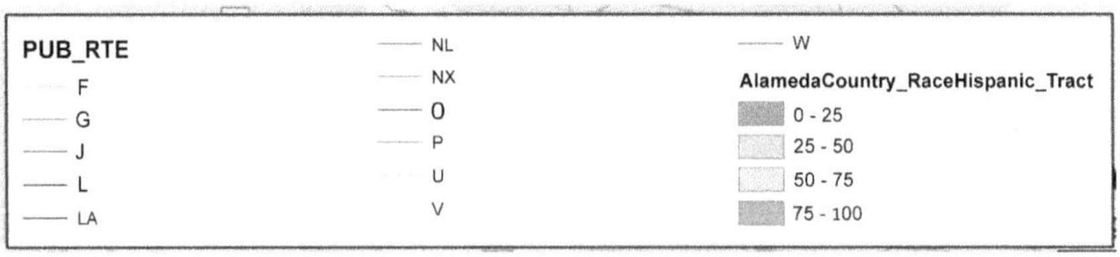

Figure 7.30: Legend with showVisibleFeatures set to True

It now shows the transbay routes, but the heading is still the field name of **PUB_RTE**, which is not very useful.

Headings in the legend cannot be changed in `LegendElement` or `LegendItem`; they must be changed in the layer itself.

The heading can be changed using what you learned earlier about the symbology of a layer. The group heading is part of the `renderer` object.

5. To change the heading, you will create a symbology variable for the bus route layer and then iterate through the groups returned from the `groups` object of the renderer, looking for the group heading `"PUB_RTE"`. When that group heading is found, you will change it to `"Transbay Routes"` in the symbology variable. Then, you will set the bus routes layer symbology to the symbology variable. Type in the following:

```
busRouteLyrSym = busRouteLyr.symbology
for group in busRouteLyrSym.renderer.groups:
    print(group.heading)
    if group.heading == "PUB_RTE":
        group.heading = "Transbay Routes"
        print(group.heading)
busRouteLyr.symbology = busRouteLyrSym
```

Run the cell. You will see the following in the **Out** cell:

```
PUB_RTE
Transbay Routes
```

The legend should be updated to look like the figure below:

Transbay Routes
F
G
J
L
LA
NL
NX
O
P
U
V
W
AlamedaCountry_RaceHispanic_Tract
0 - 25
25 - 50
50 - 75
75 - 100

Figure 7.31: Legend with updated heading

6. The census layer heading also needs to be updated, as it has no heading. `GraduatedColorsRender` does not have a `groups` object, so you will update the layer name. Type in the following:

```
censusLyr.name = "Percent Minority by Tract"
```

Run the cell. The legend should now look like the figure below:

Figure 7.32: Legend with updated heading

You may notice the difference in the font size between the heading (**Transbay Routes**) and the layer name (**Percent Minority by Tract**) in the legend. This is because you are using `AdjustColumnsAndFont`, and headings and layer names get different font sizes. You will learn more about how to ensure your legend is better set up in ArcGIS Pro to be manipulated by Python in *Chapter 12*.

It is important to set as much of your legend display as possible. Features like which legend items to show, the font, and the font size cannot be changed using ArcPy.

7. The legend can be moved using the `elementPositionX` and `elementPositionY` properties. The legend's anchor point is set to the bottom left. You will move the *x* and *y* positions to leave a 0.1-inch space between the edge of the map and the start of the legend, by setting the positions to 0.7, which allows for the 0.1 inch of space between the legend boundary and the map boundary. This is because the legend has a 0.1-inch gap for the background and border and the map edge is set with a 0.5-inch margin. In the next cell, type in the following:

```
legend.elememtPositionX = 0.7
legend.elementPositionY = 0.7
```

Run the cell. The legend has now moved to be set where you want it.

You have seen how to make many changes to the legend. You have also seen some of the limitations in what you can do with ArcPy. ArcPy can assist in automating map production, but you still need to take the time to ensure that the layout is cartographically sound and has been set up properly. This will be explored further in *Chapter 12, Case Study: Advanced Map Automation.*

North arrow, scale bar, and text

The north arrow and scale bar are both map surround elements. As such, they have limited properties that can be adjusted through ArcPy. *You cannot change the north arrow type or scale bar type through ArcPy. They must be set up with what you want in ArcGIS Pro.* However, you can move them around and change their sizes.

Be aware that if you change the size of your scale bar, you will change the scale it is showing.

Text elements have more options. In addition to being able to move text elements around the layout, you can update the text within the element and its font size. *You cannot change the font; it must be set in ArcGIS Pro.*

In this exercise, you will move the scale bar and north arrow around the layout, and change the text of the title element.

1. Continuing to work in the `LayoutElements` Notebook, in the next cell, you access the north arrow, scale bar, and title using the `listElements()` method. Type in the following:

   ```
   scalebar = layout.listElements("MAPSURROUND_ELEMENT","Scale Bar")[0]
   northArrow =
   layout.listElements("MAPSURROUND_ELEMENT","North Arrow")[0]
   title = layout.listElements("TEXT_ELEMENT","Title")[0]
   ```

   Run the cell. There will be nothing returned.

2. In the next cell, you will move the scale bar using the `elementPositionX` and `elementPositionY` properties. The scale bar's anchor point is set to the bottom right. You will move the *x* position to just inside the frame at `10.4` and the *y* position to also just inside the frame at `0.6`. Type in the following:

   ```
   scalebar.elementPositionX = 10.4
   scalebar.elementPositionY = 0.6
   ```

Run the cell. There will be nothing returned, but the scale bar should have moved over to the edge of the figure:

Figure 7.33: Moved scale bar

3. Next, you will move the north arrow to be above the middle of the scale bar. The north arrow's anchor point is the bottom middle. Since the width of the scale bar is returned in inches, you can divide that by two to get the length of half the scale bar. This value is then added to the scale bar's *x* position to find the *x* position that is the midpoint of the scale bar. You will set the north arrow's *x* position to that value. To set the north arrow's *y* position, you will use the *y* position of the scale bar, its height, and add 0.25 inches to it. This will set the north arrow *y* position to 0.25 inches above the scale bar. Type in the following:

```
northArrow.elementPositionX = scalebar.elementPositionX -
(scalebar.elementWidth/2)
northArrow.elementPositionY = scalebar.elementPositionY +
scalebar.elementHeight  + .25
```

Run the cell. There will be nothing returned, but the north arrow should have moved over the middle of the scale bar:

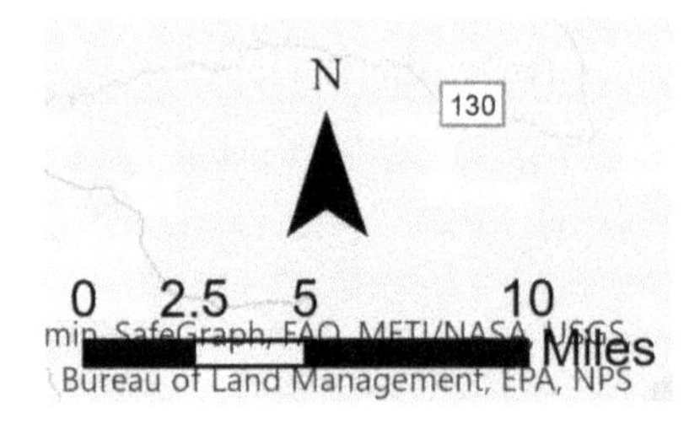

Figure 7.34: Moved north arrow

4. The title element has placeholder text. You can access that text using the `text` property of `TEXT_ELEMENT`. Since you have a placeholder here, you will just set the `text` property to a new string. Type in the following:

```
title.text = "AC Transit Trans-Bay Bus Routes and Percent Minority
in Alameda County"
```

   Run the cell. There will be nothing returned, but the title will have changed in the figure. Since the anchor point is the top center, the title stays centered:

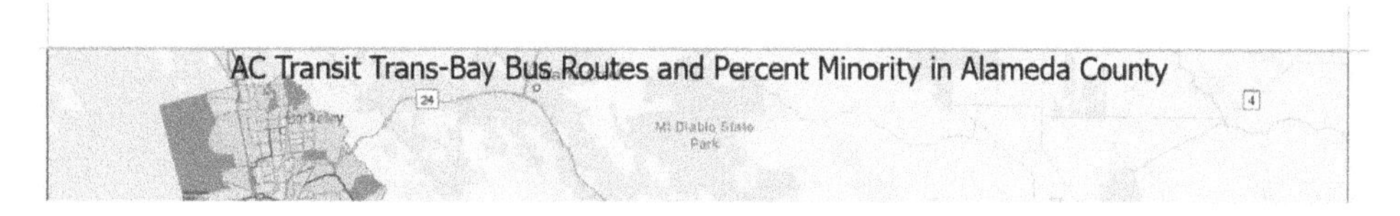

Figure 7.35: Updated title

   If you want to ensure that your title is in the correct location after changing the text, you can use `elementPositionX` and `elementPositionY` to check the anchor point is correct.

> You can set a variable to hold the title text and then use standard Python functions on it, such as `replace()`, string indexing, or any other functions that work on strings. Once you have finished creating the new text, set the `text` property equal to the variable containing your new title text.

You have now updated the symbology of the layer, updated the legend, moved the north arrow and scale bar, and updated the title. The layout is now ready for export. In the next section, you will learn how to export a layout to different formats.

Exporting layouts

Exporting layouts to different file types is a useful process to automate. Often, your `arcpy.mp` scripts will do some work on the map and layout and end with exporting to a file. Each file format has a different method for exporting. The following are all of the methods available:

- `exportToAIX`
- `exportToBMP`
- `exportToEMF`
- `exportToEPS`
- `exportToGIF`

- `exportToJPEG`
- `exportToPDF`
- `exportToPNG`
- `exportToSVG`
- `exportToTGA`
- `exportToTIFF`

All of the methods have the same required argument: the full name of the export file, including the path.

You do not need to include the file type extension; if it is not present, it will be added.

All of the methods also have an argument for `resolution`. For most, it is set to 96 **dots per inch (DPI)** by default. Beyond that, different methods have different arguments for things like **color**, **quality**, and **compression**. The most common export files you will use are **JPEG**, **PNG**, and **PDF**. In this exercise, you will export your layout to all three formats in a new Notebook.

1. Within the **Projects** tab of the **Catalog** tab, right-click on **Chapter7** and select **New** > **Notebook** to create a new Notebook.
2. Rename the Notebook to `ExportLayouts`.
3. In the first cell, you will need to import the os module, which will help you create full paths for your export files. Type in the following:

   ```
   import os
   ```

   Run the cell.
4. In the next cell, you will create a variable for the output location of the files. Type in the following:

   ```
   outputLoc = r"C:\PythonBook\Chapter7"
   ```

   Run the cell.
5. In the next cell, you will create an `ArcGISProject` object based on the current project you have open. Then, you will create a list of all the layouts within the project. Type in the following:

```
project = arcpy.mp.ArcGISProject("CURRENT")
layouts = project.listLayouts()
```

Run the cell.

6. The next cell will be for exporting to JPEG. On top of the required filename argument, the `exportToJPEG()` method has five additional optional parameters:

    - `resolution`: The DPI of the exported file. If not set, 96 is the default value.
    - `jpeg_color_mode`: This can be set to `8-BIT_GRAYSCALE` or `24-BIT_TRUE_COLOR` (default).
    - `jpeg_quality`: Sets the amount of compression applied to the JPEG; a value between 0 and 100. 100 provides the best quality but will create large files. If not set, 80 is the default value.
    - `embeded_color_profile`: A Boolean that will embed the color profile information into the JPEG's metadata. If not set, `True` is the default value.
    - `clip_to_elements`: A Boolean that will clip the layout to the smallest bounding box containing all the layout elements when set to `True`. If not set, `False` is the default value.

    You will be setting the resolution for this example. To export a layout, you will need to access the layout object within the list of layouts. You will do this by iterating through the list and using the layout name to create a name for the export file. Then, you will export the file. You will add a print statement to see the full name of the file you are exporting. Type in the following:

    ```
    for layout in layouts:
        name = layout.name
        jpgName = os.path.join(outputLoc,name+".jpg")
        print(jpgName)
        layout.exportToJPEG(jpgName, resolution=250)
    ```

    Run the cell. You should see `C:PythonBook\Chapter7\AlamedaCountyTransbayBus.jpg` printed and the file should be in that location.

7. In the next cell, you will export to PNG. On top of the required filename argument, the `exportToPNG()` method has four additional optional parameters:

    - `resolution`: The same as in `exportToJPEG()`.

- `color_mode`: Can be set to `8-BIT_ADAPTIVE_PALETTE`, `8-BIT_GRAYSCALE`, `24-BIT_TRUE_COLOR`, or `32-BIT_WITH_ALPHA` (default).
- `transparent_background`: A Boolean that can set the white page background to transparent when `True`. If not set, `False` is the default value.
- `clip_to_elements`: The same as in `exportToJPEG()`.

You will be setting the resolution for this example. The export process is the same as with `exportToJPEG()`; the only difference is the method being used and the variable names. Type in the following:

```
for layout in layouts:
    name = layout.name
    pngName = os.path.join(outputLoc,name+".png")
    print(pngName)
    layout.exportToPNG(pngName,resolution=250)
```

Run the cell. You should see `C:PythonBook\Chapter7\AlamedaCountyTransbayBus.png` printed and the file should be in that location.

8. In the next cell, you will export to PDF. For exporting to PDF, the default resolution is set to 300 DPI. The `exportToPDF` tool has many additional parameters. Some of them allow you control over the compression and quality of the image.

For a full explanation of all the parameters available, refer to the documentation here: `https://pro.arcgis.com/en/pro-app/latest/arcpy/mapping/layout-class.htm`. In most cases, you will just be using the defaults and will not need to worry about changing them.

The export process is the same as for `exportToJPEG()` and `exportToPNG()`. The only difference is the method being used and the variable names. Type in the following:

```
for layout in layouts:
    name = layout.name
    pdfName = os.path.join(outputLoc,name+".pdf")
    print(pdfName)
    layout.exportToPDF(pdfName,resolution=250)
```

Run the cell. You should see `C:PythonBook\Chapter7\AlamedaCountyTransbayBus.pdf` printed and the file should be in that location.

You have seen how to use three of the more common export methods. All three worked in the same way, and it is no different for the other methods. These examples only export one layout, but in a map with many layouts, the code would export all of them. You could create code to iterate through a folder of projects, make some updates, and export the layouts. You could also use this to create custom map series and export each page as the layout updates.

Summary

In this chapter, you learned how to reference a project and map in an ArcGIS Pro Notebook by using the `arcpy.mp` module. You also learned how to add, remove, and move layers to your maps with the `arcpy.mp` module. You explored the different options to symbolize vector and raster data using the different renderer and colorizer classes available through the module. You learned what legend properties you can adjust by using the `LegendElement` and `LegendItem` classes. You also learned how to move the scale bar and north arrow, and change the text in a text element, by accessing them through the `MapSurroundElement` class. You exported your layout to JPEG, PNG, and PDF using the export methods in the layout class. The example code you have seen can be applied to one map in a project, multiple maps in a project, or even multiple projects by adding loops to loop through all maps or projects. In *Chapter 12*, you will see more advanced examples using these skills.

It is important to remember that good cartography in layouts starts from building a layout in ArcGIS Pro; but Python can help you make updates to those elements more efficiently, especially when you are making the same changes across many maps and projects.

In the next chapter, you will learn how to integrate the `pandas` data toolkit with ArcGIS API for Python to perform geospatial analysis.

Part III

Geospatial Data Analysis

8

Pandas, Data Frames, and Vector Data

Data analysis is a popular use of Python. Since Python can read and write lots of data formats, has powerful built-in mathematical functionality, and has a large number of third-party modules written for specific analytical and statistical realms, it has gained wide popularity among analysts and scientists.

One of the most popular data analysis modules is Pandas. It has become a standard tool for data analysis and data science, and has extended into geospatial analysis and geodata science.

In this chapter, we will cover the following topics:

- What a DataFrame is
- The basics of using Pandas, including reading and writing files
- Performing data analysis and manipulation with Pandas
- Using **Spatially Enabled DataFrames (SEDFs)**

To complete the exercises in this chapter, please download and unzip the `Chapter8.zip` folder from the GitHub repository for this book: `https://github.com/PacktPublishing/Python-for-ArcGIS-Pro/tree/main/Chapter8`.

Introduction to Pandas

Pandas is a Python module used for data analysis and manipulation. It is an open-source module that can be installed and used separately from ArcGIS Pro; in fact, it is the most popular Python data analysis module. As it is so useful and well-known, it is included along with Python when ArcGIS Pro is installed.

Its origins lie in the financial world, where statistical analysis is used constantly. In 2007, needing a more powerful tool to perform quantitative analysis, a financial analyst and programmer named Wes McKinney developed the first version of Pandas. It was made open source in 2012 and was quickly recognized as a powerful and flexible data tool.

Pandas DataFrames

The basic data structure in Pandas is the Pandas **DataFrame**. A Pandas DataFrame is essentially a data table, much like ArcGIS attribute tables or Excel tables, but with a whole lot of built-in features that make it easy to manipulate and manage data.

Pandas Series

DataFrames are made up of individual data "columns." In Pandas, a data column is known as a Pandas **Series**. It is a one-dimensional array of data, of any data type, but only *one* data type. Groups of Series are combined to create a DataFrame.

Much like a Python, a Series is a set of values grouped in an array that has a direction (meaning the values are in a specific order) and only one dimension, though unlike a data list, values in a Series must all be of the same data type.

In this example, the data list `valueslist` data list is passed to the Pandas `Series` function to create a Series:

```
import pandas as pd
valueslist = [3,5,6,7,9]
a_series = pd.Series(valueslist)
```

This Series contains data values that are all integers. The data type can be implicitly passed, as in the example above, or it can be explicitly passed as the parameter `dtype`. Other optional parameters include the `name` of the Series or its `index`:

```
valueslist = [3,5,6,7,9]
b_series = pd.Series(valueslist, dtype=float, name= 'values')
```

By combining one or more Pandas Series, you can create a Pandas DataFrame:

```
df = pd.DataFrame({"a":a_series,"b": b_series})
```

Another way to combine the Pandas Series is to use the `concat` function, which will concatenate Pandas Series into a DataFrame or into one Series, depending on the `axis` orientation used. An axis of 1 will put all of the Series in the iterable list passed to `concat` into a DataFrame:

```
pd.concat([a_series, b_series],axis=1)
```

From these one-dimensional Pandas Series (think of these as the "columns"), we can create Pandas DataFrames that combine any type of data, from strings to numbers, or even binary data types like spatial data.

Read more about Pandas Series here in the Pandas API docs: `https://pandas.pydata.org/docs/reference/api/pandas.Series.html`.

The technical definition of **data frames**, of which the Pandas DataFrame is but one implementation, is lists of equal length that have a **magnitude** and **direction**. The vectors within the list are akin to columns, and the magnitude or length of the vector is the number of rows. As each column is a separate vector, they can be different data types. This allows one data frame to hold multiple data types, each in its own column.

The concept of data frames is not only found in Pandas. Other data processing libraries, such as Apache Spark and Apache Sedona, use data frames to store data in memory. The R statistical language is also built on data frames. It is a popular and powerful data concept that makes up the basis of modern data engineering.

Spatially Enabled DataFrames

Esri created the **Spatially Enabled DataFrame** (**SEDF**) object to allow you to add spatial object types to Pandas DataFrames. This object is used to perform geospatial operations within the Pandas framework. It acts just like a Pandas DataFrame, except that there is a `"SHAPE"` column that can be used to perform geospatial analysis such as buffers, clips, or spatial joins.

We will use **SEDF** objects to do some spatial **ETL** in this chapter: **extracting** data from a data source, **transforming** it into a Spatially Enabled DataFrame and performing analysis, and **loading** it into a new data source.

Read more about Spatially Enabled DataFrames here: `https://developers.arcgis.com/python/guide/introduction-to-the-spatially-enabled-dataframe/`.

Pandas can read data from CSVs, Excel files, JSON data, databases, and more, even from feature classes when ArcGIS Pro is installed, allowing it to be used in both spatial and non-spatial data engineering tasks. The freedom it offers for either one-off data analysis in Jupyter Notebooks or data processing within a Python script is unparalleled in the Python module ecosystem.

Esri has made it possible to use SEDFs either with or without the ArcPy module. If you have installed **PySHP**, **Shapely**, and **Fiona**, three open-source modules built to read, write, and edit spatial data, you can use SEDFs along with the ArcGIS API for Python module. This means that you can execute scripts or Jupyter notebooks on a MacBook, as the requirement of having the Windows-only ArcPy module on the machine is removed.

Read more about Pandas and ArcGIS here: `https://developers.arcgis.com/python/guide/part3-introduction-to-pandas/`.

Installing Pandas

Pandas is installed along with Python when ArcGIS Pro is installed, but it can also be installed using `pip`. In a command line or terminal, type this command:

```
pip install pandas
```

This will make it possible to get the latest Pandas version from the **Python Package Index** (**PyPI**).

Using the package manager **Conda** is also a popular method for installing Pandas. The pip manager is included with Python, and is more ubiquitous, but the Conda package manager is well-known within the scientific and data science communities.

Similar to `pip`, Conda can be used in the command line. If you have Conda installed, use this command to install Pandas:

```
conda install pandas
```

Getting data into (and out of) a Pandas DataFrame

While the Pandas library is worthy of a whole book or more, even just on its uses for geospatial and attribute data, we are going to focus on major functions that you can use right away to improve your geospatial data processing. The first one is getting data into a data frame from either a CSV or JSON file, or even a shapefile or feature class.

Read more details about Pandas and data science here: https://jakevdp.github.io/PythonDataScienceHandbook/03.00-introduction-to-pandas.html.

Reading data from a file

Here are some basic examples showing how to read data into a Pandas DataFrame from various sources, including CSVs, feature classes, or ArcGIS Online layers, which are among the many file read options included in Pandas. The result of all of these methods is a data frame, which is often assigned to the variable `df`, but can be assigned to any valid variable.

- Importing Pandas as the variable `pd` is a well-known shorthand. This makes it easier to write and to access the submodules and methods of the Pandas library:

  ```
  import pandas as pd
  ```

- CSVs, or comma-separated values, are common data files stored in plain text:

  ```
  df = pd.read_csv('example.csv')
  ```

- Reading JSON data is also very common. Note that the `pd.read_json` method is being discontinued in favor of the `pd.io.json.read_json` method:

  ```
  df = pd.io.json.read_json('example.json')
  ```

- Importing data from a database is common in enterprise-level code. The `read_sql` method requires a connection to talk to the database. These connections often require a second module, which in this example is psycopg2 (used to connect to PostgreSQL databases). Other possible modules include SQLAlchemy or `mysql.connector`.

  Pass a `select` statement, or the name of a specific table, to the function:

  ```
  import pandas as pd
  import psycopg2 as pg
  ```

```
engine = pg.connect("dbname='db_name' user='pguser' host='127.0.0.1'
port='5432' password='pgpass'")
df = pd.read_sql('select * from my_table', con=engine)
```

- To access spatial data, the Pandas included with the Python version in ArcGIS Pro has a `DataFrame.spatial` submodule, which has the `from_featureclass` method:

```
import pandas as pd
from arcgis.features import GeoAccessor, GeoSeriesAccessor
df = pd.DataFrame.spatial.from_featureclass('a.shp')
```

- If your computer does not have ArcPy installed, the ArcGIS API for Python requires that you log in to your ArcGIS Online account to be able to read from shapefiles. This is done either in the ArcGIS Pro Python environment using `gis = GIS("Pro")` or like this in a script:

```
from arcgis import GIS
gis = GIS("https://www.arcgis.com", "username", "password")
```

Using the ArcGIS API for Python (the `arcgis` module) to access ArcGIS Online layers is a common practice. With the module, you can add, update, or remove layers that are stored online, which can save you a lot of time and credits. In the following example, you will use the `arcgis` module to access a layer in the ArcGIS Online cloud:

1. First, import the module:

```
from arcgis import GIS
gis = GIS("Pro")
```

2. Using a layer ID, you can access a layer and turn it into a DataFrame:

```
layerid = '85d0ca4ea1ca4b9abf0c51b9bd34de2e'
geocontent = gis.content.get(layerid)
```

3. Select the layer of interest from the variable:

```
glayer = geocontent.layers[0]
```

4. Pass the selected layer to the Pandas `DataFrame.spatial` method to create the DataFrame:

```
from arcgis.features import GeoAccessor, GeoSeriesAccessor
df_layer = pd.DataFrame.spatial.from_layer(glayer)
```

These methods are often used in a common modern analysis workflow where data is stored in the cloud for web applications but pulled down to your local machine for processing, or pushed up to the cloud as a layer after creating a new dataset.

Writing data to a file

There are many write methods available for Pandas DataFrames, and only a few common ones are shown below. These methods do not depend on the read method, meaning you can (for example) read data from a database into a Pandas DataFrame and then output a CSV from the same data frame.

- To write a data frame as a CSV, use this method:

  ```
  df.to_csv('output.csv')
  ```

- Creating a JSON file from a data frame uses this method:

  ```
  df.to_json('output.json')
  ```

- Writing to a database is a bit more complex, as it requires a connection to the database. In the following example, SQLAlchemy is used to connect to a SQLite database:

  ```
  from sqlalchemy import create_engine
  engine = create_engine('sqlite://dbname')
  df.to_sql('my_table', engine)
  ```

- Spatial data formats such as shapefiles and feature classes can be created, but only from a Spatially Enabled DataFrame. These data frames have a special method, `spatial.to_featureclass`:

  ```
  df.spatial.to_featureclass('output.shp')
  ```

These are only some of the methods available for accessing data in files and loading it into Pandas DataFrames.

Exercise: From GeoJSON to CSV to SHP using Pandas

In this exercise, you will explore Pandas using an example of addresses stored in a GeoJSON file. The address file is from `openaddresses.io` and represents a county in Pennsylvania. You will use the basic features of Pandas and Notebooks to transform data from the raw format supplied by OpenAddresses into a feature class. Along the way, you will create maps displaying the address data.

Pandas can be used in a standalone Python script, but for this example we will use a Notebook, a common method.

1. Open ArcGIS Pro and start a new project, then add a new Notebook from the **Insert** tab. Rename the Notebook to `Chapter8`.
2. In the first cell, you will import `arcpy` and `arcgis`, as well as `pandas`. Type in the following:

```
import arcgis, arcpy
import pandas as pd
```

3. In the same cell, you will read the `cameron-addresses-county.geojson` file into a Pandas DataFrame. You will open the file and assign it to a data frame using the `pd.io.json.read_json` command. While this file is a GeoJSON, Pandas will treat it as a plain JSON file. However, it needs to read the JSON in a line-by-line format. To do so, you will pass the `lines=True` parameter to Pandas, so that it reads each line in the JSON file as a separate row in the resulting Pandas DataFrame. You will assign the result of the read object to a variable, `df_json`.

   Type in the following, making sure to adjust the file path to match where you downloaded the *Chapter 8* data to:

```
df_json = pd.io.json.read_json("cameron-addresses- county.geojson",
                               lines=True)
```

   Run the cell.

4. In the next cell, you will test the newly loaded DataFrame using the `.head()` method, which gets the first five rows of the DataFrame. This is helpful for checking the rows and columns of the DataFrame to make sure that they are structured as expected. Type in the following:

```
df_json.head()
```

   Run the cell. You should see the following output:

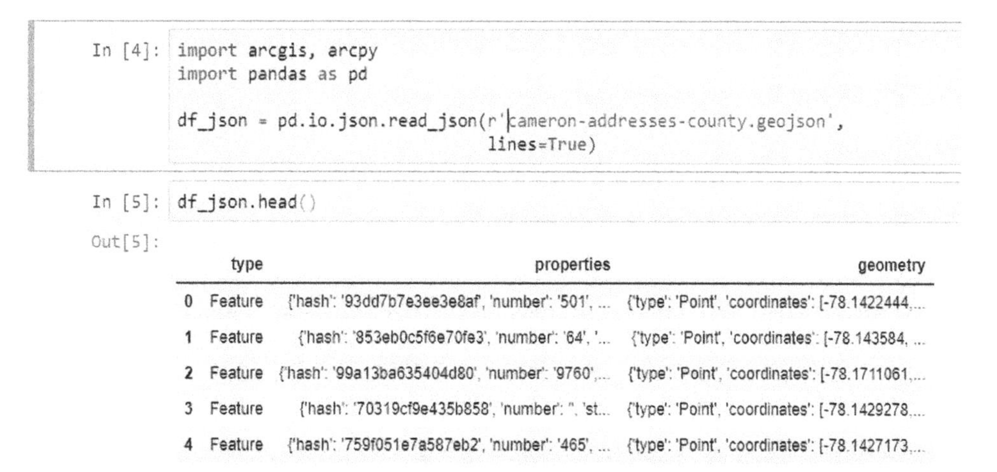

```
In [4]: import arcgis, arcpy
        import pandas as pd

        df_json = pd.io.json.read_json(r'cameron-addresses-county.geojson',
                                       lines=True)

In [5]: df_json.head()

Out[5]:
```

	type	properties	geometry
0	Feature	{'hash': '93dd7b7e3ee3e8af', 'number': '501', ...	{'type': 'Point', 'coordinates': [-78.1422444,...
1	Feature	{'hash': '853eb0c5f6e70fe3', 'number': '64', '...	{'type': 'Point', 'coordinates': [-78.143584, ...
2	Feature	{'hash': '99a13ba635404d80', 'number': '9760',...	{'type': 'Point', 'coordinates': [-78.1711061,...
3	Feature	{'hash': '70319cf9e435b858', 'number': '', 'st...	{'type': 'Point', 'coordinates': [-78.1429278,...
4	Feature	{'hash': '759f051e7a587eb2', 'number': '465', ...	{'type': 'Point', 'coordinates': [-78.1427173,...

Figure 8.1: A new Notebook in ArcGIS Pro, and the file read and df.head() used to show the first five rows

Similarly, you can use `df_json.tail()` to see the last five rows. If you pass an integer to `head()` or `tail()`, it will show you that number of rows; the default is five.

5. You can access these columns by passing the name of the column you want to the data frame in brackets. This will allow us to access only the data in that column, instead of all of the data in all of the columns in the data frame. We will need this to be able to perform specific operations on the columns of interest.

   Type the following into the next Notebook cell:

   ```
   df_json["properties"]
   ```

   Run the cell. You should only see the `properties` column of the data frame.

Now that the data has been imported into a spreadsheet, it must be flattened or **normalized** to be able to access the data within the JSON structure using Pandas tools.

Normalizing the nested JSON data

The rows shown have three columns ("type", "properties", and "geometry"). While this is correct, it's not the form that we want the data to be in for our use. Both `df_json["properties"]` and `df_json["geometry"]` contain nested JSON data, which we want to see as rows of data so that they can be processed in the data frame.

The solution is to use a Pandas function called `json_normalize`, which makes it possible to take nested JSON objects (JSON objects inside other JSON objects) and expand each JSON object into a row. We will have to do this twice, as `df_json["properties"]` and `df_json["geometry"]` both have nested JSON data that we want to expand and then join back together into the final data frame.

6. Continuing in the same Notebook, you will pass the `df_json["properties"]` column to the `pd.json_normalize` function, which creates a new data frame (`df_properties`) containing only the data from the column. Enter this code into the next Notebook cell:

```
df_properties = pd.json_normalize(df_json['properties'])
df_properties
```

In Jupyter Notebooks, if you type the name of the data frame as the last line of the code cell, the data frame will be displayed in the output window below the cell. This is useful as it allows us to see the new `df_properties` data frame, so we can confirm that the data is now correctly formatted.

Run the cell. You should see the following output:

```
In [8]: df_properties = pd.json_normalize(df_json['properties'])
        df_properties
Out[8]:
```

	hash	number	street	unit	city	district	region	postcode	id
0	93dd7b7e3ee3e8af	501	CASTLE GARDEN RD						7579
1	853eb0c5f6e70fe3	64	BELDIN DR						4502
2	99a13ba635404d80	9760	MIX RUN RD						8448
3	70319cf9e435b858								
4	759f051e7a587eb2	465	CASTLE GARDEN RD						6447
...	...	...	...	...	...	...	...	...	...
7492	ca7d22c4f71a10ce		14918 MONTOUR RD		LEIDY TWP				
7493	8d74e4cf313ffbae		14881 MONTOUR RD		LEIDY TWP				
7494	4329357ebc44aa92		14847 MONTOUR RD		LEIDY TWP				
7495	c8089ef2d8c609dc	7094	HUNTS RUN RD		PORTAGE TWP				
7496	9877bd7a32f16636	715	HAWK RD		PORTAGE TWP				

7497 rows × 9 columns

Figure 8.2: The new df_properties data frame is created and displayed

7. Now you need to repeat the normalization process for the `df_json["geometry"]` column. In the next cell, type in the following code to create a new data frame from the `df_json["geometry"]` column:

    ```
    df_geometry = pd.json_normalize(df_json['geometry'])
    df_geometry
    ```

    Run the cell. You should see the following output:

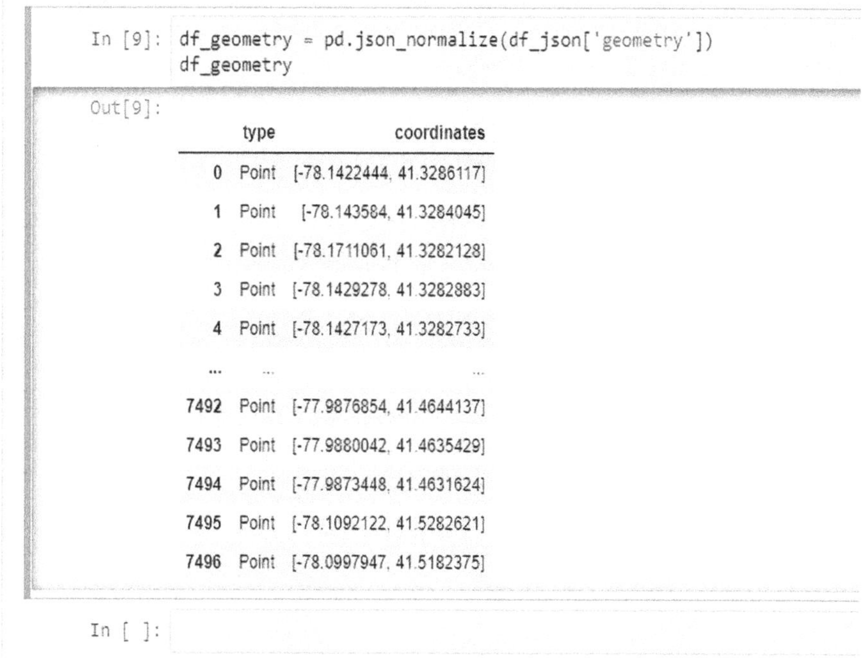

```
In [9]: df_geometry = pd.json_normalize(df_json['geometry'])
        df_geometry
Out[9]:
```

	type	coordinates
0	Point	[-78.1422444, 41.3286117]
1	Point	[-78.143584, 41.3284045]
2	Point	[-78.1711061, 41.3282128]
3	Point	[-78.1429278, 41.3282883]
4	Point	[-78.1427173, 41.3282733]
...	...	...
7492	Point	[-77.9876854, 41.4644137]
7493	Point	[-77.9880042, 41.4635429]
7494	Point	[-77.9873448, 41.4631624]
7495	Point	[-78.1092122, 41.5282621]
7496	Point	[-78.0997947, 41.5182375]

```
In [ ]:
```

Figure 8.3: The column is now normalized (the keys are now column labels and the values are rows)

This data frame will have the same number of rows as the original data frame (`df_json`), but because it is **normalized** – meaning the nested JSON keys are now column labels, and the nested JSON values are now row values – the data is easier to access and perform Pandas data operations on. The data must also be normalized to (eventually) be written out as a shapefile.

Joining data frames

Now that we have taken the original `df_json` data frame and split it into two normalized data frames (`df_properties` and `df_geometry`), we need to join the data frames back together to reunite the attribute data with the geometry data.

To accomplish this, we will use the Pandas `join` function, which makes it possible to join two data frames together into a new data frame. There are a number of optional parameters for the `join` function, but because we have the same number of rows in both of the data frames we are joining, and they are using the same index values (meaning a row at index *N* in one data frame corresponds to the row at index *N* in the other data frame), we can just pass the data frame to the `join` function and ignore the other parameters.

8. You don't need the `"type"` column in the `df_geometry` data frame, so you can just add the `df_geometry["coordinates"]` column to the `df_properties` data frame using the `join` function. In the next cell, type in:

```
df_data = df_properties.join(df_geometry['coordinates'])
df_data
```

Run the cell. The resulting data frame (`df_data`) now contains all of the attribute values from the `df_properties` data frame and the `df_geometry['coordinates']` column:

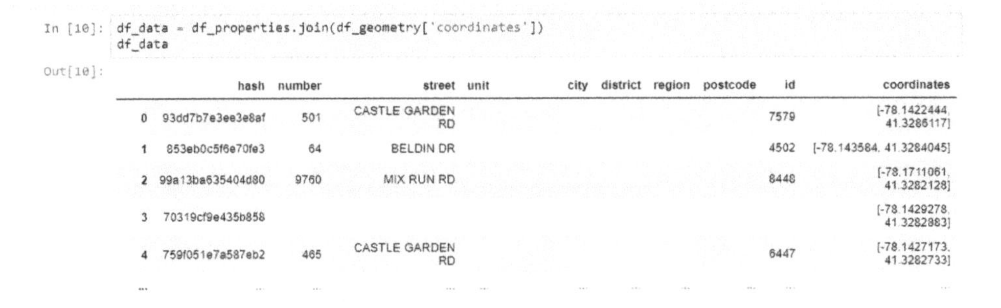

```
In [10]: df_data = df_properties.join(df_geometry['coordinates'])
         df_data
Out[10]:
```

	hash	number	street	unit	city	district	region	postcode	id	coordinates
0	93dd7b7e3ee3e8af	501	CASTLE GARDEN RD						7579	[-78.1422444, 41.3286117]
1	853eb0c5f6e70fe3	64	BELDIN DR						4502	[-78.143584, 41.3284045]
2	99a13ba635404d80	9760	MIX RUN RD						8448	[-78.1711061, 41.3282128]
3	70319cf9e435b858									[-78.1429278, 41.3282883]
4	759f051e7a587eb2	465	CASTLE GARDEN RD						6447	[-78.1427173, 41.3282733]
...	...	...	...	...	...	...	...	...	...	...

Figure 8.4: The result of the join operation

As you can see in *Figure 8.4*, the data has been normalized and joined using the two operations described above. However, you need to do one more operation to get the data into the final format.

9. There are two steps to the process of adding the latitude and longitude to the data frame. First, the `df_data['coordinates']` column is actually stored as an "object" data type. The Pandas function `to_list` allows us to convert the data from an object type to a list, and then create a new data frame with two columns (`'long'`and `'lat'`). As the `'coordinates'` column is a list type of data, you need to split the values in the list into their own columns: one for longitude and one for latitude. Luckily, Pandas can easily perform this type of operation. In the next cell, type in the following:

```
df3 = pd.DataFrame(df_data['coordinates'].to_list(),
                    columns=['long','lat'])
df3
```

Run the cell. You should see the following:

```
In [11]: df3 = pd.DataFrame(df_data['coordinates'].to_list(), columns=['long','lat'])
         df3
Out[11]:
```

	long	lat
0	-78.142244	41.328612
1	-78.143584	41.328404
2	-78.171106	41.328213
3	-78.142928	41.328288
4	-78.142717	41.328273

Figure 8.5: The new data frame with the coordinates column now split up into two new columns

10. We can then join the new `df3` data frame back to the `df_data` data frame, and reassign the result to a variable that will overwrite the original `df_data` data frame. In the next cell, type in the following:

```
df_data = df_data.join(df3)
df_data
```

Run the cell. You should see the following:

```
In [12]: df_data = df_data.join(df3)
         df_data
Out[12]:
```

	hash	number	street	unit	city	district	region	postcode	id	coordinates	long	lat
0	93dd7b7e3ee3e8af	501	CASTLE GARDEN RD						7579	[-78.1422444, 41.3286117]	-78.142244	41.328612
1	853eb0c5f6e70fe3	64	BELDIN DR						4502	[-78.143584, 41.3284045]	-78.143584	41.328404
2	99a13ba635404d80	9760	MIX RUN RD						8448	[-78.1711061, 41.3282128]	-78.171106	41.328213
3	70319cf9e435b858									[-78.1429278, 41.3282883]	-78.142928	41.328288
4	759f051e7a587eb2	465	CASTLE GARDEN RD						6447	[-78.1427173, 41.3282733]	-78.142717	41.328273

Figure 8.6: The newly joined data frame

Dropping columns

The data frame now has the new `long` and `lat` columns added, but there are extraneous columns that we should get rid of before the data is ready, as they will not be needed for the final product. Luckily, Pandas makes it easy to drop columns and create a new data frame without the columns you don't need.

11. As the `coordinates` column has been split into the `long` and `lat` columns, we can drop it. The `coordinates` column is also the wrong data type to be used to create a shapefile. Additionally, the `id` and `hash` columns provided by `openaddresses.io` are not needed for this analysis and can be dropped. Using the data frame `drop` function, you can specify the columns to drop. In the next cell, type in the following:

    ```
    df_data = df_data.drop(columns=['id','coordinates','hash'])
    df_data
    ```

    The result is a copy of the original data frame in memory when run. If you want to replace the original (and not assign it to a new variable), you can use the `"inplace=True"` parameter.

Run the cell. You should see the following output:

```
In [40]: df_data = df_data.drop(columns=['id','coordinates','hash'])
         df_data
Out[40]:
```

	number	street	unit	city	district	region	postcode	long	lat
0	501	CASTLE GARDEN RD						-78.142244	41.328612
1	64	BELDIN DR						-78.143584	41.328404
2	9760	MIX RUN RD						-78.171106	41.328213
3								-78.142928	41.328288
4	465	CASTLE GARDEN RD						-78.142717	41.328273

Figure 8.7: The extra columns are now dropped from the data frame

Creating a CSV

Now that the `df_data` data frame is in the format you want, you can use the Pandas write functionality to create an output file. While there are many different output possibilities, we'll create a CSV for now.

12. To create a CSV, use the Pandas `to_csv` function, specifying the file path you want the CSV to be saved to. In the next cell, type in the following:

```
df_data.to_csv(r'C:\Projects\output.csv')
```

Run the cell and view the resulting output; there is one issue with it. While it outputs the row index, there is no data label for the index column, which results in a malformed CSV file:

```
output.csv
1  ,number,street,unit,city,district,region,postcode,long,lat
2  0,501,CASTLE GARDEN RD,,,,,,-78.1422444,41.3286117
3  1,64,BELDIN DR,,,,,,-78.143584,41.3284045
4  2,9760,MIX RUN RD,,,,,,-78.1711061,41.3282128
5  3,,,,,,,,-78.1429278,41.3282883
6  4,465,CASTLE GARDEN RD,,,,,,-78.1427173,41.3282733
```

Figure 8.8: The output of the CSV, which has no data label (column name) for the first column

13. To remedy this, you can add a name to the index column using the `name` property of the `index` column. Type in the following in the next cell. I've chosen the name `"oid"`, but it could be any valid string:

```
df_data.index.name = 'oid'
```

14. Now that the data label (or column name) has been added, the CSV can be written again, and the column will have the correct name. Rerun the cell containing the `df_data.to_csv()` line and review the output:

```
output.csv
1 oid,number,street,unit,city,district,region,postcode,long,lat
2 0,501,CASTLE GARDEN RD,,,,,,-78.1422444,41.3286117
3 1,64,BELDIN DR,,,,,,-78.143584,41.3284045
4 2,9760,MIX RUN RD,,,,,,-78.1711061,41.3282128
5 3,,,,,,,,-78.1429278,41.3282883
6 4,465,CASTLE GARDEN RD,,,,,,-78.1427173,41.3282733
```

Figure 8.9: The correct output with the correct number of column names

Creating a Spatially Enabled DataFrame from a DataFrame

Since our data frame has the `long` and `lat` columns, we can create a Spatially Enabled DataFrame from the original Pandas DataFrame. To do this, we will use a special feature of the ArcGIS API for Python, the **GeoAccessor**, which adds the spatial column to the data frame. In this case, we want to be able to see the data on a map. The GeoAccessor makes it easy to read in data from files or data frames.

15. Continuing in the same Notebook, you are going to plot the `df_data` data frame from the last section on a map. After importing the relevant modules, you will create a new SEDF called `sdf_address`, and use the `from_xy` method of `GeoAccessor` to generate a new `SHAPE` column. This method accepts a pair of X/Y columns, which, in the case of the `df_data` data frame, are named `long` and `lat`.

You will use the `spatial.plot` method to plot the data. In the next cell, type in the following:

```
from arcgis.features import GeoAccessor
sdf_address = GeoAccessor.from_xy(df_data, 'long', 'lat')
sdf_address.spatial.plot()
```

Run the cell. You should see the plot:

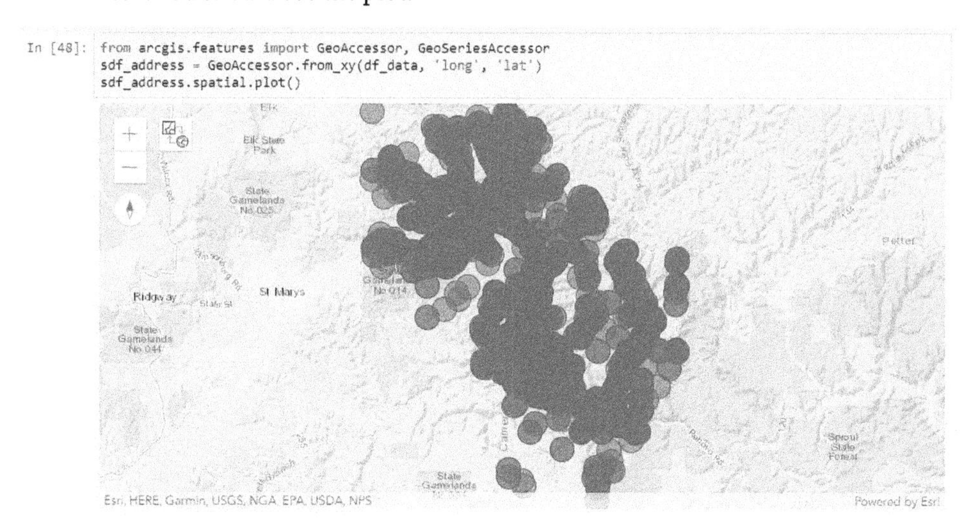

Figure 8.10: The new Spatially Enabled DataFrame plotted on a map in a Notebook

16. Next, you will create a shapefile from the new data frame with the spatial column, using the `spatial.to_featureclass` method we saw earlier. In the next cell, type in:

```
sdf_address.spatial.to_featureclass(r'C:\Projects\test.shp')
```

Run the cell. You should see the following output:

```
'C:\\Projects\\test.shp'
```

Because the new data frame is a Spatially Enabled DataFrame, it can be saved as a shapefile, feature class, or even a GeoJSON, among other formats. Specify the output type using the correct extension for that data type.

17. Once the data is saved as a shapefile, add the data to an ArcGIS Pro map using the **Add Data** interface:

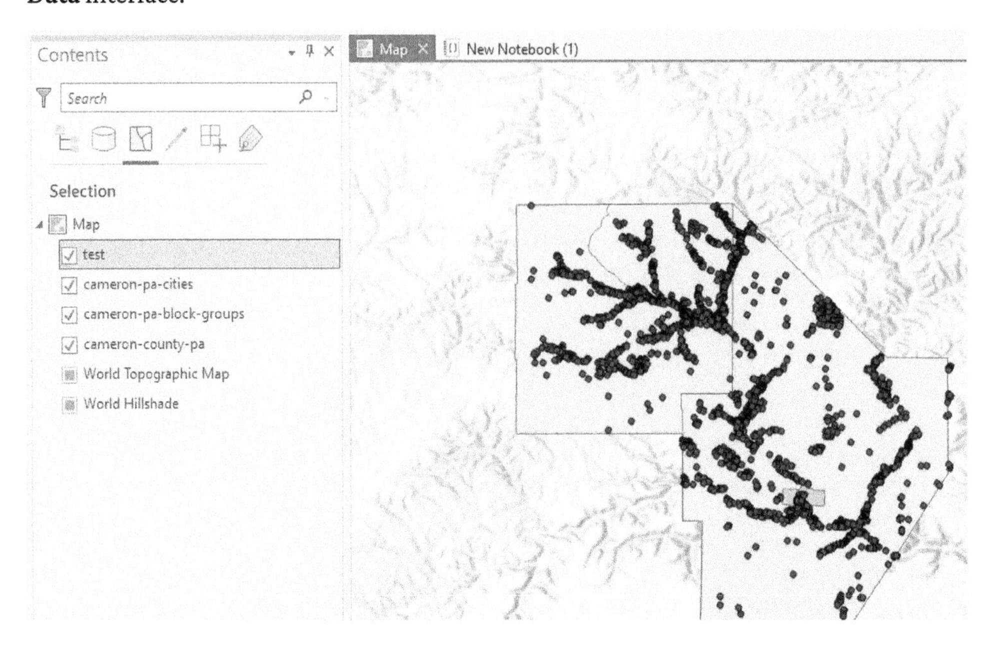

Figure 8.11: The new shapefile has been added to a map

The new Spatially Enabled DataFrame is still a Pandas DataFrame. Both spatial and non-spatial operations can be performed on these data frames.

18. For the example address data, you can use Pandas functionality to remove rows of data that don't have valid addresses. In the next cell, you will import `numpy`, another Python module, and use the Pandas `replace` function to convert empty spaces into `NaN` values (which are the `NULL` values for Pandas). Type in the following:

```
import numpy as np
sdf_address['number'] = sdf_address['number'].replace(r'^\s*$',
                                                      np.NaN,
                                                      regex=True)
sdf_address
```

Note that you are using a **regular expression (regex)** to locate all empty spaces within the `number` column of the data frame. Regexes are a complex topic best covered elsewhere, but in general they are used for pattern matching.

The regular expression and the replacement value (`np.NaN`) are passed to the `replace` function, and the result is written back into the `number` column.

Run the cell. You should see the following output:

```
In [60]: import numpy as np
         sdf_address['number'] = sdf_address['number'].replace(r'^\s*$', np.NaN, regex=True)
         sdf_address
```

3	NaN							-78.142928	41.328288	{"spatialReference": {"wkid": 4326}, "x": -78....
4	465	CASTLE GARDEN RD						-78.142717	41.328273	{"spatialReference": {"wkid": 4326}, "x": -78....
...	...	...	...	...	...	...	...	...	...	...
7492	NaN	14918 MONTOUR RD		LEIDY TWP				-77.987685	41.464414	{"spatialReference": {"wkid": 4326}, "x": -77....
7493	NaN	14881 MONTOUR RD		LEIDY TWP				-77.988004	41.463543	{"spatialReference": {"wkid": 4326}, "x": -77....
7494	NaN	14847 MONTOUR RD		LEIDY TWP				-77.987345	41.463162	{"spatialReference": {"wkid": 4326}, "x": -77....
7495	7094	HUNTS RUN RD		PORTAGE TWP				-78.109212	41.528262	{"spatialReference": {"wkid": 4326}, "x": -78....
7496	715	HAWK RD		PORTAGE TWP				-78.099795	41.518237	{"spatialReference": {"wkid": 4326}, "x": -78....

7497 rows × 10 columns

Figure 8.12: Empty strings have been replaced in the number column

Observe that all empty strings in the `number` column have been identified and replaced.

Dropping NaN values using dropna

Now that we have identified these `NaN` values, we can use a special function of Pandas: `dropna`. We don't want to process any rows with these `NULL` values in those columns, and this function is an effective way to reduce the number of rows with bad data.

19. In the next cell, you will use the `dropna` function on your data frame and specify that rows with `NaN` values in the `number` column should be dropped, using the optional `subset` parameter. Type in the following:

```
sdf_address = sdf_address.dropna(subset=['number'])
sdf_address
```

Run the cell. You should see the following output:

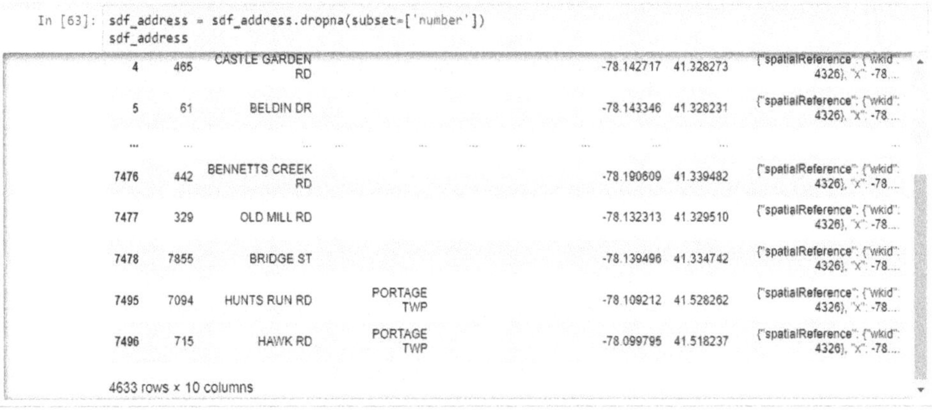

```
In [63]: sdf_address = sdf_address.dropna(subset=['number'])
         sdf_address
```

4	465	CASTLE GARDEN RD		-78.142717	41.328273	{"spatialReference": {"wkid": 4326}, "x": -78....
5	61	BELDIN DR		-78.143346	41.328231	{"spatialReference": {"wkid": 4326}, "x": -78....
...	...	...	...	...	...	...
7476	442	BENNETTS CREEK RD		-78.190609	41.339482	{"spatialReference": {"wkid": 4326}, "x": -78....
7477	329	OLD MILL RD		-78.132313	41.329510	{"spatialReference": {"wkid": 4326}, "x": -78....
7478	7855	BRIDGE ST		-78.139496	41.334742	{"spatialReference": {"wkid": 4326}, "x": -78....
7495	7094	HUNTS RUN RD	PORTAGE TWP	-78.109212	41.528262	{"spatialReference": {"wkid": 4326}, "x": -78....
7496	715	HAWK RD	PORTAGE TWP	-78.099795	41.518237	{"spatialReference": {"wkid": 4326}, "x": -78....

4633 rows × 10 columns

Figure 8.13: A reduction in the number of rows after the dropna function is used. Nearly 3000 were dropped.

The result of the operation is a reduction in the number of rows in the Pandas data frame. These can be done *in place* (meaning there is no need to assign it to a new data frame variable) using the `inplace=true` parameter. You can also use the `how` parameter to control whether a row or column will be dropped if it contains a minimum of one `NaN` value (`how='any'`) or or whether they should only be dropped if all the values in the row or column are `NaN` (`how='all'`).

Querying the data frame

Pandas makes it easy to perform queries on data frames. These queries can be used to create subsets and/or to select specific rows or columns. There are two major ways you will encounter for using a query to select specific rows of data. The first is the `query` method, a function built into DataFrames and accessed using dot notation. We will look at the second method shortly.

20. In this case, you will use the `query` method to find all addresses where the street name is equal to `"CASTLE GARDEN RD"`. Type in the following:

```
sdf_query = sdf_address.query("street == 'CASTLE GARDEN RD'")
sdf_query
```

Run the cell. The result of this query is a new `DataFrame` (`sdf_query`). This new DataFrame is still a Spatially Enabled DataFrame because it started as one, but it is limited to only rows where the street column matches the conditional passed to the query function:

Out[104]:

	OBJECTID	SHAPE	city	district	lat	long	number	postcode	region	street	unit
0	1	{"x": -8698754.85605153, "y": 5060933.31315760...			41.328612	-78.142244	501			CASTLE GARDEN RD	
3	4	{"x": -8698807.499038724, "y": 5060883.1484360...			41.328273	-78.142717	465			CASTLE GARDEN RD	
17	18	{"x": -8698868.023445867, "y": 5060812.0821932...			41.327794	-78.143261	423			CASTLE GARDEN RD	

Figure 8.14: The query results are shown for rows that meet the query condition

21. By using the DataFrame's `spatial.plot()` method, you can see the results of the query. In the next cell, type in the following:

```
sdf_query.spatial.plot()
```

Run the cell. You should see the following output (and can ignore the warning in red):

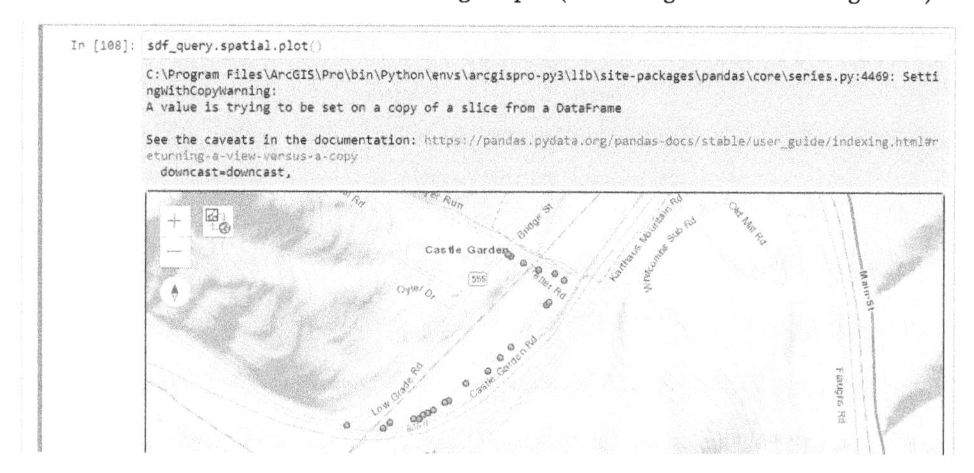

Figure 8.15: Using the spatial.plot() method to view the data frame created from a query

The query function explained above is a newer Pandas DataFrame method. The older way to do a query and return a subset is to use brackets to contain a conditional.

22. In the next cell, you will perform the same query as above with the older method, passing the column of interest to the DataFrame and specifying the condition that has to be met. Type in the following:

    ```
    sdf_query = sdf_address[sdf_address["street"] == 'CASTLE GARDEN RD']
    ```

    Run the cell. This method is related to the method of selecting a subset of columns. By passing a set of columns to the DataFrame in brackets, we can view only those records.

23. To view the first five records in the new data frame, you can use the head() method again in the next cell:

    ```
    sdf_query[["number","street"]].head()
    ```

    Run the cell. You should see the following output:

```
In [107]: sdf_query[["number","street"]].head()
Out[107]:
```

	number	street
0	501	CASTLE GARDEN RD
3	465	CASTLE GARDEN RD
17	423	CASTLE GARDEN RD
26	361	CASTLE GARDEN RD
34	303	CASTLE GARDEN RD

Figure 8.16: A subset of a DataFrame showing the selected columns only

24. Similarly, you can create a new DataFrame by specifying the columns and assigning the result to a variable. In a new cell, type in the following:

    ```
    new_sdf = sdf_query[["number","street"]]
    ```

    Run the cell.

These methods make it easy to get rid of extraneous rows or columns. Once the data has been selected and cleaned up, it can then be saved as a file, or even published to ArcGIS Online.

Read more about queries here: https://developers.arcgis.com/python/guide/working-with-feature-layers-and-features/.

Publishing the data to ArcGIS Online

Pandas DataFrames and the ArcGIS API for Python are used for pushing data into ArcGIS Online, making them available for online maps. Layers that already exist on ArcGIS Online can be called using an ID and edited locally with Pandas and the ArcGIS API for Python. Both spatial and attribute data can be edited in this fashion.

25. Continue in the same Notebook you have been working on so far. The first step is to log in to your ArcGIS Online account using the ArcGIS API for Python, as you've been doing so far. Then, a local data frame can be pushed up to ArcGIS Online using the `to_featurelayer` method, which requires a layer name as a parameter. In the next cell, type in the following:

```
from arcgis import GIS
gis = GIS("Pro")
sdf_layer = sdf_address.spatial.to_featurelayer("sdf-address")
sdf_layer
```

Run the cell. You should see the following output:

Figure 8.17: The result of publishing a layer to ArcGIS Online using the to_featurelayer method

If you log in to ArcGIS Online, you will find the new feature layer in the **Content** tab. It can be viewed on a map to confirm the results of the layer publish event:

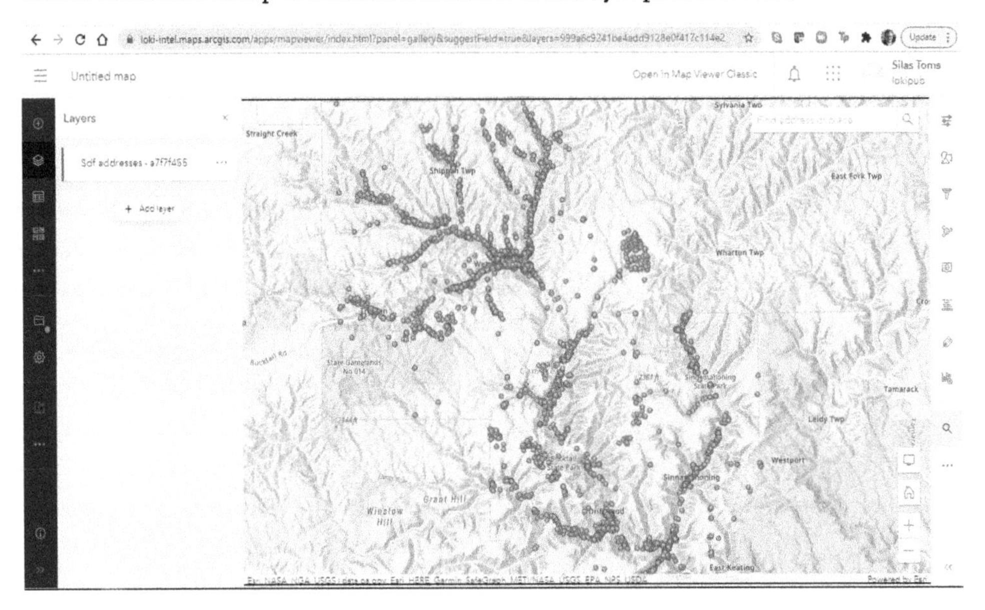

Figure 8.18: Confirming in ArcGIS Online that the SEDF has been published

26. Similarly, once the credentials have been shared, the available layers in the **Content** tab of ArcGIS Online can be called and used locally in your Notebook by referencing a layer ID and using the ArcGIS API for Python to get the layer. In the next cell, type in the following to do this with the layer we published in the previous step:

```
from arcgis.gis import GIS
gis = GIS("Pro")
layer_id = sdf_layer.id
item = gis.content.get(layer_id)
flayer = item.layers[0]
flayer
```

Run the cell. You should get the following output, which tells you that the feature layer has become a **hosted layer** in AGOL:

```
In [96]: from arcgis.gis import GIS
         gis = GIS("Pro")
         layer_id = '999a6c9241be4add9128e0f417c114e2'
         item = gis.content.get(layer_id)
         flayer = item.layers[0]
         flayer
Out[96]: <FeatureLayer url:"https://services3.arcgis.com/xnEvOtpnHkZAyTSk/arcgis/rest/services/ae3a5a/FeatureServer/0">
```

Figure 8.19: A layer stored in ArcGIS Online has been accessed

Read more about hosted layers here: `https://doc.arcgis.com/en/arcgis-online/manage-data/publish-features.htm`.

The ability to create data locally, edit it, and publish it to ArcGIS Online as a feature layer, and then later pull down the data, edit it again, and push it back to ArcGIS Online, makes it easy to control and update your data.

Converting an ArcGIS Online layer to a DataFrame

Once a layer from ArcGIS Online has been retrieved using the `arcgis` module, it can be loaded into a DataFrame using the Spatially Enabled DataFrame's `spatial.from_layer` method. This will allow you to perform normal Pandas operations on the data contained in the layer, as well as spatial operations.

27. In the next cell, you will load the feature layer from the previous step and examine the first few layers. Type in the following:

```
sdf_layer = pd.DataFrame.spatial.from_layer(flayer)
sdf_layer.head()
```

Run the cell. You should see the following rows:

```
In [95]: sdf_layer = pd.DataFrame.spatial.from_layer(flayer)
         sdf_layer.head()
Out[95]:
```

	OBJECTID	SHAPE	city	district	lat	long	number	postcode	region	street	unit
0	1	{"x": -8698754.85605153, "y": 5060933.31315760...			41.328612	-78.142244	501			CASTLE GARDEN RD	
1	2	{"x": -8698903.979641397, "y": 5060902.5976116...			41.328404	-78.143584	64			BELDIN DR	
2	3	{"x": -8701967.725798959, "y": 5060874.1798889...			41.328213	-78.171106	9760			MIX RUN RD	
3	4	{"x": -8698807.499038724, "y": 5060883.1484360...			41.328273	-78.142717	465			CASTLE GARDEN RD	
4	5	{"x": -8698877.518998435, "y": 5060876.8482161...			41.328231	-78.143346	61			BELDIN DR	

Figure 8.20: Converting a feature layer from ArcGIS Online into a DataFrame

28. Layers from ArcGIS Online can be queried in a very similar manner to data frames, even before being converted into a data frame. In the next cell, you will query the feature layer from ArcGIS Online and then convert it into a Spatially Enabled DataFrame using its `sdf` method. The `sdf` method allows you to create a data frame directly from a feature layer. Type in the following:

```
df = flayer.query(where="street = 'CASTLE GARDEN RD'").sdf
df
```

Note that there is a `where` keyword used in the conditional, unlike the `query` function for DataFrames.

Run the cell. You should see the following:

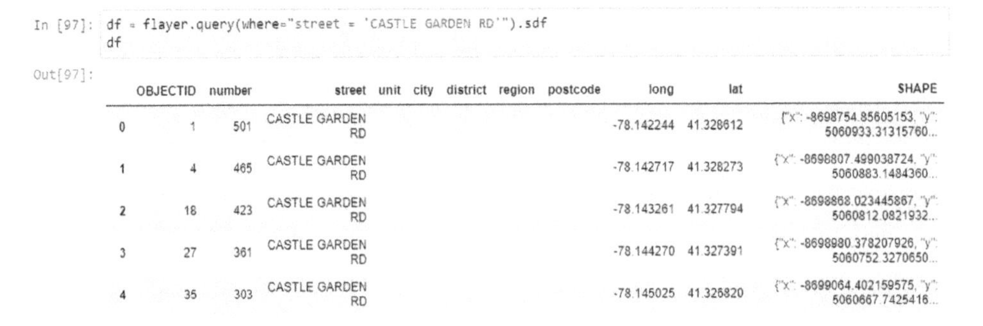

```
In [97]: df = flayer.query(where="street = 'CASTLE GARDEN RD'").sdf
         df
Out[97]:
```

	OBJECTID	number	street	unit	city	district	region	postcode	long	lat	SHAPE
0	1	501	CASTLE GARDEN RD						-78.142244	41.328612	{"x": -8698754.85605153, "y": 5060933.31315760...
1	4	465	CASTLE GARDEN RD						-78.142717	41.328273	{"x": -8698807.499038724, "y": 5060883.1484360...
2	18	423	CASTLE GARDEN RD						-78.143261	41.327794	{"x": -8698868.023445867, "y": 5060812.0821932...
3	27	361	CASTLE GARDEN RD						-78.144270	41.327391	{"x": -8698980.378207926, "y": 5060752.3270650...
4	35	303	CASTLE GARDEN RD						-78.145025	41.326820	{"x": -8699064.402159575, "y": 5060667.7425416...

Figure 8.21: Querying a feature layer and assigning it to a DataFrame

The data frame returned will only contain the data rows that meet the conditional statement.

Indexing and slicing DataFrame rows and columns

Sometimes you may need to get just one row or a subset of rows by referencing their row indices. This is known as **slicing**, and is performed in a similar manner to slicing a Python list.

Using the `.loc` function of DataFrames, you can pass row indices to a DataFrame (or Spatially Enabled DataFrame) to access those rows. For instance, `sdf_row = sdf_address.loc[100]` will give you the row at index 100.

29. In the next cell, you will get a subset of rows from the `sdf_address` DataFrame by passing start and stop row indices. Type in the following:

```
sdf_slice = sdf_address.loc[100:110]
sdf_slice
```

Run the cell. You should see the rows you selected:

```
In [116]: sdf_slice = sdf_address.loc[100:110]
          sdf_slice
Out[116]:
```

	number	street	unit	city	district	region	postcode	long	lat	SHAPE
100	275	STEAM MILL RD						-78.173693	41.456348	{"spatialReference": {"wkid": 4326}, "x": -78....
101	6709	MAY HOLLOW RD						-78.247728	41.476234	{"spatialReference": {"wkid": 4326}, "x": -78....
102	7424	LOW GRADE RD						-78.144548	41.329661	{"spatialReference": {"wkid": 4326}, "x": -78....
103	9008	MIX RUN RD						-78.184851	41.329107	{"spatialReference": {"wkid": 4326}, "x": -78....
104	9839	MIX RUN RD						-78.169372	41.328625	{"spatialReference": {"wkid": 4326}, "x": -78....
105	2578	CLEAR CREEK RD						-78.328743	41.513034	{"spatialReference": {"wkid": 4326}, "x": -78....
106	119	ROSE LN						-78.305540	41.537632	{"spatialReference": {"wkid": 4326}, "x": -78....
107	107	BO MAR DR						-78.297447	41.537670	{"spatialReference": {"wkid": 4326}, "x": -78....
108	735	ROUTE 46						-78.281283	41.537746	{"spatialReference": {"wkid": 4326}, "x": -78....
109	4271	RICH VALLEY RD						-78.327787	41.537522	{"spatialReference": {"wkid": 4326}, "x": -78....
110	1018	RICH VALLEY RD						-78.255853	41.521735	{"spatialReference": {"wkid": 4326}, "x": -78....

Figure 8.22: Slicing a data frame using the loc function to get a set of rows

30. Similarly, these slicing and access by index methods can be combined with column selection to get a section of rows for only specific columns. In the next cell, you will select the `"number"`, `"street"`, `"long"`, and `"lat"` columns for the row numbers specified in the previous step. Type in the following:

```
sdf_slice = sdf_address.loc[100:110][['number','street','long','lat']]
sdf_slice
```

Run the cell. You should see the following output:

```
In [117]: sdf_slice = sdf_address.loc[100:110][['number','street','long','lat']]
          sdf_slice
Out[117]:
```

	number	street	long	lat
100	275	STEAM MILL RD	-78.173693	41.456348
101	6709	MAY HOLLOW RD	-78.247728	41.476234
102	7424	LOW GRADE RD	-78.144548	41.329661
103	9008	MIX RUN RD	-78.184851	41.329107
104	9839	MIX RUN RD	-78.169372	41.328625
105	2578	CLEAR CREEK RD	-78.328743	41.513034
106	119	ROSE LN	-78.305540	41.537632
107	107	BO MAR DR	-78.297447	41.537670
108	735	ROUTE 46	-78.281283	41.537746
109	4271	RICH VALLEY RD	-78.327787	41.537522
110	1018	RICH VALLEY RD	-78.255853	41.521735

Figure 8.23: This slice is also a subset of columns

As you can see, slicing and accessing specific rows is made easy by the `loc` function. Accessing specific columns is also made easy by passing a list of column names to the DataFrame in square brackets.

31. Using the shapefile called `cameron-county-pa.shp`, you will create a Spatially Enabled DataFrame. You will then perform slicing using `loc` to access the `"SHAPE"` row of the SEDF. In the next cell, type in the following:

    ```
    sdf_county = pd.DataFrame.spatial.from_featureclass(r"cameron-
    county-pa.shp")
    sdf_county.loc[0]['SHAPE']
    ```

    Run the cell. You should see the following output:

    ```
    In [44]: sdf_county = pd.DataFrame.spatial.from_featureclass(r"cameron-county-pa.shp")
             sdf_county.loc[0]['SHAPE']
    Out[44]:
    ```

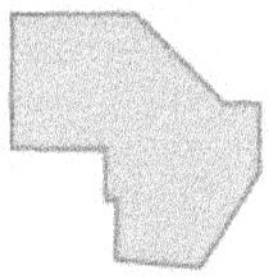

    Figure 8.24: Using loc to get the first row from the county data frame

The `spatial.plot` method will also work to view the data, which is one of the best things about Notebooks, as the data frame and the visualization are available in the same space. In the next cell, type in the following:

```
sdf_county.spatial.plot()
```

Run the cell. You should see the data visualized on the map:

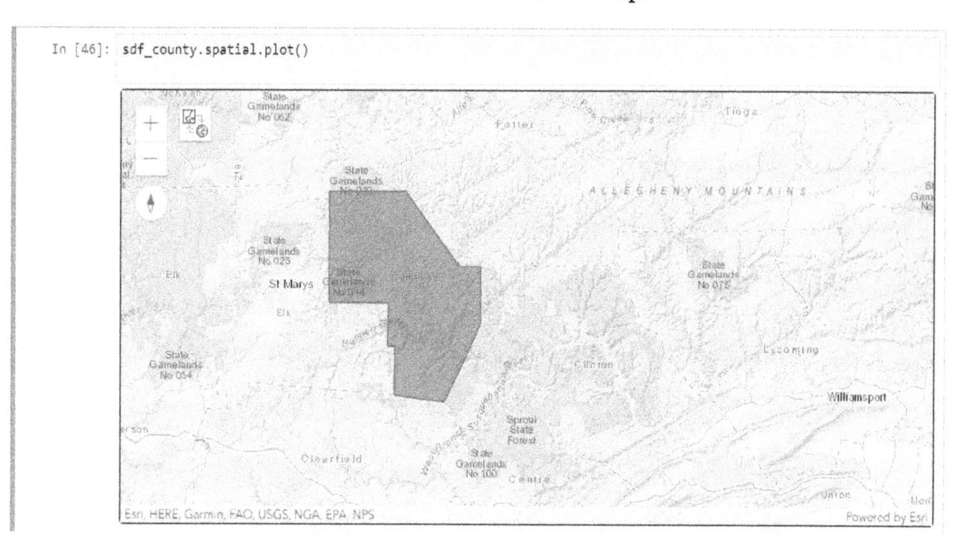

Figure 8.25: Visualization of a polygon using Spatially Enabled DataFrames

Read more about ArcGIS API for Python visualizations and symbology here: `https://developers.arcgis.com/python/guide/visualizing-data-with-the-spatially-enabled-dataframe/`.

Summary

In this chapter, you were introduced to Pandas and DataFrames. You were also introduced to the ArcGIS concept of the Spatially Enabled DataFrame. You explored reading data into Pandas, using multiple methods designed to read from different file types. You also looked at manipulating data using splicing and column manipulation. You learned how to query data in two different ways, and how to output new datasets from the transformed data.

In the next chapter, we will explore using raster data with ArcPy and ArcGIS API for Python.

9

Raster Analysis with Python

Both the arcpy and arcgis modules have raster tools included in their classes and methods. For arcpy, the tools exposed are available in the Spatial Analyst toolbox, as well as some unique methods of performing Map Algebra known as **operators**. For the arcgis module, the tools are available in the arcgis.raster class.

In this chapter, we will use a Notebook in ArcGIS Pro and a **digital elevation model** (**DEM**) TIF file to explore how to use Python with raster data. We will cover the following topics:

- Raster data objects and their properties
- ArcPy raster tools: The Spatial Analyst toolset and Map Algebra
- Using arcgis.raster

Each section covers a distinct component and does not require the completion of the preceding component.

To complete the exercises in this chapter, please download and unzip the Chapter9.zip folder from the GitHub repository for this book: https://github.com/PacktPublishing/Python-for-ArcGIS-Pro/tree/main/Chapter9

Raster data objects

Raster data can be read and written using ArcPy. The module allows you to create a new raster object, add data to it, and save it as a raster dataset, or to read an existing raster into memory to perform analysis on it.

ArcPy also allows you to use an existing raster to create a new raster object with the same schema as the original.

A raster is referred to as a **raster object** when it has been read into memory using Python. A raster object can be *read-only*, meaning it can't be overwritten, or it can have *read-and-write permissions* so that the data in the original raster can be changed when the raster object is saved.

In this section, you will use a Notebook to explore the available raster tools in the `arcpy` module. Open ArcGIS Pro and start a new project, and then add a new Notebook from the **Insert** tab. Rename the Notebook to `Chapter9`. Using the Notebook will allow you to view the output of using these tools.

Creating a new blank raster

When a new raster is needed, a few parameters must be supplied to create the output raster dataset. These include basics such as the output folder and filename, cell size, number of bands, pixel type (meaning available shades of color), and spatial projection (which defaults to the spatial reference set in the ArcGIS environment), as well as other required and optional parameters.

This example shows required and optional parameters passed to the ArcPy tool called `CreateRasterDataset`. There are many other values available to be passed to the tool and we have chosen these only to demonstrate their order and use.

1. In a Notebook cell, enter the following lines and adjust the filepaths to match your system:

```
import arcpy
out_path = r'C:\projects\'
out_name = 'raster1.tif'
cell_size = '20'
pixel_type = '16_BIT_UNSIGNED'
spatial_ref = arcpy.SpatialReference(2227)
num_of_bands = '1'
config_keyword ='' # optional
pyramids = '' # optional
tile_size = '256' # optional
compression = 'NONE' # optional
pyramid_origin = '' # optional

out_raster = arcpy.management.CreateRasterDataset(out_path,out_name,
cell_size, pixel_type, spatial_ref, num_of_bands, config_keyword,
pyramids, tile_size, compression, pyramid_origin)
```

The parameters must be passed to the CreateRasterDataset function in the order as shown. As for the file type, the extension chosen will define the type of raster file created. In the example above, we have chosen to create a TIF raster, with a cell size of 20. The spatial reference system chosen is in feet, so the cell size represents 20 feet.

Read more about the required and optional parameters here: https://pro.arcgis.com/en/pro-app/latest/tool-reference/data-management create-raster-dataset.htm

Reading and copying raster properties

To capture the details from another raster when creating a new raster, use the getRasterInfo method of the existing raster object. This method simplifies the process of creating a new raster, as the cell size, spatial reference, and all other details are copied from the existing raster into the new raster object. *The data is not copied, however.*

To use this method, first read the existing raster into memory using the Raster tool to create a raster object. Then, use the raster object and pass its getRasterInfo property to the Raster tool to create a new raster.

2. Type the following code into a Notebook cell, adjusting the filepath to match an existing raster on your system (the raster we just created, for example):

   ```
   from arcpy import Raster
   orig_raster = Raster(r'C:\project\oldraster')
   orig_raster_info = orig_raster.getRasterInfo()
   new_raster = Raster(orig_raster_info)
   ```

   Run the cell.

Check out more information here about the raster object: https://pro.arcgis.com/en/pro-app/latest/arcpy/classes/raster-object.htm

Creating a raster object from an existing raster

By passing a string filepath to the arcpy.Raster class, we can access the data, including its metadata and the array data itself.

The raster `readOnly` property is set to `True` by default, to avoid accidentally overwriting the raster or its cells, but it can be set to `False` when creating the raster object.

Here is an example of accessing an existing raster and creating a raster object. We'll use the USGS DEM data (which is a TIF) included in the chapter's ZIP file.

3. Enter the following into a Notebook cell, being sure to adjust the filepath to match where you copied the TIF:

```
import arcpy
data = r'USGS_13_n38w123_20210301.tif'
raster_obj = arcpy.Raster(data)
raster_obj
```

Running the cell should give you the following output:

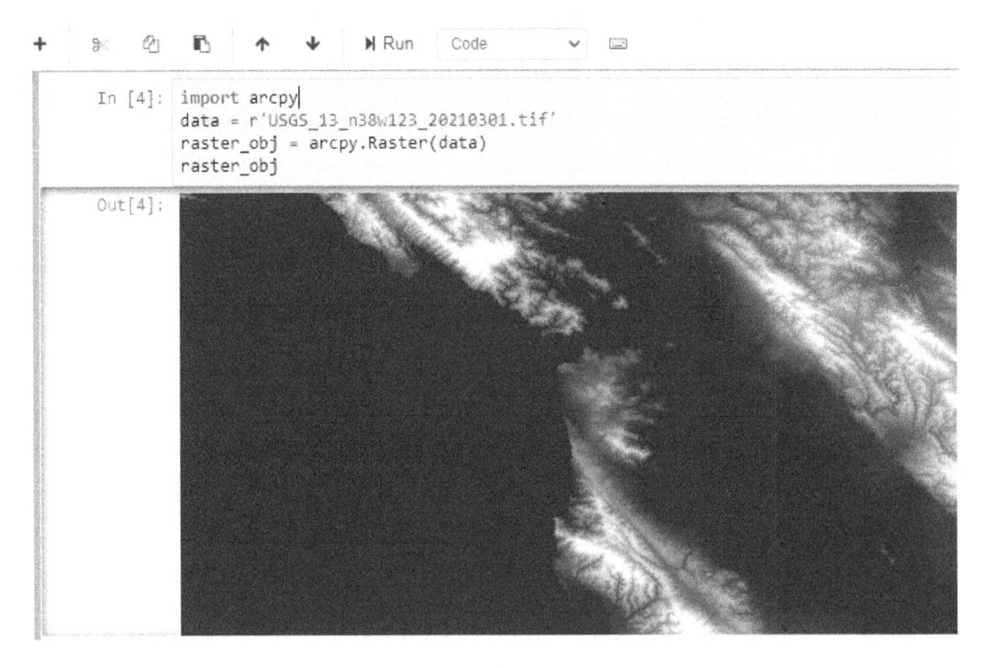

Figure 9.1: The resulting raster object

As a quick reminder, the final line in the screenshot above (`raster_obj`) creates an output in the **Out** cell. In this case, the output is the raster object visualization. It can be thought of as the equivalent of a `print` statement of a script.

Saving a raster

All of the work done on a raster object will be performed in memory, which is not made permanent unless it is saved to your hard drive. Even if a raster is added to the **Map** component of your project in ArcGIS Pro, unless it is saved explicitly it will not be available the next time you open your project, and you will see the dreaded red exclamation point next to your layers in the **Map**:

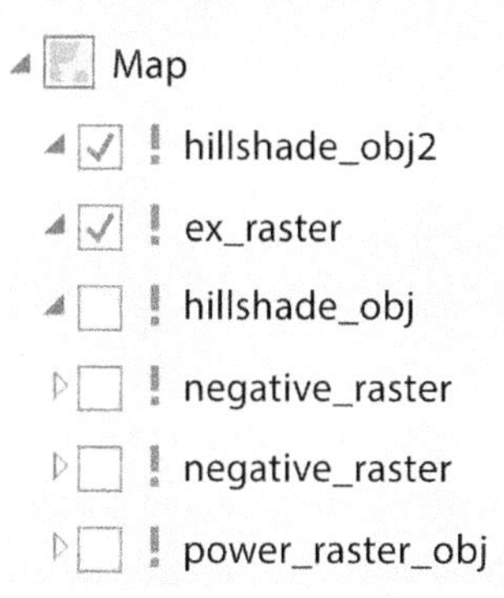

Figure 9.2: Red exclamation points are displayed when a dataset isn't available

To save the raster object to the hard drive, use the save method of the raster object.

4. In a Notebook cell, type the following, adjusting the filepath to where you want to save to:

```
import arcpy
data = r'USGS_13_n38w123_20210301.tif'
raster_obj = arcpy.Raster(data)
raster_obj.save(r"C:\Projects\raster_save.tif")
```

   Run the cell.

Once the raster object is saved, it can be added again to a **Map** or read back into memory in a Notebook. This will allow you to explore the raster properties.

Accessing the raster properties

Once you have the raster object in memory, you can explore the properties of the raster. These properties include geographic details such as the spatial coordinate system and the spatial extent, or raster-specific properties such as the number of bands or the NoData value. Some properties are about the raster itself and some are about the cell values contained within the raster. Some of these methods and properties are read-only, while others are adjustable.

With the built-in `dir` function, you can explore the available methods and properties available with a raster object. This will allow you to use the methods and properties in your analysis or when assessing a dataset.

5. In the next cell, type in the following to read in this chapter's raster object and explore its properties:

```
import arcpy
data = r'USGS_13_n38w123_20210301.tif'
raster_obj = arcpy.Raster(data)
dir(raster_obj)
```

The output of the `dir` function is shown below. All of the functions of the `Raster` class are listed, starting with internal functions that start with (and end with) double underscores. The public functions come after these internal functions, and each allows you to access a property of a raster object (such as the `extent` property) or a built-in function (such as the `save` function). These functions are accessed using dot notation, where a period is placed after the raster object and then the function is written:

```
In [20]: dir(raster_obj)

Out[20]: ['RAT', '__abs__', '__add__', '__and__', '__bool__', '__class__', '__delattr__', '__delitem__', '__dict__',
'__dir__', '__divmod__', '__doc__', '__eq__', '__floordiv__', '__format__', '__ge__', '__getattribute__', '__
getitem__', '__gt__', '__hash__', '__iadd__', '__iand__', '__ifloordiv__', '__ilshift__', '__imod__', '__imul
__', '__init__', '__init_subclass__', '__invert__', '__ior__', '__ipow__', '__irshift__', '__isub__', '__iter
__', '__itruediv__', '__ixor__', '__le__', '__lshift__', '__lt__', '__mod__', '__module__', '__mul__', '__ne_
_', '__neg__', '__new__', '__or__', '__pos__', '__pow__', '__radd__', '__rand__', '__rdivmod__', '__reduce_
_', '__reduce_ex__', '__repr__', '__rfloordiv__', '__rlshift__', '__rmod__', '__rmul__', '__ror__', '__rpow_
_', '__rrshift__', '__rshift__', '__rsub__', '__rtruediv__', '__rxor__', '__setattr__', '__setitem__', '__siz
eof__', '__str__', '__sub__', '__subclasshook__', '__truediv__', '__weakref__', '__xor__', '_repr_png_', 'add
Dimension', 'appendSlices', 'bandCount', 'bandNames', 'bands', 'blockSize', 'catalogPath', 'compressionType',
'exportImage', 'extent', 'format', 'functions', 'getBandProperty', 'getColormap', 'getDimensionAttributes',
'getDimensionNames', 'getDimensionValues', 'getHistograms', 'getProperty', 'getRasterBands', 'getRasterInfo',
'getStatistics', 'getVariableAttributes', 'hasRAT', 'hasTranspose', 'height', 'isInteger', 'isMultidimensiona
l', 'isTemporary', 'maximum', 'mdinfo', 'mean', 'meanCellHeight', 'meanCellWidth', 'minimum', 'name', 'noData
Value', 'noDataValues', 'path', 'pixelType', 'properties', 'read', 'readOnly', 'removeVariables', 'renameBan
d', 'renameVariable', 'save', 'setBandProperty', 'setColormap', 'setHistograms', 'setProperty', 'setStatistic
s', 'setVariableAttributes', 'slices', 'spatialReference', 'standardDeviation', 'uncompressedSize', 'variable
Names', 'variables', 'width', 'write']
```

Figure 9.3: The result of using the built-in dir function

For example, you can get the extent of the raster, and its constituent minimum and maximum values (known as `XMin`, `YMin`, `XMax`, and `YMax`), using the `extent` function. You might then use those properties to create an ArcPy `Point` object from these inputs if you need to create a vector representation of the raster extent.

6. To do this, in the next cell, type in the following:

```
lowerLeft = arcpy.Point(raster_obj.extent.XMin, raster_obj.extent.YMin)
print(lowerLeft)
```

Run the cell. The raster object's `extent` properties, `XMin` and `YMin`, are accessed directly and passed to the `arcpy.Point` function.

Accessing raster and cell value properties

Many properties of rasters are accessed to assess the cell values contained within the raster. These include properties such as mean, minimum, and maximum, and the NoData value. In the example raster used in this chapter, the cell values represent elevation data, so understanding the maximum or minimum value is very useful.

7. In the next cell, the minimum and maximum cell values are called using the respective properties:

   ```
   import arcpy
   data = r'USGS_13_n38w123_20210301.tif'
   raster_obj = arcpy.Raster(data)
   print(raster_obj.minimum, raster_obj.maximum)
   ```

   Running this cell should give you:

   ```
   -103.81411743164062 985.8399047851562
   ```

8. The mean value of all the raster cells is also easy to obtain using the built-in `mean` method:

   ```
   print(raster_obj.mean)
   ```

   This gives us:

   ```
   72.00359745334588
   ```

9. For rasters, there are often cells that have no data. Different raster formats use various values to represent these NoData cells. Sometimes, there is even more than one NoData value used within a single raster.

   To access the specific NoData values, use the `noDataValue` or `noDataValues` method:

   ```
   print(raster_obj.noDataValue)
   ```

   This raster uses a float as a NoData value:

   ```
   -3.4028234663852886e+38
   ```

10. As there are sometimes multiple values used as NoData values, the `noDataValues` method exists, and will return a tuple (unless there is a `NoneType` used as the NoData value):

```
print(raster_obj.noDataValues)
```

This gives us the following:

```
(-3.4028234663852886e+38,)
```

11. The height and width of the raster object represent the number of cells in the *y* and *x* dimensions, respectively, from some geographic origin point. A raster does have to be a rectangle but does not have to be a square. This means that the height and width will not always be equal:

```
print(raster_obj.height,raster_obj.width)
```

In this case, running the cell shows us the height and width are the same:

```
10812 10812
```

12. Other raster information is available using the `properties` method:

```
raster_obj.properties
```

Run the cell. This is the output for the example raster:

```
{'KIND': 'IMAGE', 'BAND_COUNT': 1, 'HAS_TABLE': False, 'HAS_XFORM':
False, 'DataType': '*'}
```

The value of any specific cell can be accessed directly using the `GetCellValue` function of ArcPy. You do not have to create a raster object first, but instead access the raster directly by passing the raster filepath as the first parameter. The second parameter is the cell location in *x*/*y* notation, and the third (optional) parameter is the band parameter, which can be used to get the value of the cell in either all bands when left blank or only in specific bands when included.

13. In the next Notebook cell, type in the following to access the value of a specific raster cell. The *x* and *y* value (the second parameter) must be in the same spatial reference system as the raster:

```
result = arcpy.GetCellValue_management(r'USGS_13_n38w123_20210301.
tif', "-122.45 37.767", "1")
cellvalue = int(result.getOutput(0))
```

Running this gives us the value at the specific cell:

```
'87.660217'
```

These tools make it easy to assess your raster and to access the data each cell contains. The properties of the raster will determine the allowable operations, so accessing them is an important component of each analysis you will perform.

Geographic properties

There are a number of geographic properties used to help understand the area on Earth that the raster represents. In the digital elevation dataset used in this chapter, the raster represents the San Francisco Bay Area from Marin County in the west to Oakland in the east.

14. The `extent` of a raster represents the geographic space that the raster occupies. Type the following into the next cell:

    ```
    print(raster_obj.extent)
    ```

    Run the cell. The extent will be reported in the coordinates of the spatial reference system of the raster. In this case, the raster uses a geographic coordinate system based on the North American datum of 1983, so it is reported in longitude and latitude:

    ```
    -123.000555555794 36.999444440379 -121.999444440278 38.000555555895
    NaN NaN NaN NaN
    ```

15. To access the spatial reference system itself, the `spatialReference` method is used:

    ```
    raster_obj.spatialReference
    ```

    Running this gives you a table that looks like the following:

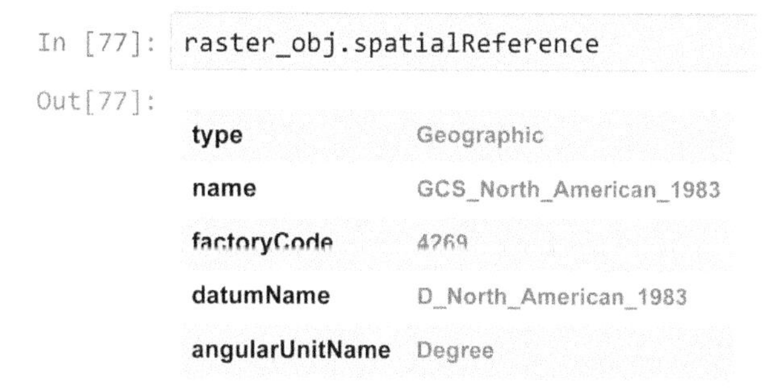

Figure 9.4: The components of the spatialReference property

Raster properties are important for understanding your datasets. ArcPy makes it easy to access these properties, and even to adjust them as well.

ArcPy Raster tools

Now that you understand how to create a raster object, let's explore the use of raster objects with raster tools. These tools are the same ones available in ArcToolbox and are executed by passing the same parameters as you do when using the user interface through ArcGIS Pro.

However, by using Python, we can automate the analysis and run it as a script, or run these tools in the Jupyter Notebook environment in ArcGIS Pro, as shown below.

In this section, you will use a digital elevation model to explore the tools that are available. These tools include Slope, Hillshade, and Conditionals, to name just a few. Some of these tools can be used without a Spatial Analyst license, but most advanced raster tools in ArcGIS Pro require Spatial Analyst. You can use the same Notebook you created at the start of the chapter for this code exploration.

The Spatial Analyst toolset and the sa module

The Spatial Analyst toolset enables advanced spatial modeling and analysis. It is represented as the `sa` module within ArcPy. We were already introduced to this module in *Chapter 2, Basics of ArcPy*, and we will recap and extend what we know in this section. A Spatial Analyst extension license is required to access these tools.

To enable the license, you may need to log in to `arcgis.com` and go to the **Licenses** menu in your account:

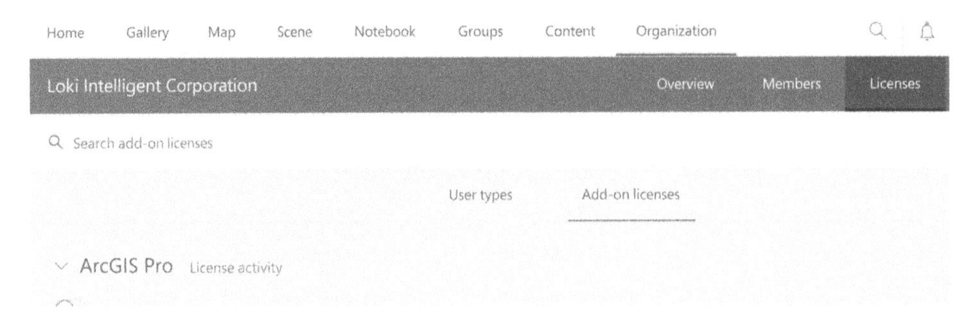

Figure 9.5: Organization page of arcgis.com, with the Licenses section at the upper right

Navigate down to the Spatial Analyst license section and click **Manage**. Then, click the on/off switch to enable the license.

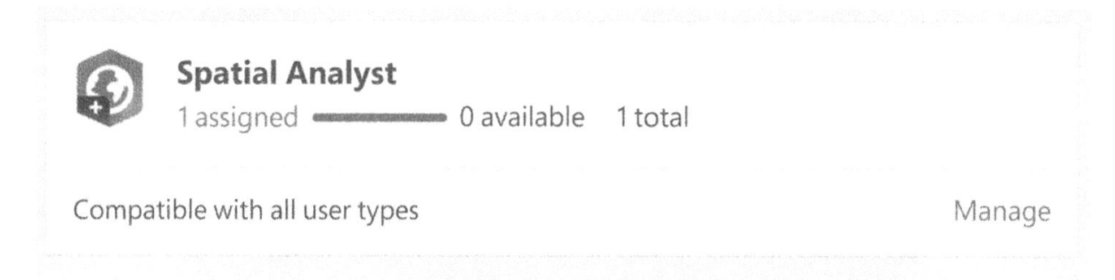

Figure 9.6: The Spatial Analyst license card in ArcGIS Online

Once the license has been checked out, the extension can be used successfully. Using ArcPy's `CheckOutExtension` method, check out the extension in a Notebook:

```
import arcpy

arcpy.CheckOutExtension("Spatial")
```

The `'CheckedOut'` message returned confirms that the extension is licensed and available for use.

Generating a raster object

All of the tools in ArcToolbox can be used with in-memory raster objects.

1. For instance, we can create a `Slope` raster from the raster object:

    ```
    import arcpy
    data = r'USGS_13_n38w123_20210301.tif'
    raster_obj = arcpy.Raster(data)
    slope_raster = arcpy.sa.Slope(raster_obj)
    slope_raster
    ```

The result is a Slope dataset, which can be viewed directly in your Notebook by running the cell:

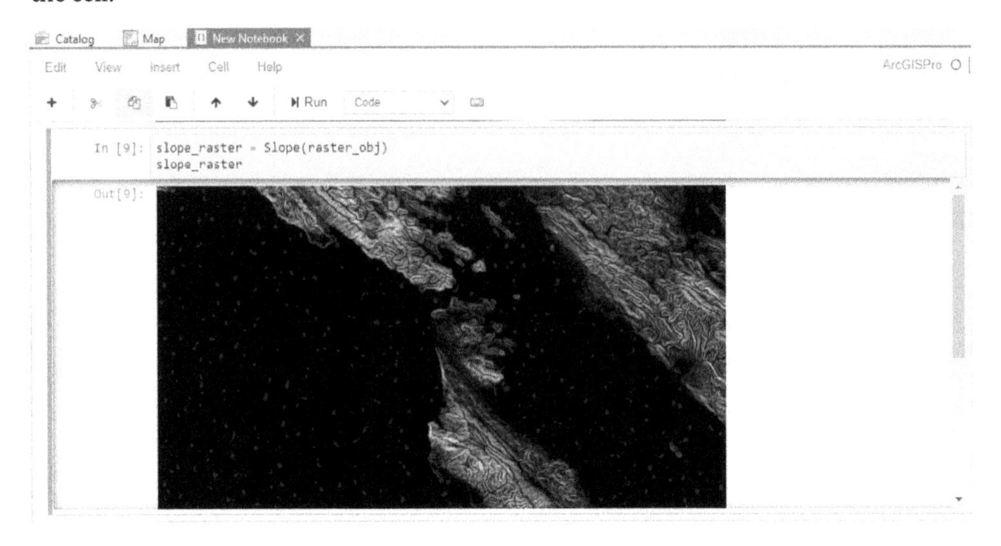

Figure 9.7: The resulting slope object

The Slope dataset is also added to the **Map** component automatically:

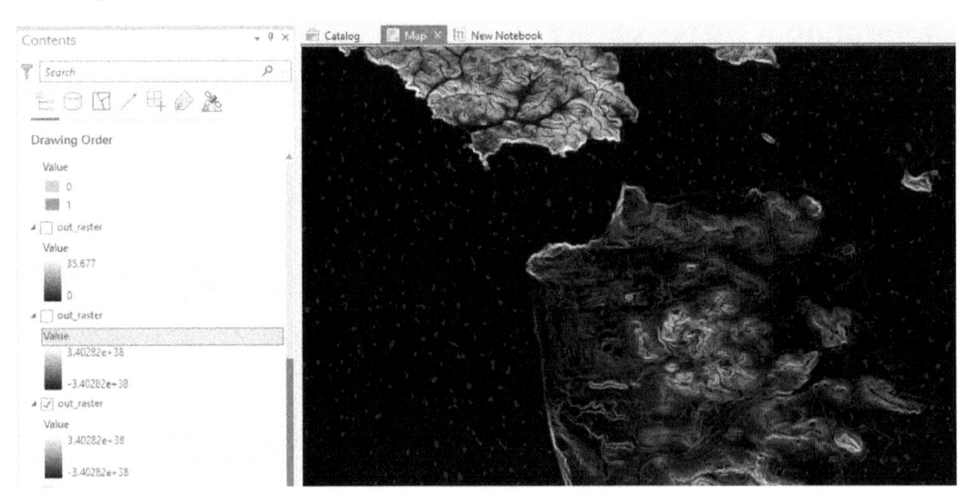

Figure 9.8: The Map component and its Table of Contents

Remember that these are objects in memory, and not saved anywhere. To save these types of raster objects, you need to use the save method of the object, as we've been doing throughout:

```
slope_raster.save(r"C:\Projects\rast_slope_obj.tif")
```

Statistical raster creation tool

Sometimes you will need to create a raster for statistical analysis with either a **constant** value, a **distributed** value, or a **random** value, and ArcPy allows for that as well. These functions will produce rasters with one value assigned to every cell, rasters where a normal Gaussian distribution is applied, or random rasters, which will assign a random value to all cell values in the output raster. These are useful for adding other raster values to a blank raster canvas or for creating random values for testing purposes.

The output spatial reference system (as well as other parameters) is controlled by environmental variables, or by the spatial reference of the map view. If there is no known spatial coordinate system, the spatial reference of the output rasters will be set to `Unknown`.

You will need to check out a Spatial Analyst license, as explained above.

2. For instance, the Create Constant Raster tool allows you to generate a raster of a specific data type (floating point data or integer data), and a constant value.

   Cell size and extent are optional parameters, which can otherwise be set using environmental variables, but are included here. In the next cell, type in the following:

   ```
   import arcpy
   const_raster = arcpy.sa.CreateConstantRaster(13, "INTEGER", 2,
   Extent(0, 0, 500, 500))
   const_raster.save("C:/projects/constant_raster")
   ```

   Run the cell.

3. A raster with a normal Gaussian distribution is created using the Create Normal Raster tool. This raster will assign a floating-point value to all cells. In the next cell, type in the following:

   ```
   from arcpy.sa import Extent, CreateNormalRaster
   normal_raster = CreateNormalRaster(1, Extent(0, 0, 200, 200)) * 3.7 + 24
   normal_raster.save(r"C:\arcpy\normal_raster")
   ```

   Run the cell.

4. For a random raster, use the Create Random Raster tool. The parameters (seed value, cell size, and extent) are optional. In the next cell, type in the following:

   ```
   from arcpy.sa import Extent, CreateRandomRaster
   outRandRaster = CreateRandomRaster(45, 10, Extent(0, 0, 250, 250))
   outRandRaster.save(r"C:\arcpy\random_raster")
   ```

Run the cell.

Learn more about this method of creating statistical rasters here: https://pro.arcgis.com/en/pro-app/latest/tool-reference/spatial-analyst/an-overview-of-the-raster-creation-tools.htm

Conditionals

If you want to select cell values using a conditional statement, use the `ExtractByAttributes` tool instead. Using the `Value` keyword, we can select cells that meet the condition, and the resulting raster will assign a NoData value to all cells that do not meet the condition.

5. In this example, we'll select all areas that are below 200 meters in the elevation model:

```
extract_raster = arcpy.sa.ExtractByAttributes(raster_obj,
                                               "Value <= 200")
extract_raster
```

Run the cell. The result of the code shows a raster object where only cells with values below 200 meters are retained. All other cells (the white cells) are NoData cells:

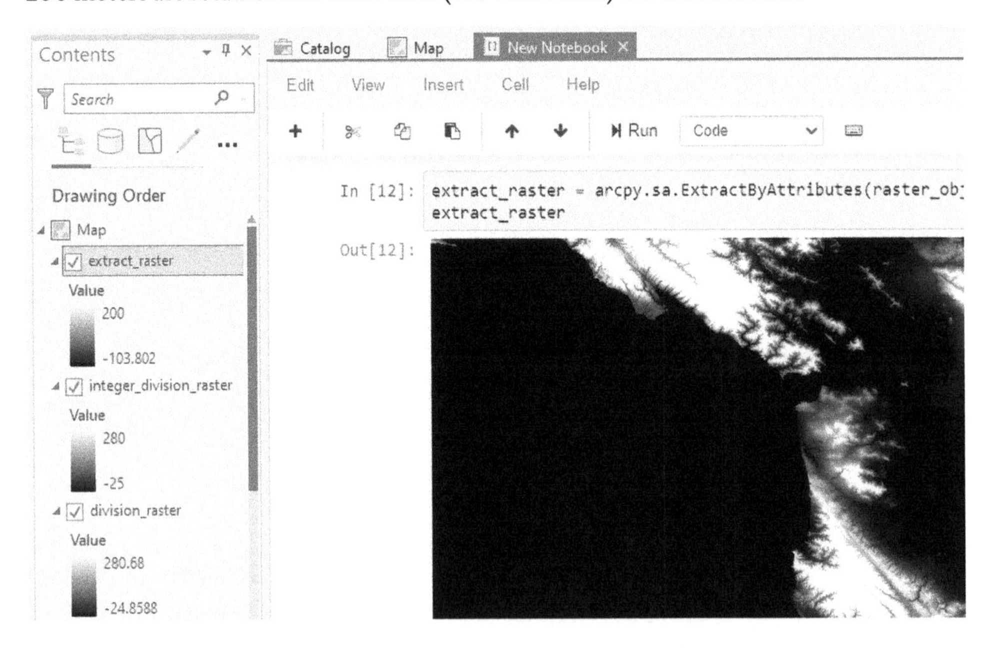

Figure 9.9: Cells that are not white represent values below 200 meters

The screenshot below shows what information is returned when you click on a white cell on the `extract_raster` in the **Map** component.

The original raster in the Table of Contents represents the same cell as a value of 227.95 meters in the info popup:

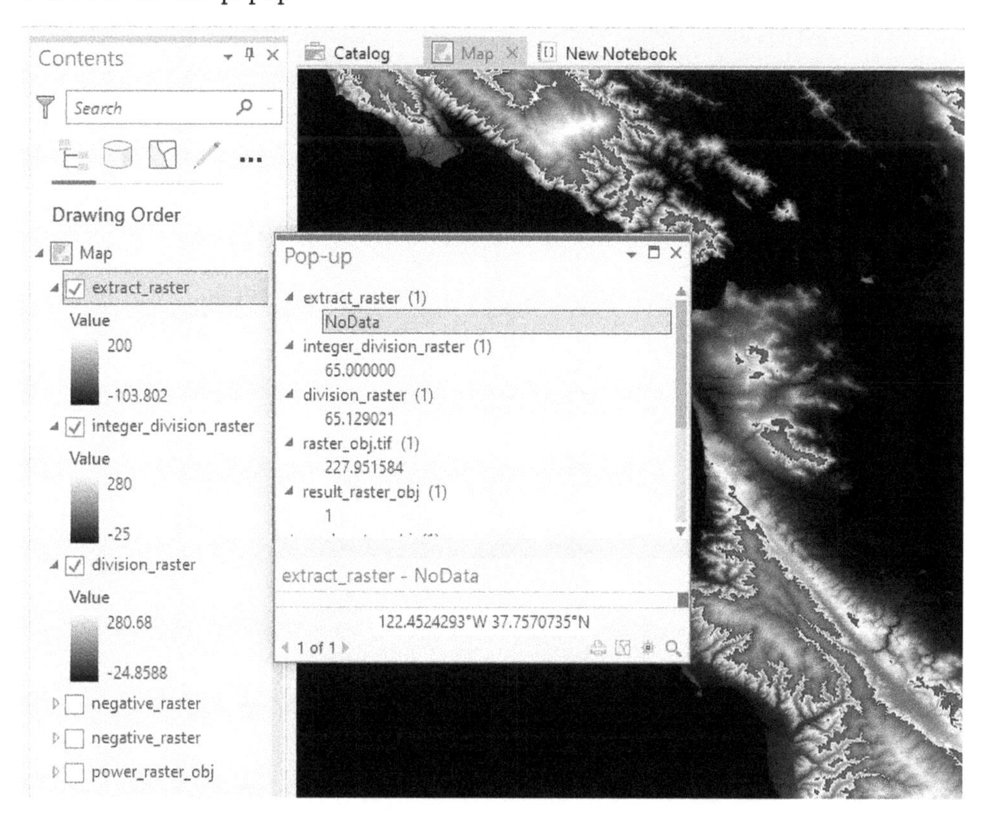

Figure 9.10: Clicking on the Map shows the value of a cell on the extracted raster to be NoData where the original (raster_obj.tif) was above 200

The Hillshade tool

Another often-used tool within the Spatial Analyst toolset is the Hillshade tool. It requires an input raster object and a number of optional parameters, including the azimuth (angle of the light source), the altitude angle of the light source, the model shadow type, and the Z factor (the number of ground units in each surface z-unit or elevation unit). The output is a `Hillshade` object, which can be saved to disk and used for creating terrain models.

In the example below, we will create a `Hillshade` from a raster object and a set of parameters.

6. In the next cell, type in the following:

```
import arcpy
azimuth = 200
altitude = 45
model_shadows = 'NO_SHADOWS'
z_factor = 1
hillshade_obj = arcpy.sa.Hillshade(raster_obj, azimuth, altitude, 
model_shadows, z_factor)
hillshade_obj
```

Run the cell. This screenshot shows the result you should obtain:

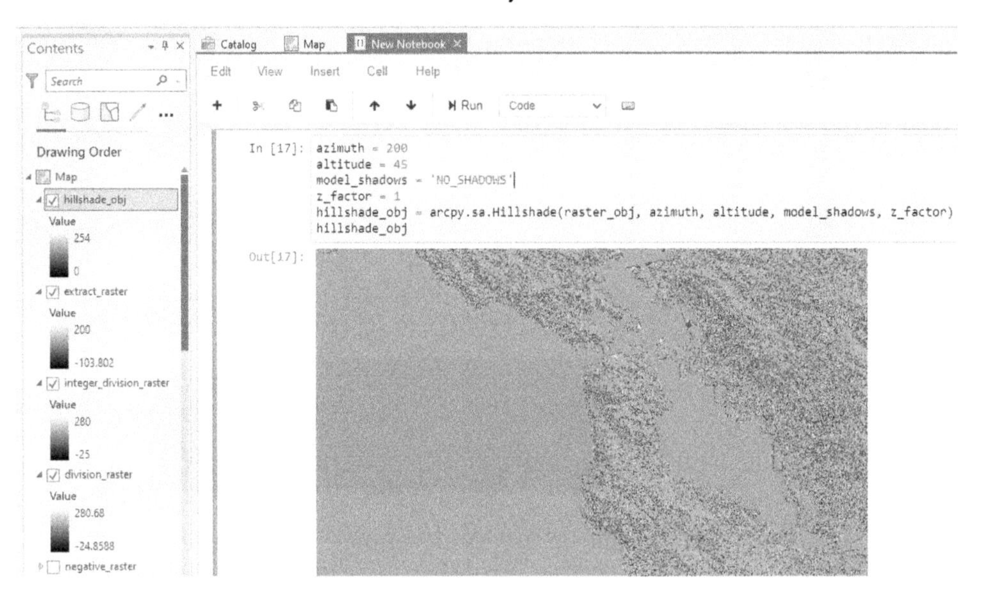

Figure 9.11: The resulting hillshade object produced by the Hillshade tool

When running this tool, the results are sometimes ugly because the data below zero is also turned into a "hill."

It results in an odd and unwanted Hillshade result, as shown below:

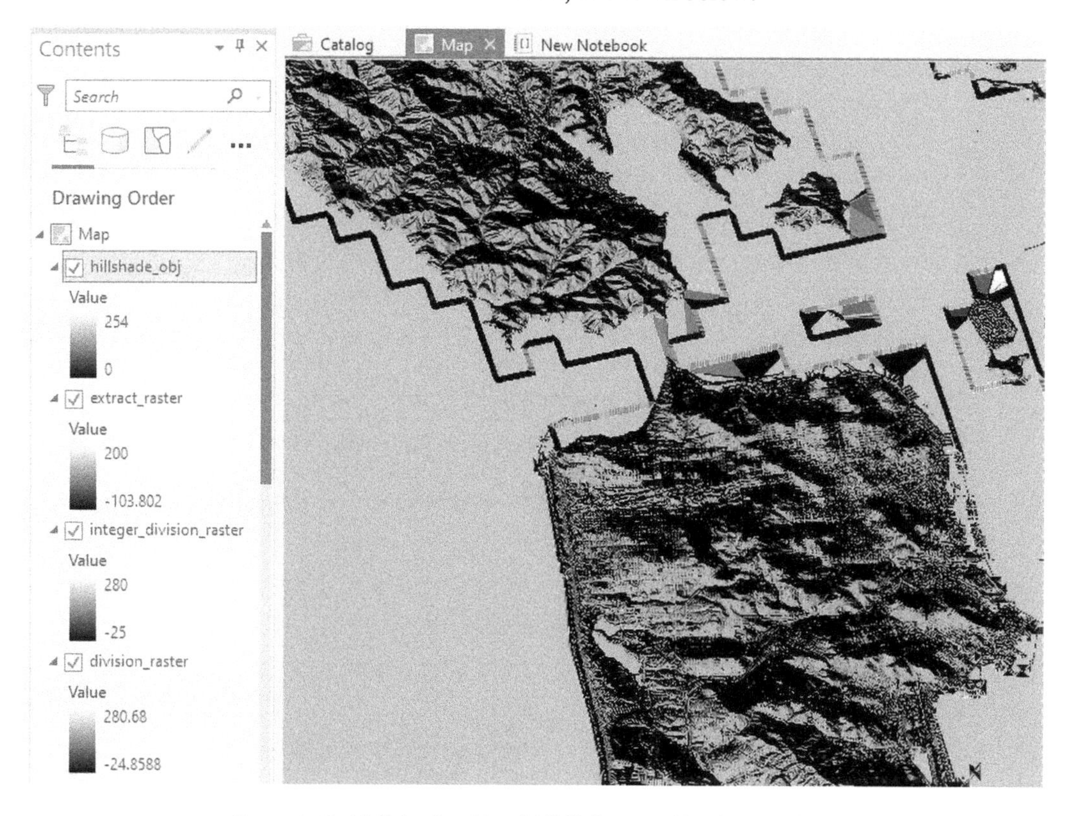

Figure 9.12: A hillshade with odd "hills" created by data artifacts

To avoid this, you can use the tool in combination with the `ExtractByAttributes` tool to perform the Hillshade operation only on the cells that meet the conditional. Since the `ExtractByAttributes` tool returns an output raster, it can be added to the Hillshade tool as an input.

7. In this code below, we see that two operations are combined into one line. The `ExtractByAttributes` tool is executed first. All cell values above 1 are extracted, and the resulting raster object is passed to the `Hillshade` object:

```
from arcpy.sa import Hillshade, ExtractByAttributes
azimuth = 200
altitude = 45
model_shadows = 'NO_SHADOWS'
z_factor = 1
hillshade_obj2 = Hillshade(ExtractByAttributes(raster_obj,
                                        "Value >= 1"),
                    azimuth, altitude,model_shadows, z_factor)
hillshade_obj2
```

Run the cell. The result of both operations (`hillshade_obj2`) is added to the Table of Contents of the **Map** component of the project, as shown in the screenshot below:

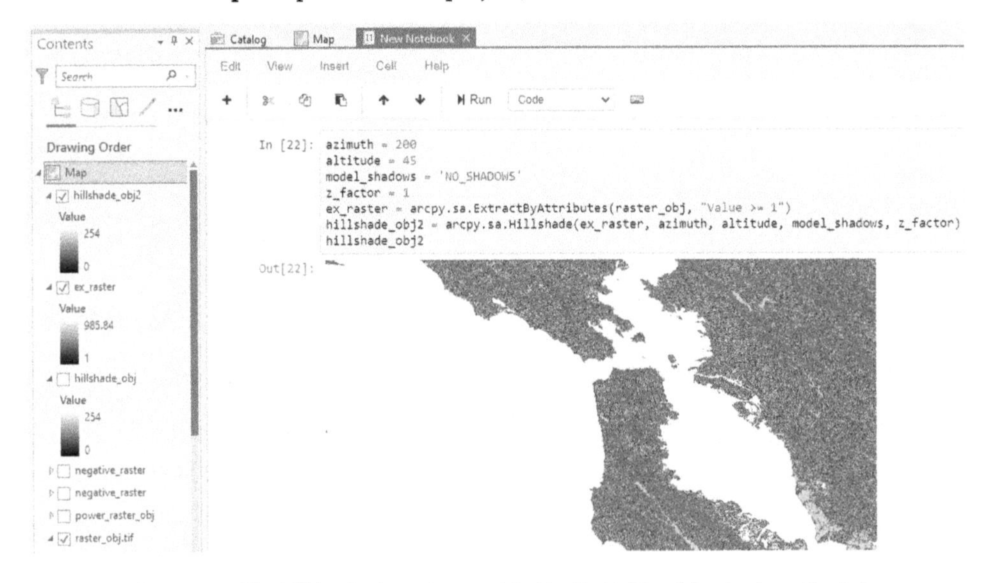

Figure 9.13: The hillshades have been added to the Table of Contents on the left

Remember that the layers added to the **Map** component are only in memory and will not be saved to disk unless you explicitly use the raster object's `save` method.

This will result in a much prettier hillshade:

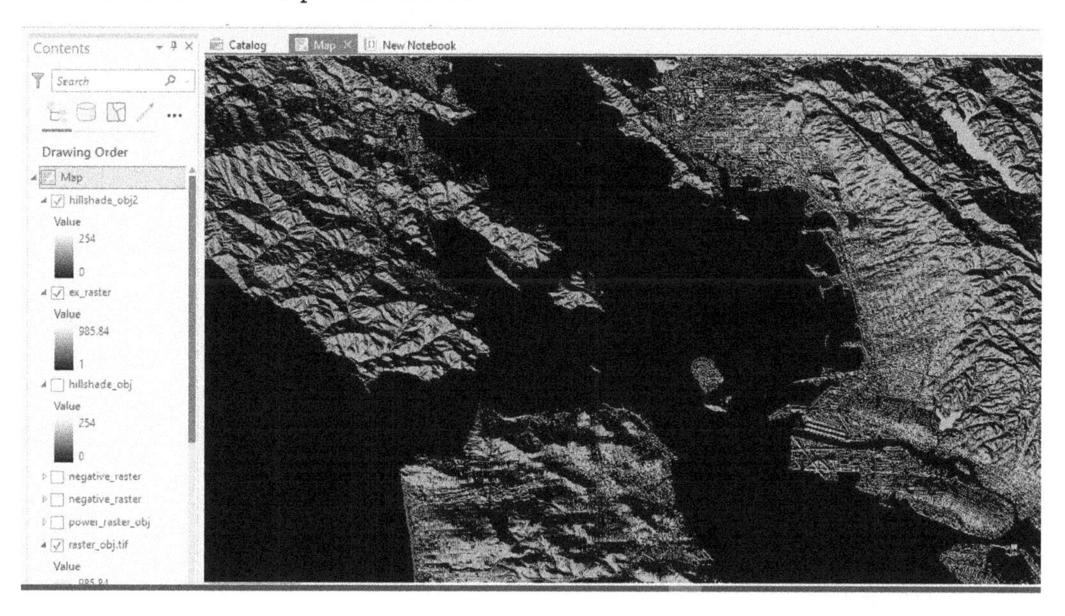

Figure 9.14: A prettier hillshade produced by using only elevations above 1 meter

The Conditional tool

Another very useful tool in the `sa` module is the Con or Conditional tool, which we briefly saw in action in *Chapter 2, Basics of ArcPy*. It allows you to perform an if/else evaluation of a raster's cells, and to assign a value when the condition is true, as well as an optional value for when the condition is false.

For instance, if a cell has a value of 5 and the conditional states that all values below 10 are to be replaced with the value 3, then the condition is true and the cell value will be adjusted to 3. However, if the cell value is 11 for the same conditional, the condition is false, and either nothing will happen, or it will be replaced with the value designated for false conditions.

There are two ways to organize this tool. One method mirrors how the tool is filled out using the ArcToolbox interface, and the other uses a Map Algebra-like statement. The second method avoids the conditional's reliance on a default field name like `VALUE`, as sometimes the field name may not be known to you when you are performing the analysis and may differ from the default.

8. Here is the first way to do it, where the first parameter is the input raster object, and the second is the value of the raster object cells, the third is a value to be used for a cell if the conditional is false, and the last is the conditional.

If you wanted to select all elevations above 200 feet, you could use this statement:

```
from arcpy.sa import Con
from arcpy import Raster
data = r'USGS_13_n38w123_20210301.tif'
input_raster = Raster(data)
result = Con(input_raster, input_raster, 0, "VALUE > 200")
```

Run the cell.

9. Here is the second way, known as the Map Algebra form. Instead of using a `where` conditional with the keyword `VALUE`, a **greater than** operator (see next section) is used against the raster object itself to create the condition. Cells that meet the condition are populated with values from the original input raster. Values that do not meet the condition are assigned a NoData value:

```
result2 = Con(input_raster > 200, input_raster)
```

Run the cell.

The Con tool can be combined with other tools such as `IsNull`, which will allow you to assess a raster for Null cell values and replace them or ignore them as needed.

10. In this example, the `Null` values are replaced with the value zero if the condition is true (in other words, the cell values are `Null`). Otherwise, the cell values are not replaced:

```
result2 = Con(IsNull(input_raster), 0, input_raster)
```

Run the cell.

Similarly, the Con tool can be used inside another Con tool and combined with other operators, creating a complete analysis flow in one statement. If you need to select areas that are near sea level or near the top of hills, but avoid areas in between, you can use this type of process. It will allow for multiple comparisons to happen in order to replace values and then reassess the result.

11. To see this in action, type in the following:

```
result3 = Con(input_raster1 < 34,1, Con((input_raster1 >= 34) &
(input_raster1 < 37),2, Con((input_raster1 >= 37) & (input_raster1 <
45),3, Con(input_raster1 >= 45,4))))
```

Run the cell. In this code, the conditional statements work together to create the output. The first value supplied to the main, or outside, Conditional tool is the result of the operation **`input_raster1 < 34`**, meaning all cell values below 34 are assigned the value 1, and all cells equal to or above 34 are assigned the result of the next conditional. In this next conditional, values that are above or equal to 34 but below 37 are assigned the value 2; all others are assigned the value of the next conditional. In this third conditional, those cell values above or equal to 37 and below 45 are assigned the value 3. The final Con tool says that all cell values above or equal to 45 are assigned the value of 4. In this way, the DEM is divided into 4 "bins" where each bin is associated with a band of elevation.

The Conditional tool is even more useful when combined with Map Algebra operators, which we will discuss next.

Map Algebra

ArcPy makes it easy to use Map Algebra on rasters. Map Algebra operators allow for mathematical or conditional operations on the values of the cells of the rasters, including multiplying the values or selecting specific cells based on those values and a conditional statement.

To explore how we'll use Map Algebra within the Notebook, let's take a look at the following example, where a new raster object is created from an existing raster by applying a conditional value against the existing raster object. This might apply when you need to select only high elevation areas within a DEM, as part of a raster elevation workflow.

12. In the next cell, type in the following:

    ```
    new_raster_obj = raster_obj > 30
    new_raster_obj
    ```

Run the cell. The result of this operation is a Boolean, showing True or False areas that do or do not meet the condition, as shown in this screenshot of the resulting raster:

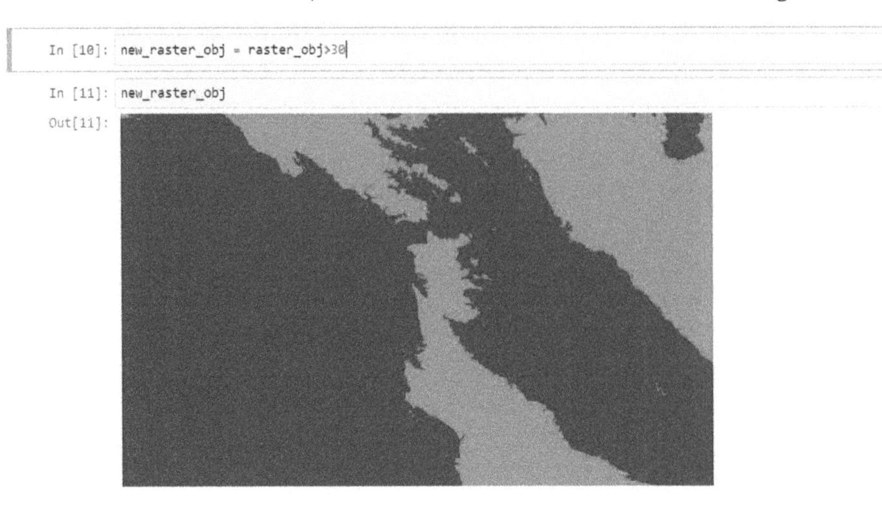

Figure 9.15: This Boolean result shows areas above (light) or below (dark) 30 meters

Similarly, a mathematical operation could be directly applied against a raster object, such as multiplying the raster object by 4. This would be useful for creating an exaggerated DEM or to scale a raster to match other rasters:

13. In the next cell, type in the following:

```
times_raster = raster_obj * 4
times_raster
```

Run the cell. The screenshot below contains an info popup that shows the value of the same cell:

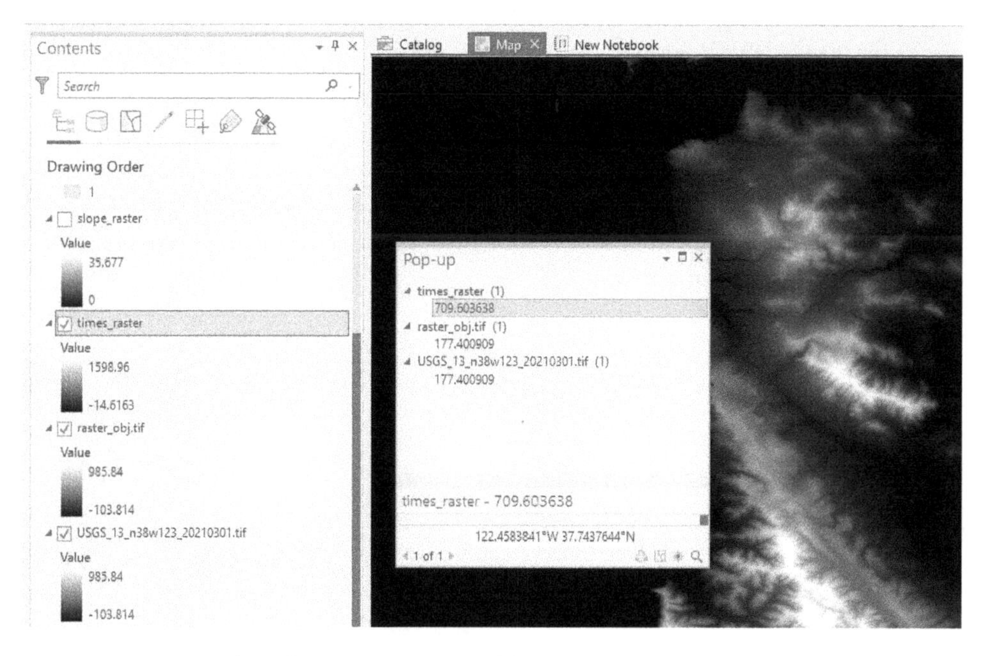

Figure 9.16: The cell queried shows two different values: in the original raster and the raster object in memory, it is ~117, and the cell in the result of the multiplication operation is 4 times that value

Another example would be adding two raster objects together. In this manner, the resulting raster object would represent the addition of all of the cell values, using this format. Imagining that there are two rasters (`raster1` and `raster2`) in your project, you could do:

```
raster_obj_addition = Raster("raster1") + Raster("raster2")
```

Let's look at an example with actual rasters:

14. First, we create a Boolean raster that represents a `True` value for all areas above 100 within the original raster object:

```
high_raster_obj = raster_obj > 100
high_raster_obj
```

Run the cell. The resulting Boolean raster has two distinct areas: cells that are above 100 meters and cells that are below (in other words, the cells where the statement `raster_obj > 100` is true or false):

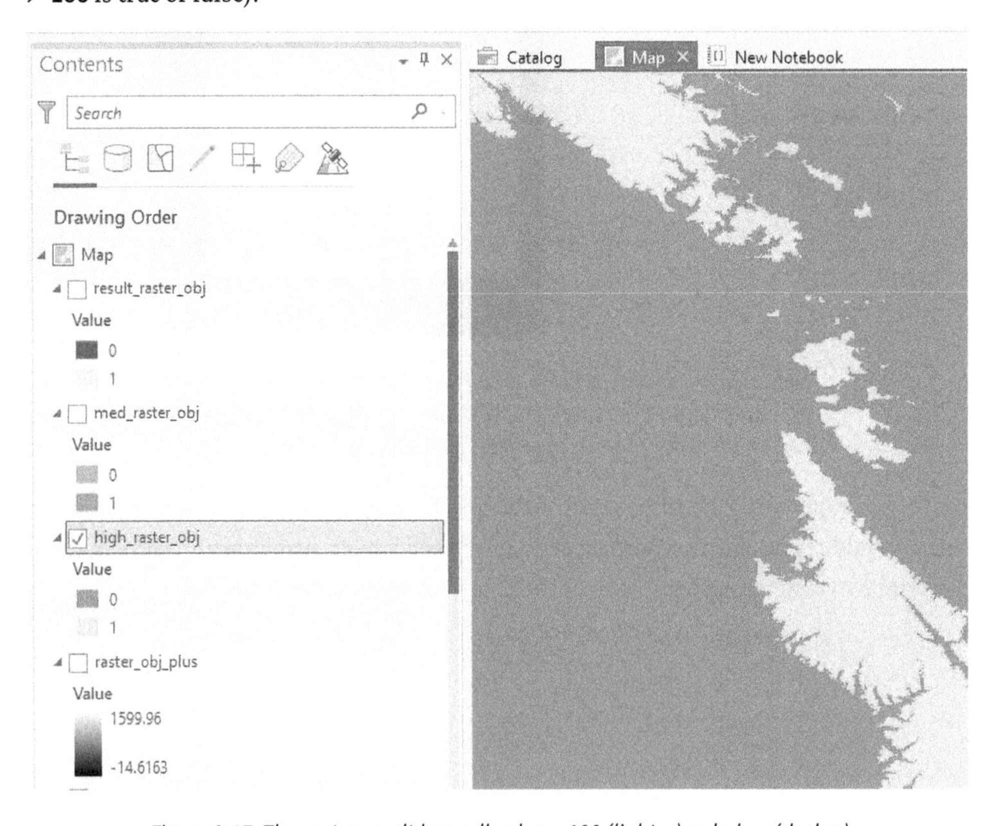

Figure 9.17: The raster result has cells above 100 (lighter) or below (darker)

15. Then, we create a raster object that is `True` if the value of the cell is below 70:

```
med_raster_obj = raster_obj < 70
med_raster_obj
```

Run the cell. This results in a Boolean with two areas, representing cells that meet the statement conditions and those that do not:

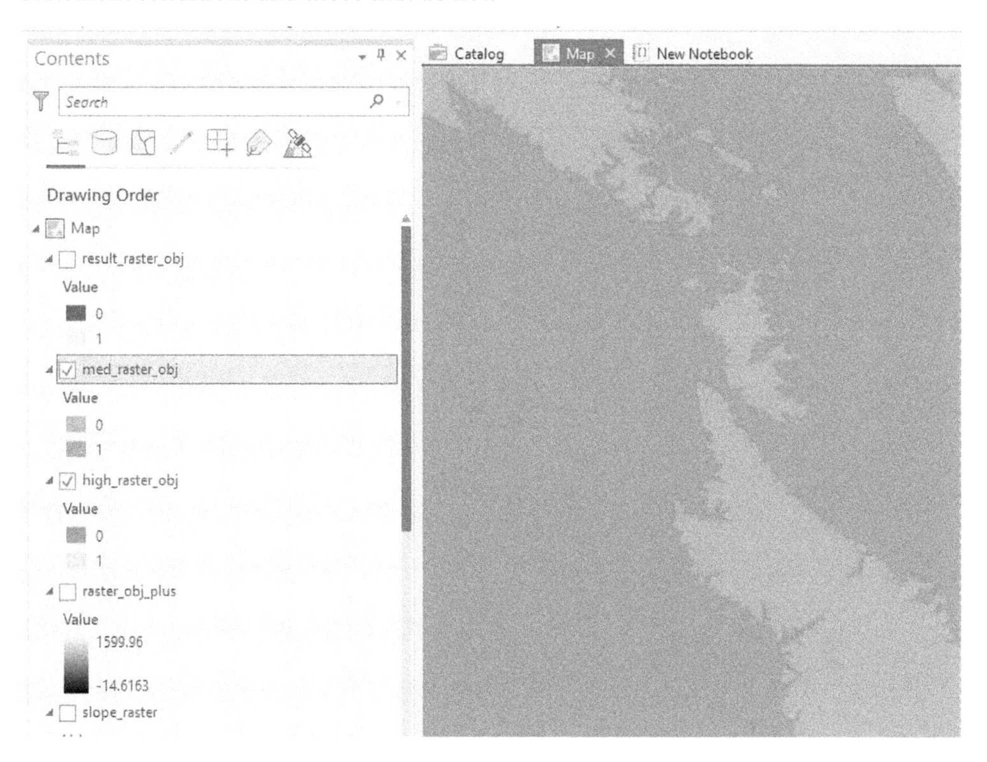

Figure 9.18: The raster result has areas above 70 (true) or below (false)

16. Finally, we add them together:

```
result_raster_objs = med_raster_obj + high_raster_obj
result_raster_objs
```

Run the cell. The resulting raster object is not Boolean, although it contains cells with values of 1 or 0. Where the raster cells are 1, the zero areas of the high raster object were added to the 1 value areas of the medium raster object.

Where the raster cells are 0, there were only zeros in both parent raster objects. It creates a kind of outline, defined as the areas between 70 and 100 in the original DEM:

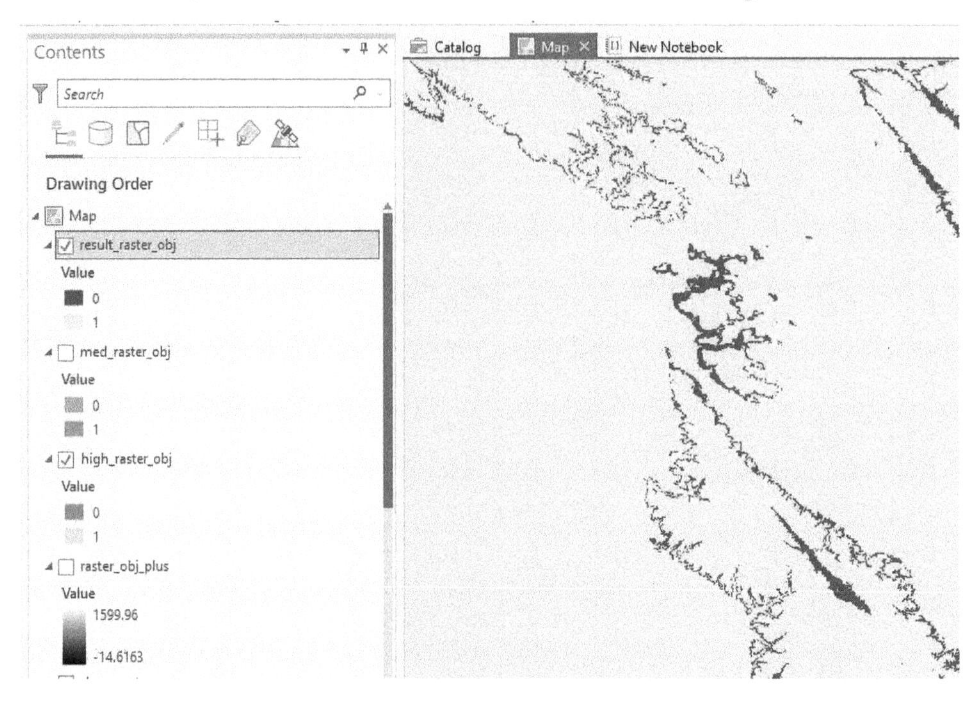

Figure 9.19: The result of the addition, where areas between 70 and 100 are given the value 1

Shorthand operators for Map Algebra

There are a number of shorthand **operators** that we can use when working with raster objects. These operators allow inline code to execute many different kinds of Map Algebra operations on a raster object. We saw above the use of multiplication using the asterisk (*), but there are many more.

17. For instance, to raise a raster to the power of a number, *N*, you would use two asterisks (**). This makes it easy to exaggerate raster elevation heights or to differentiate between cell value classes:

    ```
    power_raster_obj = raster_obj ** 2
    power_raster_obj
    ```

    Run the cell. The result of the operation is a raster where each cell value has been squared.

The following screenshot shows the cell value at a selected cell, with the cell value of the original raster object below it:

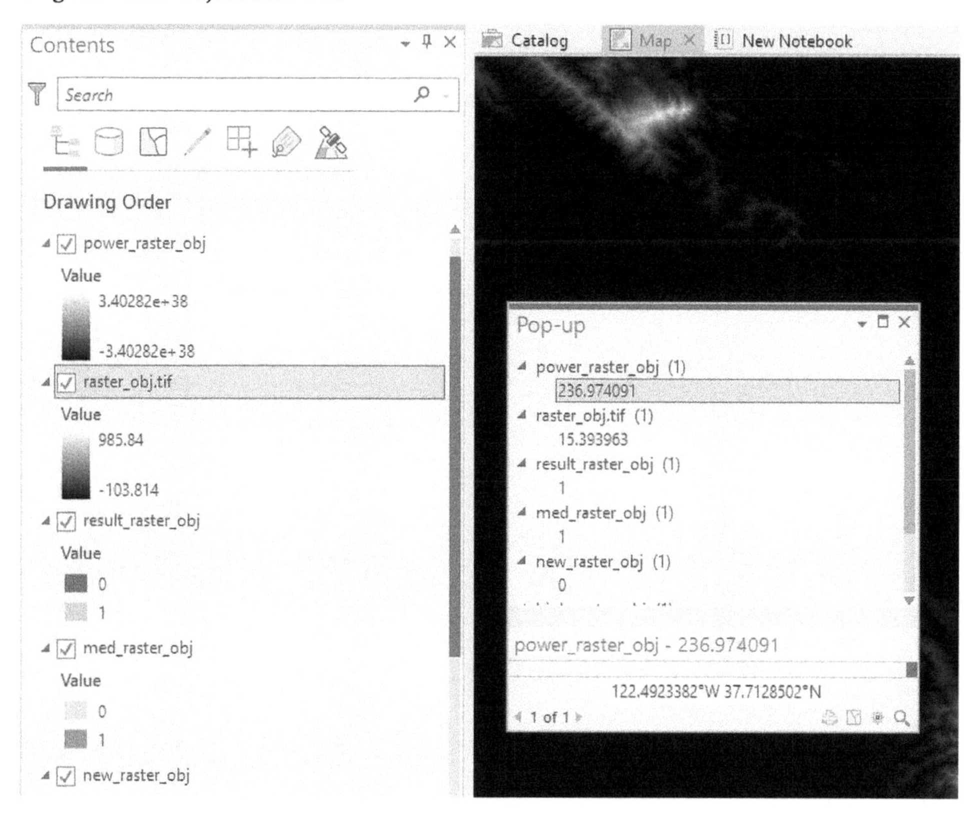

Figure 9.20: The selected cell has a value of 15.393963 to the power of 2

Addition and subtraction are performed using the expected plus (+) and minus (-) signs. This can be used with another raster object or with an integer or float value. When used with another raster, as we saw above, the value of the cell in the second raster object will be added or subtracted from the initial raster. The result is then assigned to an output raster object, which is temporary and must be saved to be committed to disk.

Negative operator

One cool feature of the operators is that some of them can be placed in front of the raster object to create a new raster object. The minus operator can be used in this fashion to create a negative of the original raster object.

This will allow you to add a positive and negative raster together in an analysis, or use the raster to create an inverse DEM.

18. In the next cell, type in the following:

```
negative_raster = -raster_obj
negative_raster
```

Run the cell. The resulting raster shows the mountain tops as valleys (where the dark areas are negative):

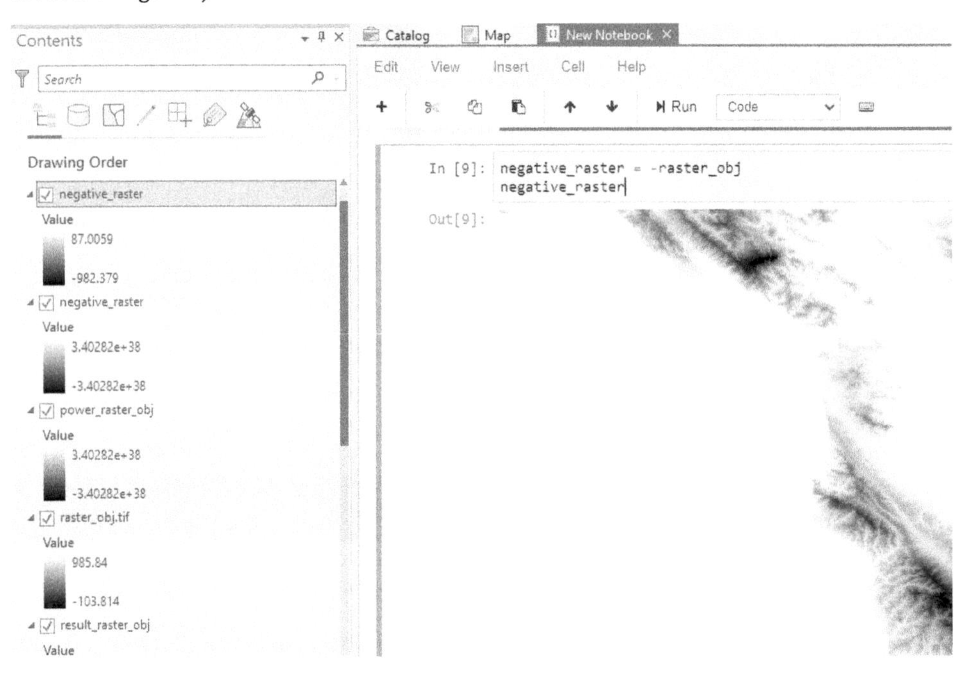

Figure 9.21: A negative raster where the highest value is on the sea floor

Division operators

Division is performed using the forward slash (/), but to achieve integer division (where the operation produces integer cell values), you'll need to use the double forward slash instead (//). This will allow you, for example, to reduce the height of hillshades.

19. In this example, we'll divide the values in each cell by 3.5 using the forward slash operator:

```
division_raster = raster_obj / 3.5
```

This gives us:

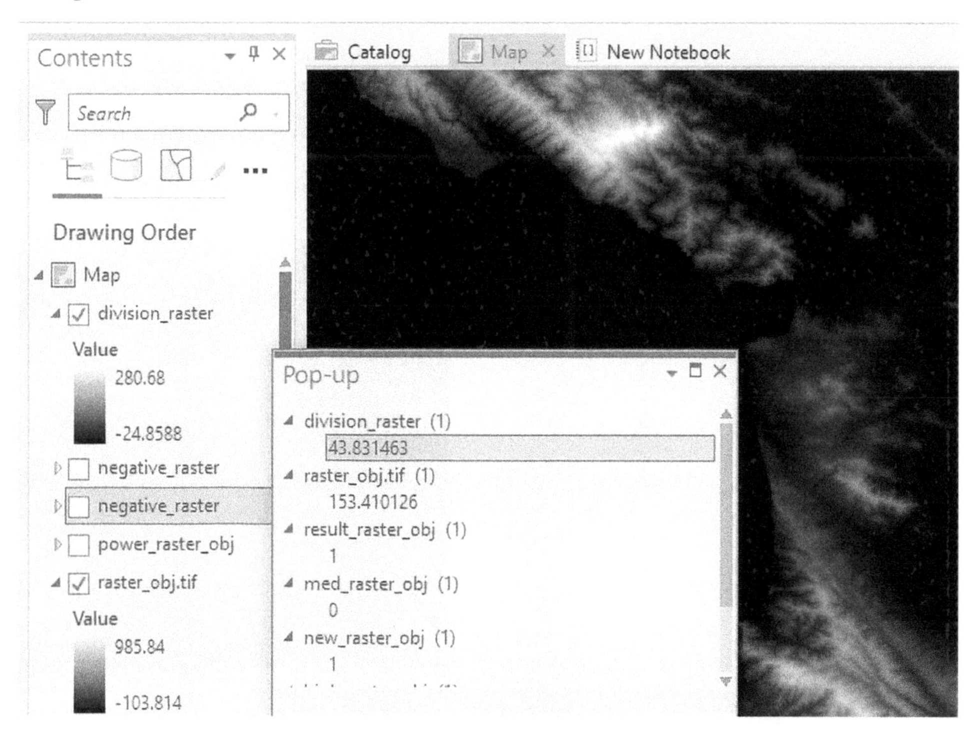

Figure 9.22: The cell value of the raster is 3.5x smaller than the original raster cell value

20. Instead of using the division operator, in this next example, we'll use the **integer division** operator. This makes the resulting cell values into integers, even when divided by a float. You may need this for simplifying raster addition analyses or for creating cell "bins":

```
integer_division_raster = raster_obj // 3.5
```

Run the cell. The resulting raster object has only integer values, as shown in the Table of Contents:

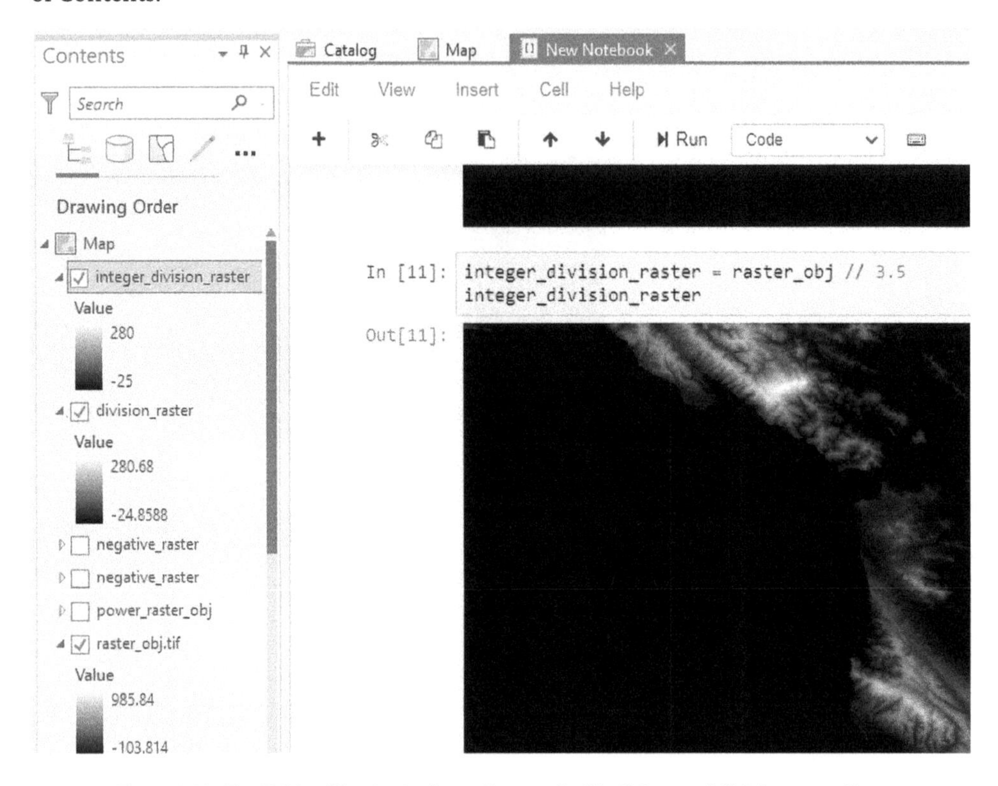

Figure 9.23: The Table of Contents shows the result of both types of division operations

Boolean operators

Another useful set of operators are Boolean (`True`/`False`) operators, which allow you to perform Boolean operations on raster cell data. You may need to find areas that meet a set of conditions and evaluate them as true or false, and then add them together.

To perform a Boolean OR, use the pipe symbol (|). The result of this operation will contain only 1 or 0 (`True` or `False`) cell values, or a NoData value:

```
raster_or_output = Raster("raster1") | Raster("raster2")
```

In the above state, the rasters would be combined in such a way that matching raster cells where there is a value above zero (on either raster) will be assigned `True` (1), cells where both rasters are equal to zero will be assigned `False` (0), and values where there is a NoData value on either raster will be assigned NoData.

A Boolean AND is performed using an ampersand (&):

```
raster_and_output = Raster("raster1") & Raster("raster2")
```

Like the OR operation, the result of this operation will contain only 1 or 0 (`True` or `False`) cell values or a NoData value. However, in this case, it assigns `True` (1) where both rasters, when overlaid, contain a value above 0; `False` (0) where the overlaid cells both have a value of 0 or where one raster has a cell of 0; and NoData where at least one raster has a cell value of NoData.

The greater than and less than operators use the (>) and (<) symbols, respectively, but remember that these are Boolean operators, and the result will be an output raster object that splits the cell values into `True` (1) or `False` (0), depending on the condition set. This may not be the desired output, so if you want to select cell values using a conditional statement, use the `ExtractByAttributes` tool from the Spatial Analyst module instead.

Read more about working with the Map Algebra operators here: `https://pro.arcgis.com/en/pro-app/latest/help/analysis/spatial-analyst/mapalgebra/working-with-operators.htm`

Using arcgis.raster

ArcGIS API for Python (the `arcgis` module) also contains useful tools for raster layers and imagery layers. Used in combination with `arcpy`, it can allow you to store rasters or imagery in the cloud, pull down the rasters, perform analysis or a process on the rasters, and then put the rasters back into ArcGIS Online.

Much like `arcpy`, there are a ton of built-in functions for `arcgis` that enable you to interact with raster data. These functions allow you, for example, to get statistical properties of the raster such as mean or maximum or minimum cell value, or to perform mathematics on the cell values. Most of these functions are part of the raster subclass of `arcgis` and are available at `arcgis.raster.functions`.

Working with imagery layers

In this example, we will load an imagery layer from the web into a map and save it to disk.

You may carry on in the same Notebook as the previous section.

1. The first step is to create a GIS object to access the satellite image from an imagery service using a URL. In a new cell, type in the following:

```
from arcgis.gis import GIS
from arcgis.raster import *
gis = GIS("pro")
map_obj = gis.map()
imagery_obj = ImageryLayer("https://sentinel-cogs.s3.us-
west-2.amazonaws.com/sentinel-s2-l2a-cogs/1/C/CV/2018/10/
S2B_1CCV_20181004_0_L2A/B02.tif", gis=gis)
imagery_obj
```

Run the cell. The result of the call to the URL is an imagery object:

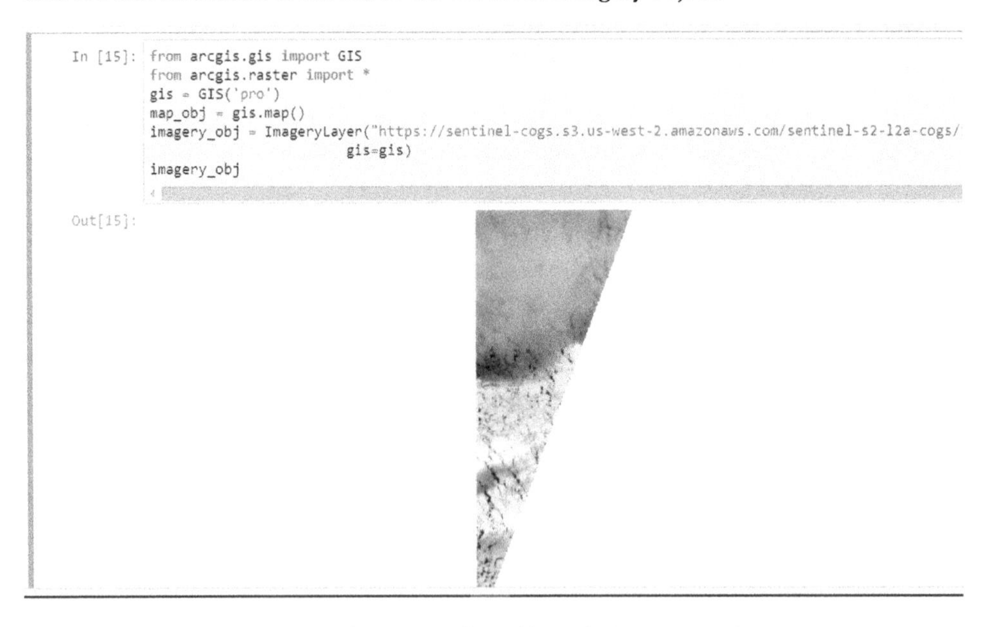

Figure 9.24: The image selected from the imagery service

2. To add the layer to the map, use the `add_layer` function:

```
map_obj.add_layer(imagery_obj)
```

Run the cell.

3. The imagery layer can be saved to disk using the `export_image` method:

```
imagery_obj.export_image(size=[1400, 600],
                          export_format="tiff",
                          f="image",
                          save_folder=r"C:\my\save_folder",
                          save_file="raster.tif")
```

   Run the cell. The output format can also be JSON, KMZ, an image, or even a NumPy array. In this example, it is an image type (`f="image"`). Other optional parameters include compression and aspect ratio.

   Much like our exploration of the `arcpy` raster tools, this is just scratching the surface of the available tools in the `arcgis` module. For instance, imagery statistics can be created using built-in functions.

4. The extent of the image can be used to understand where on Earth the image represents (which may not be obvious), and is accessed via the `extent` property:

```
imagery_obj.extent
```

   Run the cell. This is the result:

```
{
  "xmin": 300000,
  "ymin": 1890220,
  "xmax": 409800,
  "ymax": 2000020,
  "spatialReference": {
    "wkid": 32701,
    "latestWkid": 32701
  }
}
```

You can use the extent JSON to create a new JSON object for calculating a histogram. It reveals the WKID (Well Known ID), a spatial reference identifier, along with the bounding box extent envelope coordinates.

Plotting a histogram

A histogram is a summary of the distribution of the pixel values of the image, combined into bins.

It gives you a statistical insight into the images available from the imagery service or from a raster object.

The histogram values must be calculated using the `compute_histograms` function. It requires a geometry object (a Polygon or Envelope), passed as a JSON object. This JSON uses the same WKID as the extent JSON.

5. In the next cell, type in the following to use the extent in the previous step to compute a histogram:

```
hist =imagery_obj.compute_histograms({
  "xmin": 300000,
  "ymin": 1890220,
  "xmax": 301800,
  "ymax": 1900020,
  "spatialReference": {
    "wkid": 32701,
    "latestWkid": 32701
  }
})
hist
```

   Run the cell. The result of the operation is shown below:

```
In [23]: hist =imagery_obj.compute_histograms({
           "xmin": 300000,
           "ymin": 1890220,
           "xmax": 301800,
           "ymax": 1900020,
           "spatialReference": {
             "wkid": 32701,
             "latestWkid": 32701
           }
         })
         hist
```


```
Out[23]: {'histograms': [{'size': 256, 'min': 1, 'max': 14352, 'counts': [41, 0, 0, 0, 0, 0, 0, 0, 0, 70, 560, 1081, 1192, 917, 1006, 75
9, 549, 447, 381, 323, 274, 267, 519, 693, 536, 393, 343, 325, 290, 286, 233, 227, 254, 222, 199, 163, 170, 215, 230, 236, 216,
202, 206, 193, 224, 238, 248, 225, 194, 199, 155, 189, 149, 131, 122, 125, 118, 115, 96, 96, 100, 95, 81, 79, 89, 82, 79, 74, 8
4, 70, 70, 72, 60, 70, 57, 51, 50, 62, 60, 64, 47, 60, 65, 65, 52, 60, 59, 49, 60, 56, 60, 64, 58, 47, 57, 58, 51, 80, 62, 69,
72, 54, 63, 74, 65, 63, 63, 62, 54, 62, 73, 76, 65, 44, 55, 83, 69, 74, 57, 59, 68, 87, 67, 69, 66, 84, 67, 59, 90, 71, 57, 69,
77, 80, 94, 72, 85, 66, 72, 87, 54, 69, 71, 101, 82, 103, 76, 105, 65, 83, 81, 90, 92, 95, 105, 101, 79, 137, 77, 103, 115, 11
2, 121, 130, 107, 138, 150, 149, 175, 216, 190, 251, 277, 335, 329, 374, 415, 516, 444, 597, 755, 730, 850, 1003, 1121, 1180, 1
362, 1714, 1597, 1267, 2111, 2026, 2124, 1341, 1845, 2061, 1661, 1478, 1428, 1396, 1168, 1118, 1163, 876, 1007, 978, 1095, 961,
1273, 1280, 1348, 1369, 1898, 2350, 2624, 2830, 3659, 4458, 4044, 5639, 5868, 6300, 5379, 5368, 6403, 5880, 6609, 4047, 4594, 3
259, 3189, 2628, 2017, 1136, 1393, 1130, 1283, 1191, 1444, 1358, 1085, 1157, 922, 648, 429, 341, 163, 78, 25, 11, 2, 4, 0, 0,
0, 1]}]}
```


Figure 9.25: The result of the histogram calculation

6. Using the Matplotlib library, which is included with the Python installation for ArcGIS Pro, the histogram statistics can be plotted to reveal the concentration of pixel values.

In the next cell, type in the following:

```
import matplotlib.pyplot as plt
%matplotlib inline
plt.hist(hist['histograms'][0]['counts'],
         len(hist['histograms'][0]['counts']),
         density=True,
         histtype='bar',
         facecolor='b',
         alpha=0.5)
plt.show()
```

Run the cell. This code will pass the histogram data to the library (with no "binning") and create a bar graph of the data:

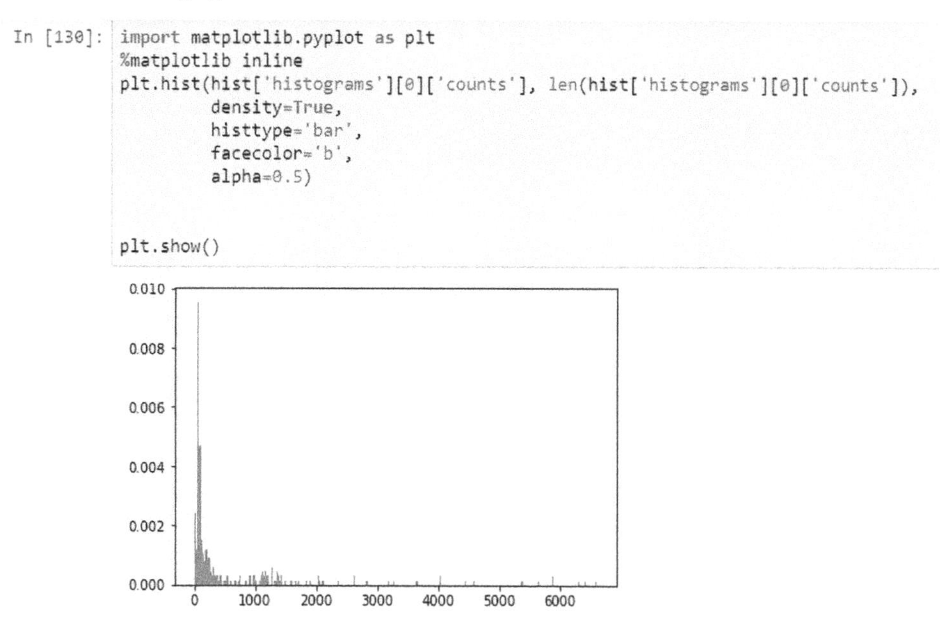

Figure 9.26: The histogram plotted with Matplotlib

This is one quick way to assess the statistics of the imagery data or raster you are working with, using the included tools for ArcGIS API for Python and the useful `matplotlib` library, which cannot be fully covered here.

Working with raster layers

The `arcgis.raster.Raster` function creates a raster object, in a similar manner to the `arcpy.Raster` function:

7. In the next cell, type in the following to generate a raster object from the DEM:

   ```
   araster_obj = arcgis.raster.Raster('USGS_13_n38w123_20210301.tif')
   ```

   Run the cell.

8. The extent of the raster can be accessed using the `extent` function:

   ```
   araster_obj.extent
   ```

   This is the result after running the cell:

   ```
   {'xmin': -123.00055555579371, 'ymin': 36.999444443707034, 'xmax':
   -121.99944444360563, 'ymax': 38.00055555589506, 'spatialReference':
   {'wkid': 4269, 'latestWkid': 4269}}
   ```

9. The raster statistics can be accessed using the `get_statistics` method:

   ```
   stats = araster_obj.get_statistics()
   ```

   This is the result, showing the minimum, maximum, and mean cell values, among other details:

   ```
   [{'min': -103.81411743164, 'max': 985.83990478516, 'mean':
   72.003597453322, 'standardDeviation': 141.66796857683, 'skipX': 1,
   'skipY': 1, 'count': 0.0}]
   ```

10. Save the raster into a folder using the `save` method:

    ```
    araster_obj.save(output_name='C:/projects/test_raster.tif',gis=gis)
    ```

    The raster can also be written to a cloud service using this `save` method, by specifying that the `gis` object is writing to ArcGIS Online instead of a local filesystem.

These methods can be combined with the ArcPy methods for raster objects to create custom raster workflows. This chapter was limited in length and could only touch on a selection of the available raster tools in these modules.

There are many other functions available for processing, and accessing the properties of, the rasters in both modules.

Read more about using the `arcgis` module for rasters here: `https://developers.arcgis.com/python/api-reference/index.html`

Summary

In this chapter, you learned how to use ArcPy to read rasters into memory or create new rasters, and to save the result. You learned how to read the raster properties for both the cell values and the geographic values. You also learned how to use the `sa` module to access Spatial Analyst tools, and to perform operations on the rasters to create Slope or Hillshade rasters. You also reviewed the uses of `arcgis.raster` to get data from the web and save it locally.

In the next chapter, you will learn about the NumPy module and how it is used with arrays and Notebooks to process raster data quickly for statistical analysis.

10

Geospatial Data Processing with NumPy

Data processing tools are often limited to the pre-built tools discussed in other chapters, or open-source tools such as Shapely, Rasterio, or GDAL. These tools can be limited in terms of processing speed and flexibility. When creating geospatial data workflows, you will often have to create custom tools to process data quickly, and those other libraries can be limiting.

NumPy offers a third way. Used for scientific computing, it is an incredibly fast and powerful module written in C, with a Python code "wrapper" so it can be used in your existing Python environment. It is built to read, analyze, and write multidimensional arrays of data.

Esri has included easy tools to convert rasters into NumPy arrays and back, which makes it easy to add custom NumPy functions into your existing pipelines. Selecting or clipping areas of a raster, performing array math, creating new arrays and populating them with data, processing specific raster bands or cell values; these are all suitable processes to perform with the NumPy module. The results can then be written back as raster output and used in other ArcGIS Pro workflows.

In this chapter, we will cover the following topics:

- The advantages of NumPy arrays in Python processing
- Importing rasters into arrays
- Replacing raster processing tools using NumPy tools
- Saving arrays as rasters
- Mathematical operations using arrays

- Exercise: Statistical analysis of raster data using NumPy
- Creating charts from NumPy arrays using Matplotlib

To complete the exercises in this chapter, please download and unzip the `Chapter10.zip` folder from the GitHub repository for this book: `https://github.com/PacktPublishing/Python-for-ArcGIS-Pro/tree/main/Chapter10`.

Introduction to NumPy

NumPy is an important data processing module included with ArcGIS Pro. NumPy was originally written by Travis Oliphant, who also went on to develop the Anaconda project. It is an open-source Python library based on two competing numeric structure libraries known as Numeric and Numarray.

NumPy was written to be able to handle large arrays of data and also to extend the functionality of Python for mathematical and scientific processing. This makes it a very useful library for writing code to read, analyze, and write raster data.

For ArcGIS Pro users and code writers, using NumPy directly allows you to create custom functions and tools that are not available in the basic tools included in ArcGIS Pro. NumPy's speed and mathematical capabilities open up new ways to perform analyses and create data workflows.

These custom data workflows will often use Pandas, and NumPy is at the heart of Pandas. Pandas is built on NumPy and its array structure, known as **ndarray** (or **N-dimensional array**), is what is used for the Pandas Series data structure. As we learned in *Chapter 8*, a Pandas Series is akin to a row of data and will often be called a row in this chapter.

Advantages of NumPy arrays

NumPy ndarrays are **homogenous** (all data is of the same data type) and **multidimensional**. This makes ndarrays very useful for many data types used in GIS, including continuous geospatial data such as rasters, which often have multiple dimensions, known as *bands*.

The core code of NumPy is written in the C programming language. This allows for better memory management of large arrays of data and quicker processing of data. Python, as an interpreted language, must be converted into byte-code to be executed, which can make the code run more slowly for some applications. Owing to that C code core, NumPy is more efficient and faster.

NumPy arrays versus Python lists

NumPy arrays are often compared to the built-in Python `list` data type. Both lists and ndarrays are ordered data structures that are mutable and enclosed in square brackets. However, there are some major differences between them.

A Python list can contain a combination of different types of data, while NumPy arrays can only contain one type of data per array. This means that there is no type checking on each object in the list. NumPy arrays are also faster when accessing data, and are more compact in memory use.

Data objects in an array are stored with much less metadata describing the object (size, reference count, object type, and object value are stored for Python lists). Python lists therefore balloon the memory usage, making it much slower to access or iterate each object.

Read more about NumPy here: `https://numpy.org/`.

Read the basic introduction to NumPy here: `https://numpy.org/doc/stable/user/absolute_beginners.html`.

Importing NumPy

The first step toward using NumPy is to import the module. Open a new Notebook in ArcGIS Pro and type the following into the first cell:

```
import numpy as np
```

Run the cell. Using the variable `np` to represent the `numpy` module is not required, but it is an accepted shorthand for Python code that makes it easier to access the submodules and properties of NumPy.

Basics of NumPy for rasters

Using NumPy for rasters is very straightforward. Rasters are data organized into regular rows and columns, and may have multiple bands of data. These data behaviors can be precisely recreated using NumPy arrays, which can have any number of rows or columns, as well as multiple dimensions.

Creating an array

Often in GIS you must create rasters for analyses. These arrays may need to be blank, allowing you to accumulate values from inputs to a continuous surface based on location; all one value to create a constant raster; or merged with vector data inputs such as GeoJSON files or shapefiles. All of these are possible with NumPy arrays.

There are many ways to create a NumPy array. Some of these are built-in tools, and some are methods to derive an array from an existing dataset such as a raster, CSV file, or JSON data, as seen in *Chapter 8*'s exploration of Pandas. Data can also be read from a vector file such as a shapefile or feature class, or even a raw text file.

To create an array of any shape where every value is 0, there is a built-in tool: `numpy.zeros`. Instead of passing in a set of values, a tuple containing only the shape of the array is passed, where the shape represents the number of rows and number of columns respectively.

This example will create a uniform array where the values are 0, with 4 rows and 6 columns:

```
nparray_zeros = np.zeros((4,6))
```

The resulting array looks like this:

```
In [96]: # one dimensional array where all values are either 0 or 1.
         # Pass a tuple with the number of rows/series (4 here) or number of columns (6 here)

         nparray_zeros = np.zeros((4,6))
         nparray_zeros

Out[96]: array([[0., 0., 0., 0., 0., 0.],
                [0., 0., 0., 0., 0., 0.],
                [0., 0., 0., 0., 0., 0.],
                [0., 0., 0., 0., 0., 0.]])
```

Figure 10.1: Creating an array of zeroes

Similarly, you can create an array where all values are 1. In this example, the result will be an array with 6 rows, each with 7 elements. We want an array with more than a single row, so we use the `numpy.ndarray.ones` function, for which the default data type is float:

```
nparray_ones = np.ones((6,7))
```

The array looks like this:

```
In [97]: # np.ones will create an array with all values = 1. The data type defaults to float
         nparray_ones = np.ones((6,7))
         nparray_ones

Out[97]: array([[1., 1., 1., 1., 1., 1., 1.],
                [1., 1., 1., 1., 1., 1., 1.],
                [1., 1., 1., 1., 1., 1., 1.],
                [1., 1., 1., 1., 1., 1., 1.],
                [1., 1., 1., 1., 1., 1., 1.],
                [1., 1., 1., 1., 1., 1., 1.]])
```

Figure 10.2: Creating an array of ones

In this next example, an array is created from a Python list with 4 elements:

```
onelist = [2,4,6,8]
nparray = np.array(onelist)
```

You can create a random array using the `randint` function. It will produce an array of the specified shape (the parameter called `size`, confusingly) and will populate the array with random values up to the first parameter. The `seed` function is used to "seed" the random generator (as "random" generators are not truly random, and rely on a seed value to create the generated output):

```
np.random.seed(0)
nparray_3d = np.random.randint(10, size=(3, 4, 5))
```

The array will look something like:

```
In [100]: # Create a random array
          np.random.seed(0)
          nparray_3d = np.random.randint(10, size=(3, 4, 5))
          nparray_3d

Out[100]: array([[[5, 0, 3, 3, 7],
                  [9, 3, 5, 2, 4],
                  [7, 6, 8, 8, 1],
                  [6, 7, 7, 8, 1]],

                 [[5, 9, 8, 9, 4],
                  [3, 0, 3, 5, 0],
                  [2, 3, 8, 1, 3],
                  [3, 3, 7, 0, 1]],

                 [[9, 9, 0, 4, 7],
                  [3, 2, 7, 2, 0],
                  [0, 4, 5, 5, 6],
                  [8, 4, 1, 4, 9]]])
```

Figure 10.3: Creating a random array

Ranges can be used to create an array as well, using the NumPy `arange` method. The following line of code will create an array with 1 row, containing the numbers 0-9:

```
nparray3 = np.arange(10)
```

As you have seen, there are many different ways to create arrays in NumPy.

Reading a raster into an array

To read data from a raster into a NumPy array, ArcPy has a built-in method, `RasterToNumPyArray`. This allows you to read any type of raster into an array and perform analysis using custom NumPy processing.

The input to `RasterToNumPyArray` can be either a raster object or a string that is the path on your computer to the raster. In the example below, you will read a raster into a raster object and pass the raster object into `RasterToNumPyArray`.

1. In your Notebook, add the following lines to a new cell. Adjust the file path variable to match where you have placed the elevation TIF from *Chapter 9*:

   ```
   import arcpy
   file_path = r'USGS_13_n38w123_20210301.tif'
   raster_obj = arcpy.Raster(file_path)
   raster_obj
   ```

   Run the cell. You should see the following output:

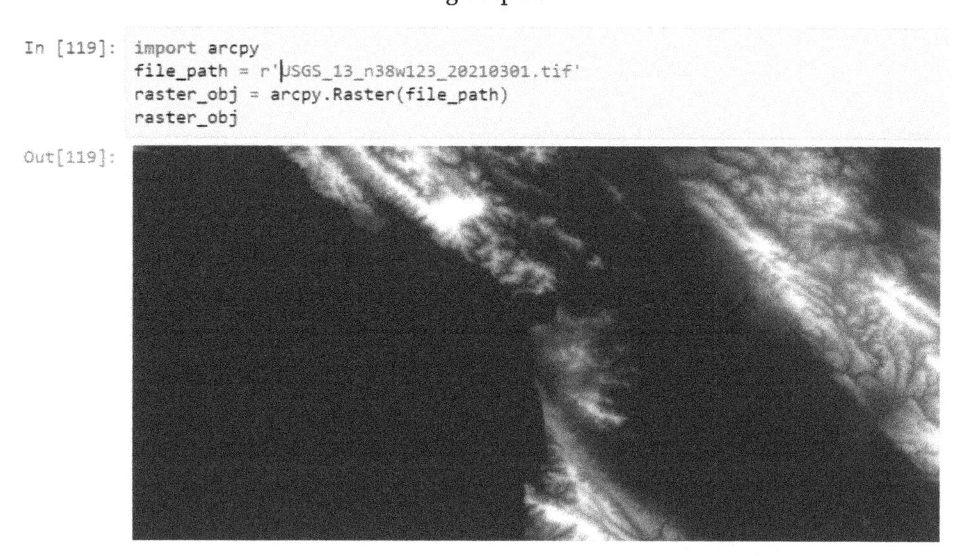

Figure 10.4: Elevation data as a raster object

2. Now that you have the raster object, it can be converted into a NumPy array:

   ```
   nparr_raster = arcpy.RasterToNumPyArray(raster_obj)
   nparr_raster
   ```

Run the cell. You should see the NumPy array in the output cell:

```
In [120]: nparr_raster = arcpy.RasterToNumPyArray(raster_obj)
          nparr_raster

Out[120]: array([[ 8.5578552e+01,  8.5281036e+01,  8.5320587e+01, ...,
                   3.2745525e+01,  3.2705288e+01,  3.2640228e+01],
                 [ 8.9238808e+01,  8.8557518e+01,  8.8136261e+01, ...,
                   3.2848251e+01,  3.2886658e+01,  3.2931221e+01],
                 [ 9.2715370e+01,  9.2176201e+01,  9.1481804e+01, ...,
                   3.2800846e+01,  3.2930138e+01,  3.3099201e+01],
                 ...,
                 [-9.9999900e+05, -9.9999900e+05, -9.9999900e+05, ...,
                   7.8969254e+01,  7.9099106e+01,  7.9647850e+01],
                 [-9.9999900e+05, -9.9999900e+05, -9.9999900e+05, ...,
                   8.3939728e+01,  8.3800331e+01,  8.4412811e+01],
                 [-9.9999900e+05, -9.9999900e+05, -9.9999900e+05, ...,
                   8.9324646e+01,  8.9062607e+01,  8.9005020e+01]], dtype=float32)
```

Figure 10.5: The Notebook representation of the array after being converted from a raster

With the raster now converted into an array, NumPy tools can be used on the raster data. We will cover these more later in the chapter.

If you want to explore more about the parameters of the `RasterToNumpyArray` function, check out the official documentation here: `https://pro.arcgis.com/en/pro-app/latest/arcpy/functions/rastertonumpyarray-function.htm`.

Now that we can create an array or import an array, let's review the properties of arrays using built-in NumPy functions.

Array properties

NumPy arrays all have a size, shape, dimension, and data type. These behaviors influence their use and behaviors.

Size

The **size** of an array refers to the number of data elements within it, which can be determined using the total number of rows multiplied by the total number of columns multiplied by the number of dimensions. The size of an array is called using the `size` property of an array, `numpy.ndarray.size`.

For example, consider the size of an array with 2 rows, each with 4 columns, with 1 dimension:

```
datalist = [[2,4,6,8], [1,3,5,7]]
nparray = np.array(datalist)
print(nparray.size)
```

The result of the print statement will be 2 * 4 * 1 = 8.

Shape

The `shape` property of an array refers to a tuple that contains up to 3 values; however, there may only be 2. Instead of the total number of elements, which is the size, the shape describes the array's structure.

If there is only 1 dimension (or *band*, in raster terminology) to the array, the shape will describe the number of rows and the number of columns that the array contains. If the number of dimensions is more than 1, the `shape` array will include a third value, with the number of dimensions first, then rows, then columns.

If we look at the array from above, created from a list of lists, it has a shape of *(2,4)*:

```
datalist = [[2,4,6,8], [1,3,5,7]]
nparray = np.array(datalist)
nparray.shape
```

This is the result:

```
(2, 4)
```

In the following example, an array with a shape of 3 dimensions, 4 rows, and 5 columns is created and populated with random integer elements:

```
np.random.seed(0)
nparray_3d = np.random.randint(10, size=(3, 4, 5))
```

Below we see the output:

```
array([[[5, 0, 3, 3, 7],
        [9, 3, 5, 2, 4],
        [7, 6, 8, 8, 1],
        [6, 7, 7, 8, 1]],
```

```
        [[5, 9, 8, 9, 4],
        [3, 0, 3, 5, 0],
        [2, 3, 8, 1, 3],
        [3, 3, 7, 0, 1]],

        [[9, 9, 0, 4, 7],
        [3, 2, 7, 2, 0],
        [0, 4, 5, 5, 6],
        [8, 4, 1, 4, 9]]])
```

Note that the `randint` function confusingly uses the parameter `size` to describe the shape, but the resulting array's shape and size properties are reported as expected:

```
print("shape: ",nparray_3d.shape,"\n", "size:", nparray_3d.size)
```

The output:

```
shape:  (3, 4, 5)
size: 60
```

You can see above the 3 dimensions of the array, each of which has 4 rows of 5 columns. As 3 * 4 * 5 is 60, the size of the array is 60.

Data type

The `dtype` property allows you to confirm the data type of the array. As it is an array, and not a Python list, all of the data in the array is the same type, so only 1 value will be returned. All arrays have a `dtype` method, no matter the data type. Let's check the `dtype` of the `nparr_raster` array from above:

```
print(nparr_raster.dtype)
```

In this case, the `dtype` is `float32`. The data type of the array will determine what kinds of mathematical operations can be performed on the array. If you are going to perform Map Algebra on environmental rasters, for instance, you want to be sure that the data types can be "added" together.

The data type can be specified using the `dtype` parameter for most array creation functions. Use the `dtype` parameter like this:

```
onelist = [2,4,6,8]
nparray = np.array(onelist, dtype=float)
```

It works the same for more dimensions:

```
twolist = [[2,4,6,8], [1,2,3,4]]
nparray = np.array(twolist, dtype=float)
```

The results look like this:

```
In [23]: onelist = [2,4,6,8]
         nparray = np.array(onelist,dtype=float)
         nparray

Out[23]: array([2., 4., 6., 8.])

In [29]: twolist = [[2,4,6,8],[1,2,3,4]]
         nparray = np.array(twolist,dtype=float)
         nparray

Out[29]: array([[2., 4., 6., 8.],
                [1., 2., 3., 4.]])
```

Figure 10.6: Specifying the float data type for two arrays

Accessing specific elements

Updating a specific element value of an array, or a set of elements, is made simple using NumPy. Finding a portion of a raster and using only that smaller subsection of the raster for analysis is a very useful function that NumPy makes easy.

To get to the specific element, you can use **indexing**. If the shape of the array has more than 1 row, the indexing is more specific. This is useful for updating continuous rasters such as precipitation.

In the following example, the indexing of the array acts much like the indexing on a Python list. The result is the last two values in the array, which have a `dtype` of `float`:

```
onelist = [2,4,6,8]
nparray = np.array(onelist,dtype=float)
print(nparray[2], nparray[3])
```

This is the output:

```
6.0 8.0
```

In the next example, the array has a shape of *(2,4)*. To access data elements, you must specify an index that includes the row number (starting from 0) and then the element in the row. Below, we print the array's shape, the fourth element of the second row, and the third element of the first row:

```
twolist = [[2,4,6,8],[1,2,3,4]]
nparray = np.array(twolist,dtype=float)
print(nparray.shape, nparray[1,3], nparray[0,2])
```

The code in a Notebook cell looks like this:

```
In [6]: twolist = [[2,4,6,8],[1,2,3,4]]
        nparray = np.array(twolist,dtype=float)
        print(nparray.shape, nparray[1,3], nparray[0,2])

        (2, 4) 4.0 6.0
```

Figure 10.7: Getting the shape of an array and accessing elements

For arrays with more than one dimension (or band for rasters), as well as multiple rows and columns, the indexing requires three values: a dimensional index, a row index, and a column index:

```
nparray = np.random.randint(10, size=(3, 2, 5))
```

This is the result:

```
array([[[8, 1, 1, 7, 9],
        [9, 3, 6, 7, 2]],

       [[0, 3, 5, 9, 4],
        [4, 6, 4, 4, 3]],

       [[4, 4, 8, 4, 3],
        [7, 5, 5, 0, 1]]])
```

We can use indexing to get the data from specific cells by listing the dimension, row, and column:

```
print(nparray[0,0,1], nparray[0,1,0], nparray[1,0,0])
```

This is the result of the indexing for this particular random array:

```
1 9 0
```

Using indexes to access elements is helpful for updating specific values, one at a time. To access or update a subset of rows, use array slicing.

Accessing a subset of the array

Instead of just accessing one element, it's often important to select a *subset* of the array. For instance, when processing a raster, it's useful to select a portion of the raster instead of working on the whole raster, which might be very large. NumPy arrays make this easy using the power of **slicing**, making it possible to perform analysis on just the subset selected.

The slicing mechanism matches the shape of the array. If there is one row, you need to pass one index "set" or description of the indexes to select. If there is more than one row, you will need two index sets, and if there is more than one dimension (think rasters with multiple bands) you will need to include three index sets.

Slicing an array with one row is simple and mirrors the slicing of a Python list. Using indexing, you select the subset of values in the array as a new array, and often assign them to a new variable:

```
onelist = [2,4,6,8]
nparray = np.array(onelist)
print(nparray[1:4])
```

This is the result of our slicing operation:

```
[4 6 8]
```

Slicing an array with more than one row (which describes most rasters) is a bit more complex. It requires two separate selection sets, which indicate the rows and the columns to select.

The first set includes the indexes of the rows to be selected, and the second set includes the indexes of the columns. These indexes can be one value (if only one row or column is to be selected) but generally more than one row or column is selected at a time. This is achieved by using the colon to separate the start and end indexes.

This example demonstrates how to select the last 4 columns of the first 2 rows. First, let's make the array:

```
nparray_2d = np.random.randint(10, size=(6, 6))
```

Our array looks like this:

```
array([[5, 9, 3, 0, 5, 0],
       [1, 2, 4, 2, 0, 3],
```

```
        [2, 0, 7, 5, 9, 0],
        [2, 7, 2, 9, 2, 3],
        [3, 2, 3, 4, 1, 2],
        [9, 1, 4, 6, 8, 2]])
```

Let's slice the last 4 columns of the first 2 rows:

```
nparray_2d[0:2, 2:6]
```

We get:

```
array([[1, 1, 2, 7],
       [9, 5, 0, 4]])
```

Slicing arrays with more than one dimension is similar, but you need to include a third set to describe the dimensions that you want to select. Let's make a random array with 3 dimensions, with 3 rows each, containing 9 elements each (the columns):

```
nparray_3d = np.random.randint(10, size=(3, 3, 9))
```

This is the array:

```
array([[[3, 0, 0, 6, 0, 6, 3, 3, 8],
        [8, 8, 2, 3, 2, 0, 8, 8, 3],
        [8, 2, 8, 4, 3, 0, 4, 3, 6]],

       [[9, 8, 0, 8, 5, 9, 0, 9, 6],
        [5, 3, 1, 8, 0, 4, 9, 6, 5],
        [7, 8, 8, 9, 2, 8, 6, 6, 9]],

       [[1, 6, 8, 8, 3, 2, 3, 6, 3],
        [6, 5, 7, 0, 8, 4, 6, 5, 8],
        [2, 3, 9, 7, 5, 3, 4, 5, 3]]])
```

To select the subset, the first index set (1) indicates that only the second band should be selected from, and the second (:) indicates that all values from the first index onward should be included (in other words, every row). The third set (2:7) indicates that only the elements from the selected columns should be included:

```
nparray_3d[1,:,2:7]
```

This is the result:

```
array([[0, 8, 5, 9, 0],
       [1, 8, 0, 4, 9],
       [8, 9, 2, 8, 6]])
```

Slicing a raster

Performing these slice operations on an array generated from a raster can be very useful, as it can replace raster clipping operations in a workflow, meaning you can choose to only work on a section of the raster. In the following example, we'll read a raster into an array, slice it, assess it, and save it.

1. First, let's import the raster and convert it to a NumPy array:

```
import arcpy
file_path = r'USGS_13_n38w123_20210301.tif'
raster_obj = arcpy.Raster(file_path)
nparr_raster = arcpy.RasterToNumPyArray(raster_obj)
print(nparr_raster.shape, nparr_raster.size)
```

   Running the cell allows us to see its shape and size:

```
(10812, 10812) 116899344
```

2. Next, we'll select a subset:

```
nparray_subset = nparr_raster[6000:, 6000:]
nparray_subset.shape
```

   Run the cell. You should see:

```
(4812, 4812)
```

   Notice that the selected data is in the lower-right quadrant of the original raster array, which had a shape of (*10812, 10812*). The new array has a shape of (*4812, 4812*) after the slicing operation.

3. After selecting the subset, the resulting array can be converted back into a raster to be viewed. In the next cell, type in:

```
raster_nparray = arcpy.NumPyArrayToRaster(nparray_subset)
raster_nparray
```

This is the resulting raster:

```
In [53]: raster_nparray = arcpy.NumPyArrayToRaster(nparray_subset)
         raster_nparray
```

Out[53]:

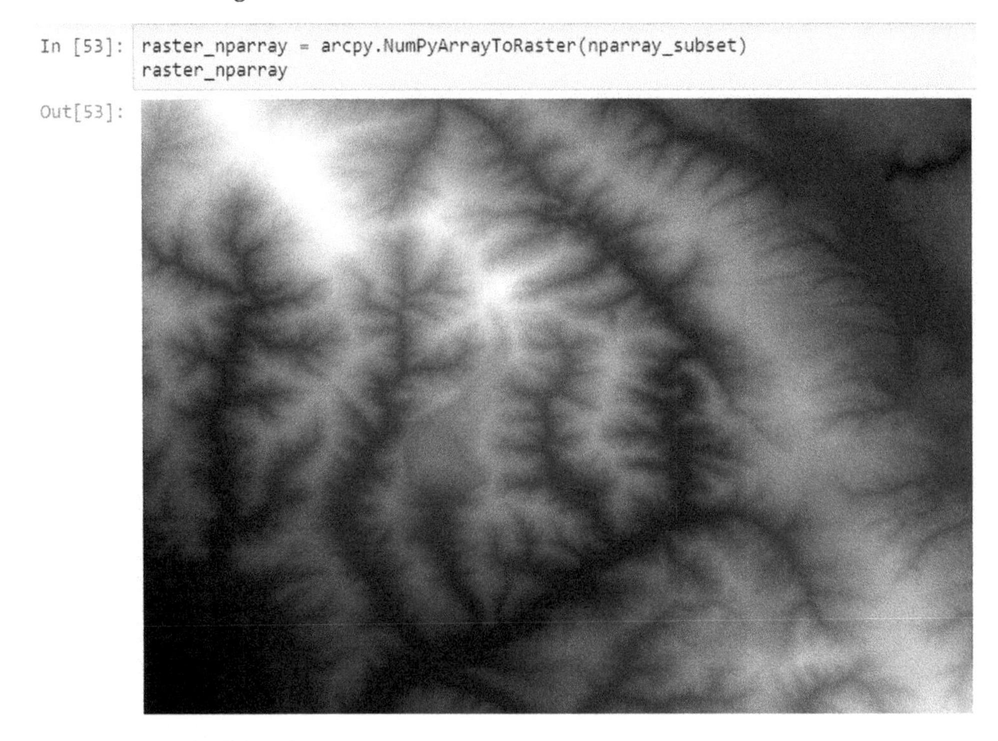

Figure 10.8: The "sliced" raster, representing a 4812x4812 area of the original raster

4. To save the subset of the raster array, use the `save` method. Give the raster output a name as well as an extension to make it a valid file:

   ```
   raster_nparray.save("C:/Projects/subset.tif")
   ```

   Run the cell.

Slicing is very common, especially when you need to clip the raster to process just one part of it. When you need to join the arrays back together, use the `concatenate` tool.

Concatenating arrays

Combining arrays of a similar shape is easy using the `numpy.concatenate` function. The `concatenate` function allows you to adjust the "axis" of the concatenation, meaning the method of joining the two arrays. The default axis (0) allows you to maintain the shape of the columns while adding the second array as new rows. Let's look at an example:

```
onelist = [[2,4,6,8],[1,4,6,7]]
nparray = np.array(onelist)
twolist = [[3,5,1,9],[2,3,8,5]]
nparray2 = np.array(twolist)

np.concatenate((nparray, nparray2), axis=0)
```

With an axis of `0`, this is the result of the concatenation:

```
array([[2, 4, 6, 8],
       [1, 4, 6, 7],
       [3, 5, 1, 9],
       [2, 3, 8, 5]])
```

Using an axis of `1` will increase the number of columns instead:

```
np.concatenate((nparray, nparray2),axis=1)
```

This is the new result:

```
array([[2, 4, 6, 8, 3, 5, 1, 9],
       [1, 4, 6, 7, 2, 3, 8, 5]])
```

Let's explore this concept with raster data.

1. First, enter the following code to create the subset array from the elevation raster into a new cell. This subset array starts at the 6,000th row and the 6,000th column of the raster array:

   ```
   file_path = 'USGS_13_n38w123_20210301.tif'
   raster_obj = arcpy.Raster(file_path)
   nparr_raster = arcpy.RasterToNumPyArray(raster_obj)
   nparray_subset = nparr_raster[6000:, 6000:]
   ```

   Run the cell.

2. Next, create a second subset array:

   ```
   nparray_subset2 = nparr_raster[:6000, 6000:]
   raster_nparray2 = arcpy.NumPyArrayToRaster(nparray_subset2)
   ```

   Run the cell. The second subset array starts at row 0 (meaning the first row of the array) and goes until row 5,999, and includes the same columns of those rows (from column 6,000 until the last column).

3. Now, we can concatenate these columns. Note that the second subset is added first to the parameters, as it's "above" the first subset:

```
joined_array_axis0 =
np.concatenate([nparray_subset2,nparray_subset],axis=0)
raster_joined_axis0 = arcpy.NumPyArrayToRaster(joined_array_axis0)
raster_joined_axis0
```

Run the cell. Here is the result:

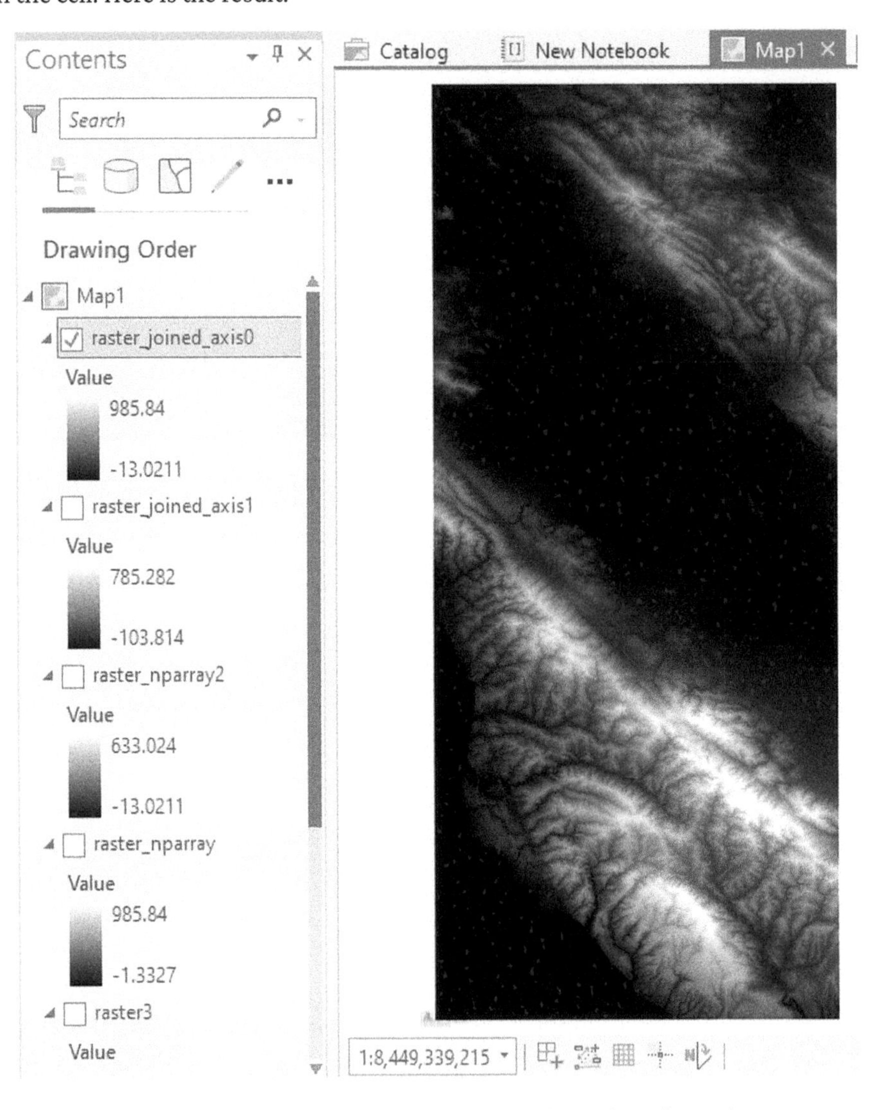

Figure 10.9: The concatenation using axis 0 joins along the y axis

The resulting array is 12,000 rows of data, each with 6,000 columns. This produces an array that stretches along the *y* axis, meaning that the columns were concatenated.

Similarly, using an axis of 1 would allow you to concatenate rows instead of columns.

4. Create a new subset using a different slice:

```
nparray_subset3 = nparr_raster[:4000, 2000:6000]
raster3 = arcpy.NumPyArrayToRaster(nparray_subset3)
```

Run the cell. In this slice, the rows go from 0 to 3,999 and the columns start at 2,000 and go to 5,999.

5. In the next slice, the columns will go from 6,000 until the last column:

```
nparray_subset4 = nparr_raster[:4000, 6000:]
raster4 = arcpy.NumPyArrayToRaster(nparray_subset4)
```

Run the cell.

6. To concatenate these two subsets, use axis 1:

```
joined_array2 = np.concatenate([nparray_subset3,
                                nparray_subset4],axis=1)
raster_joined_axis1 = arcpy.NumPyArrayToRaster(joined_array2)
```

Run the cell. This is the result:

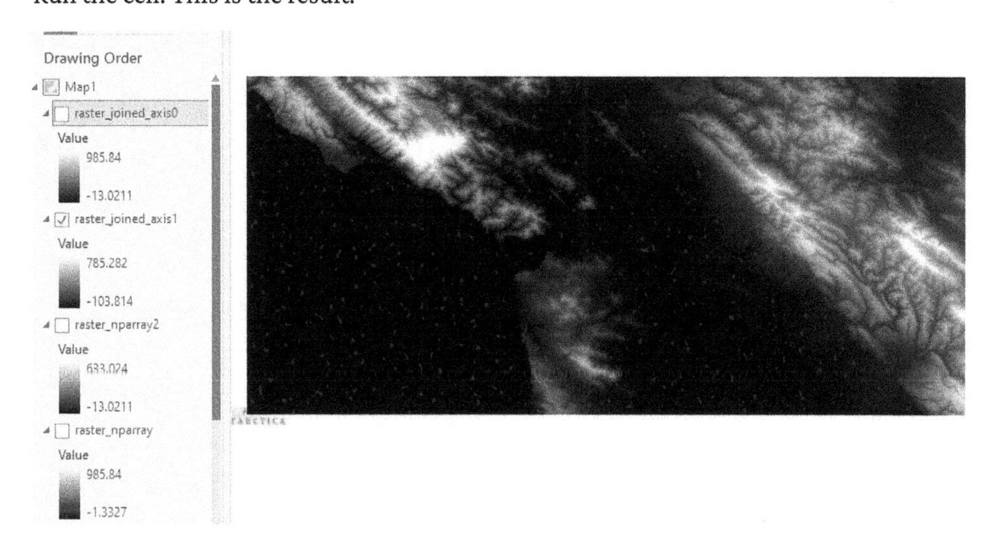

Figure 10.10: The concatenation using axis 1 joins along the x axis

The concatenation allows the array to be "mended" after splitting, and should serve as useful for combining arrays when processing data in big data pipelines.

Creating a raster from a NumPy array

Once the data has been processed, use the `NumPyArrayToRaster` function to create a new raster object, which can then be saved to your hard drive. Make sure to adjust the output file path to match your desired folder:

```
raster_nparray = arcpy.NumPyArrayToRaster(nparray_subset)
raster_nparray.save("subset.tif")
```

This ability to go from raster to raster object to array and back will make it possible to create custom tools that take advantage of the speed increases available when using NumPy for processing geospatial data.

Mathematical operations with NumPy

The structure of NumPy arrays allows for unique mathematical operations. For instance, multiplication, addition, or subtraction operations on an array can be performed using another array, or even a constant value that will add or subtract or multiply each element of an array by that same value.

You may need to create DEMs with values multiplied so that the resulting Hillshades are more extreme, or use raster math to add two rasters together. With NumPy, it is as simple as a few lines of code in a Notebook or script, as shown below. An example with addition will help explain it.

1. First, create an array with a shape of *(3,3)* using `numpy.arange` and `reshape`:

   ```
   arr1 = np.arange(9.0).reshape((3,3))
   arr1
   ```

   Run the cell to view the array.

2. To it, you will add a second array, with a shape of *(1,3)*, which means it has 1 row and 3 columns in that row. Create the second array now:

   ```
   arr2 = np.arange(3.0)
   arr2
   ```

   Run the cell to view the array.

3. Now add the two arrays together:

```
numpy.add(arr1, arr2)
```

Run the cell. The result of the addition operation is that the first column of the original array is increased by the value in the first column of the second array; the second column is increased by the value in the second column of the second array; finally, the third column is increased by the third element in the second array. Below is the output for all three cells:

```
In [39]: arr1 = np.arange(9.0).reshape((3, 3))
         arr1

Out[39]: array([[0., 1., 2.],
                [3., 4., 5.],
                [6., 7., 8.]])

In [40]: arr2 = np.arange(3.0)
         arr2

Out[40]: array([0., 1., 2.])

In [41]: numpy.add(arr1,arr2)

Out[41]: array([[ 0.,  2.,  4.],
                [ 3.,  5.,  7.],
                [ 6.,  8., 10.]])
```

Figure 10.11: Adding two arrays together

4. Another addition operation would be to add one value to every element in the array. This is done by using the `numpy.add` function and supplying a value as the second parameter, instead of another array:

```
numpy.add(arr1, 5)
```

Run the cell. You should see the following output, with all elements of `arr1` (defined in *step 1*) increased by 5:

```
In [54]: numpy.add(arr1,5)

Out[54]: array([[ 5.,  6.,  7.],
                [ 8.,  9., 10.],
                [11., 12., 13.]])
```

Figure 10.12: Adding 5 to an array

5. Another way to perform an addition operation (between arrays or between arrays and a single value) is to use this intuitive shorthand:

```
arr1 + 5
```

Run the cell. You should see the same output as that of the previous cell:

```
In [62]: arr1 + 5

Out[62]: array([[ 5.,  6.,  7.],
                [ 8.,  9., 10.],
                [11., 12., 13.]])
```

Figure 10.13: Adding 5 to an array – the shorthand way

6. Similarly, the `numpy.subtract` function allows for subtraction operations. If an array is supplied as a second array, it will subtract from each element according to the matching column index. If the second parameter is an integer or float, that value will be subtracted from each element in the array. Type the following in the next cell:

```
numpy.subtract(arr1,5)
```

Run the cell. This is the result:

```
array([[-5., -4., -3.],
       [-2., -1.,  0.],
       [ 1.,  2.,  3.]])
```

7. If you create a new array with the same number of columns and use it to subtract from the original array, each column will be subtracted by the value in that column of the second array:

```
arr4 = np.arange(1,4.0)
numpy.subtract(arr1,arr4)
```

Run the cell. You should see:

```
array([[-1., -1., -1.],
       [ 2.,  2.,  2.],
       [ 5.,  5.,  5.]])
```

8. As with addition, a shorthand that can be used is to just use the normal subtraction symbol directly:

```
arr1 - arr4
```

Run the cell. You should see:

```
array([[-1., -1., -1.],
       [ 2.,  2.,  2.],
       [ 5.,  5.,  5.]])
```

To perform multiplication using NumPy arrays, the `numpy.multiply` function is used. Much like the add and subtract operations, you can perform a multiplication of two arrays (or more) or an array and a value.

In these examples, the multiplication operation operates on every element in the first array. The second parameter can be either another array or a value.

1. First, view the initial arrays:

```
arr1, arr4
```

Running the cell gives us:

```
(array([[0., 1., 2.],
        [3., 4., 5.],
        [6., 7., 8.]]),
 array([1., 2., 3.]))
```

2. Let's see a multiplication operation between the arrays:

```
np.multiply(arr1, arr4)
```

Run the cell. We get:

```
array([[ 0.,  2.,  6.],
       [ 3.,  8., 15.],
       [ 6., 14., 24.]])
```

3. Now, the same multiplication operation between the arrays using shorthand:

```
arr1 * arr4
```

Run the cell. We see:

```
array([[ 0.,  2.,  6.],
       [ 3.,  8., 15.],
       [ 6., 14., 24.]])
```

A similar operation can be used to raise array elements to the power of the array or value provided.

1. In the next cell, raise all elements to the power of 2:

   ```
   arr1 ** 2
   ```

   Run the cell. Your output should be:

   ```
   array([[ 0.,  1.,  4.],
          [ 9., 16., 25.],
          [36., 49., 64.]])
   ```

2. Now, raise all elements of the same array to the power of 3:

   ```
   arr1 ** 3
   ```

   Run the cell. You should see:

   ```
   array([[  0.,   1.,   8.],
          [ 27.,  64., 125.],
          [216., 343., 512.]])
   ```

3. Using the two arrays, you can raise one to the power of the other:

   ```
   arr1 ** arr4
   ```

   Run the cell. You should see:

   ```
   array([[  0.,   1.,   8.],
          [  3.,  16., 125.],
          [  6.,  49., 512.]])
   ```

Other available operations include `numpy.sqrt`, which can be used to get the square root of all elements; `numpy.sin`, `numpy.cos`, or `numpy.tan`, which will get the sine, cosine, or tangent respectively; and `numpy.min` and `numpy.max`, which get the minimum or maximum element respectively.

Read more about NumPy mathematical operations here: `https://numpy.org/doc/stable/reference/routines.math.html`.

Array queries

Arrays can be queried to find specific elements that meet a condition. These queries can be performed in a few different ways: using a built-in numpy.where function or using a shorthand operation.

The numpy.where tool uses a condition to process an array. For example, let's apply it to a simple array of numbers from 0 to 9:

```
import numpy as np
arr1 = np.arange(10)
np.where(arr1 < 3, arr1, -1)
```

This is the result:

```
array([ 0.,  1.,  2., -1., -1., -1., -1., -1., -1., -1.])
```

This function will evaluate a condition (the first parameter) upon the array (the second parameter) and replace all elements that meet the condition with the third parameter. You can see that all the elements greater than or equal to 3 have been replaced with -1 values.

Check out more details on the function here: https://numpy.org/doc/stable/reference/generated/numpy.where.html.

The shorthand operation, on the other hand, uses square brackets to contain a conditional. If the shorthand conditional operation is assigned to a variable, it will produce a new array containing the elements of the original array that meet the condition. Otherwise, the operation can instead be used to populate the array with a new value that replaces all elements that meet the operation.

Consider this example below. The array has a conditional statement passed to it using square brackets, and those elements that meet the condition are replaced with the None value used by NumPy (which is known as numpy.nan):

```
arr1[arr1 < 5]
```

The result is an array containing only those elements of arr1 less than 5:

```
array([0., 1., 2., 3., 4.])
```

For instance, in the DEM file we have been using throughout this chapter, there are cells that have a value below 0, meaning they are below sea level. If we wanted to change the value of those elements below sea level to a NoData or None value, it's very easy using NumPy:

```
import arcpy
file_path = r'USGS_13_n38w123_20210301.tif'
raster_obj = arcpy.Raster(file_path)
nparr_raster2 = arcpy.RasterToNumPyArray(raster_obj)
nparr_raster2[nparr_raster2 < 0] = None
nparr_raster2
```

The code produces the following output:

```
In [104]: nparr_raster2 = arcpy.RasterToNumPyArray(raster_obj)
          nparr_raster2[nparr_raster2<0] = None

In [105]: nparr_raster2

Out[105]: array([[85.57855 , 85.28104 , 85.32059 , ..., 32.745525, 32.705288,
                  32.64023 ],
                 [89.23881 , 88.55752 , 88.13626 , ..., 32.84825 , 32.886658,
                  32.93122 ],
                 [92.71537 , 92.1762  , 91.481804, ..., 32.800846, 32.930138,
                  33.0992  ],
                 ...,
                 [      nan,       nan,       nan, ..., 78.96925 , 79.099106,
                  79.64785 ],
                 [      nan,       nan,       nan, ..., 83.93973 , 83.80033 ,
                  84.41281 ],
                 [      nan,       nan,       nan, ..., 89.324646, 89.06261 ,
                  89.00502 ]], dtype=float32)
```

Figure 10.14: Raster with elements < 0 changed to nan

Another useful operation would be to select specific elements and perform a multiplication operation. In the next example, the elements above 30 are multiplied by 3:

```
nparr_raster = arcpy.RasterToNumPyArray(raster_obj)
nparr_raster[nparr_raster > 30] * 3
```

Below is the expected output:

```
In [102]: nparr_raster = arcpy.RasterToNumPyArray(raster_obj)
          nparr_raster[nparr_raster>30] * 3

Out[102]: array([256.73566, 255.84311, 255.96176, ..., 267.97394, 267.1878 ,
                 267.01508], dtype=float32)
```

Figure 10.15: Raster with elements above 30 multiplied by 3

Note that the shape of the array remains the same; calling `nparr_raster.shape` gives us `(10812, 10812)`.

Exercise: Statistical analysis of raster data using NumPy

In the *Chapter 10* folder of the GitHub repo, you will find a set of rasters that represent pollution over New York City. This data covers 10 years of annual average pollution for a variety of pollution types. You will use the Nitrous Oxide files for this section. The files go from 2009 (`"aa1_no300m"`) to 2018 (`"aa10_no300m"`) and are at a resolution of 300 meters.

You'll use them to explore the statistical methods available using NumPy, including mean, median, and standard deviation. You'll also create histograms and charts depicting the reduction in pollution data over the 10-year monitoring period.

The data was downloaded from this dataset: `https://catalog.data.gov/dataset/nyccas-air-pollution-rasters`.

1. To start, create a new cell in your Notebook and make sure you have the filepath for the raster pollution data for 2009. You'll need to convert the raster data into a NumPy array:

```
import arcpy
file_path = r'AnnAvg1_10_300mRaster\aa1_no300m'
arcpy.Raster(file_path)
```

Run the cell. You should see the following:

```
In [131]: file_path01 = r'AnnAvg1_10_300mRaster\aa1_no300m'
          arcpy.Raster(file_path01)

Out[131]:
```

Figure 10.16: The raster object shows the pollution throughout New York City for 2009

2. Next, you will use the `RasterToNumPyArray` tool to create an array:

```
raster_adf = arcpy.Raster(file_path)
 air_array = arcpy.RasterToNumPyArray(raster_adf)
air_array
```

This is the result:

```
array([[-3.4028231e+38, -3.4028231e+38, -3.4028231e+38, ...,
        -3.4028231e+38, -3.4028231e+38, -3.4028231e+38],
       [-3.4028231e+38, -3.4028231e+38, -3.4028231e+38, ...,
        -3.4028231e+38, -3.4028231e+38, -3.4028231e+38],
       [-3.4028231e+38, -3.4028231e+38, -3.4028231e+38, ...,
        -3.4028231e+38, -3.4028231e+38, -3.4028231e+38],
       ...,
       [ 1.0976446e+01,  1.0357784e+01,  1.0420952e+01, ...,
        -3.4028231e+38, -3.4028231e+38, -3.4028231e+38],
       [-3.4028231e+38,  1.0228572e+01,  1.0283316e+01, ...,
        -3.4028231e+38, -3.4028231e+38, -3.4028231e+38],
       [-3.4028231e+38, -3.4028231e+38,  1.0180457e+01, ...,
        -3.4028231e+38, -3.4028231e+38, -3.4028231e+38]],
dtype=float32)
```

Now that you have an array, you can use some of the statistical methods built into NumPy to understand the data. To do so, you will need to query the array, and also make sure that the negative values (which represent NoData) are ignored using the NumPy nan value, which is similar to the Python None value. It allows NumPy to ignore the value of that cell when calculating statistics.

3. In this step, you will use the NumPy where function to get only values above 0. As a reminder, the function accepts a condition, the values to apply if the condition is True, and a value to apply if the condition is False. In the next cell, type in:

   ```
   no_nan_array = np.where(air_array > 0, air_array, numpy.nan)
   ```

   Run the cell. In this case, the array values are kept if the condition is True, and NumPy.nan values are used if it is False.

4. Now that the incorrect values are removed, let's explore the average value of the data. To generate the mean value of the nitrous oxide pollutant, you can use the np.nanmean function, which ignores the nan values. In the next cell, type in:

   ```
   np.nanmean(no_nan_array)
   ```

   Run the cell. This is the result for the first year of the study:

   ```
   22.480715
   ```

5. To generate the median value of the nitrous oxide pollutant, we can use the np.nanmedian function. In the next cell, type in:

   ```
   np.nanmedian(no_nan_array)
   ```

   Run the cell. This is the result for the first year of the study:

   ```
   21.17925
   ```

6. To generate the standard deviation of the nitrous oxide pollutant, use the np.nanstd function:

   ```
   np.nanstd(no_nan_array)
   ```

   Run the cell. This is the result for the first year of the study:

   ```
   8.013502
   ```

You will use these basic statistics in a graph, which will show you a comparison between the first year of the study, the 5th year, and the 10th year.

Creating charts from NumPy arrays using Matplotlib

Using the Matplotlib module, which is included in the Python installation and we touched on briefly in *Chapter 9*, you can create charts of your data. It's a convenient and powerful module that can't be captured here fully but deserves more investigation.

Read more about Matplotlib here: `https://matplotlib.org/`.

A histogram will give you a good idea about the distribution of your data in the raster array. In this example, you will use Matplotlib to generate both the histogram and the chart for the histogram, which are separate things.

7. In the next cell, you will call the Matplotlib `pyplot` tool and use it to generate the histogram and chart. Note that the `%matplotlib inline` line will allow the charts to be generated in the Notebook:

```
import numpy as np
import matplotlib.pyplot as plt
# Keep the chart in the Notebook
%matplotlib inline
num_bins = 5
n, bins, patches = plt.hist(no_nan_array, num_bins,
facecolor='blue', alpha=0.5)
plt.show()
```

Run the cell. This will result in the following histogram for the nitrous oxide levels in 2009. The *x* axis represents the cell value, and the *y* axis represents the number of times the cell value occurs:

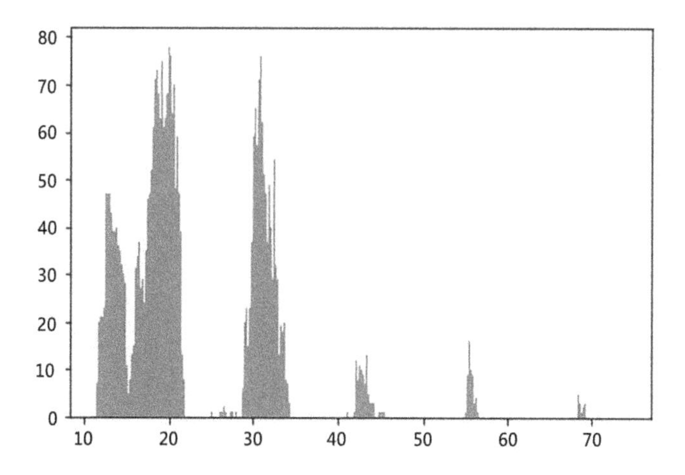

Figure 10.17: Histogram showing nitrous oxide levels in 2009

The parameters passed to the `plt.hist` method include the array, the number of "bins" (meaning how many sections the values are divided into, which in this case is 5), and those controlling the color and opacity of the result. Then, the `plt.show()` method is called to finally generate the chart.

8. For the histograms of the years 2013 and 2018, which can be created in the exact same manner, you can see the reduction in the count for higher-value readings. Rerun the code in the steps above, replacing the raster with the filepath for the raster for 2013 (`AnnAvg1_10_300mRaster\aa5_no300m`) and then the raster for 2018 (`AnnAvg1_10_300mRaster\aa10_no300m`) to produce the mean, median, and standard deviation for those years.

In 2013, there are fewer high values (in other words, values above 50):

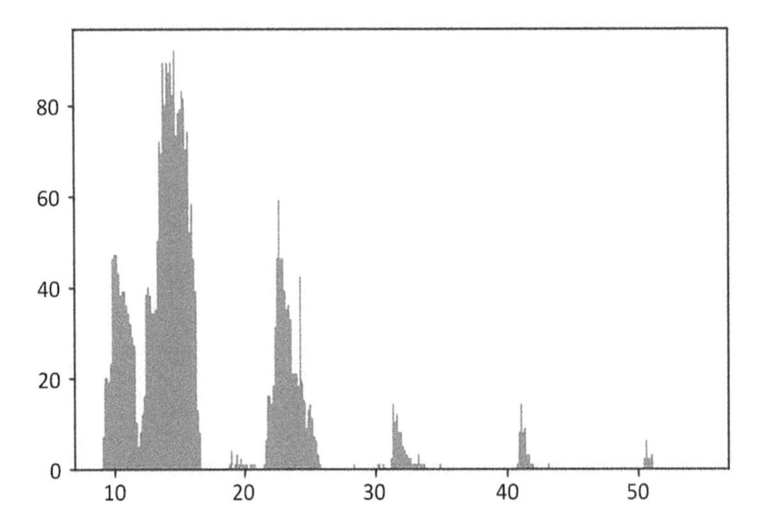

Figure 10.18: Histogram showing nitrous oxide levels in 2013

In 2018, the values have been reduced even further:

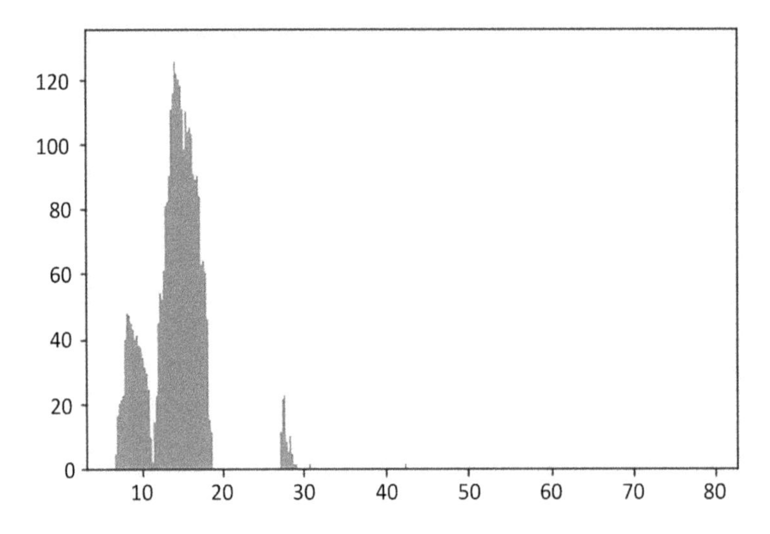

Figure 10.19: Histogram showing nitrous oxide levels in 2018

This means that there has been a reduction in nitrous oxide pollution over the 10-year monitoring period.

To be able to compare the values across the years, you will need to read three different rasters into arrays and generate their statistics. The code below will perform the statistical analysis and generate the bar chart.

9. First, import the necessary modules and make sure that the charts will appear in the Notebook:

   ```
   import arcpy
   import numpy as np
   import matplotlib.pyplot as plt
   %matplotlib inline
   ```

   Run the cell.

10. Then, create NumPy arrays from the rasters, which must first become raster objects:

    ```
    file_path10 = 'AnnAvg1_10_300mRaster\aa10_no300m' # year 10
    file_path05 = 'AnnAvg1_10_300mRaster\aa5_no300m'  # year 5
    file_path01 = 'AnnAvg1_10_300mRaster\aa1_no300m'  # year 1

    arrays =[arcpy.RasterToNumPyArray(arcpy.Raster(file_path01)),
              arcpy.RasterToNumPyArray(arcpy.Raster(file_path05)),
               arcpy.RasterToNumPyArray(arcpy.Raster(file_path10))]
    ```

    Run the cell.

11. Now, query each array to get only values above 0, and pass the resulting array to a new list:

    ```
    nan_arrays = []
    for r_array in arrays:
        nan_arrays.append(np.where(r_array>0,r_array,np.nan))
    ```

    Run the cell.

12. Then, in a new cell, generate the average (mean) value of each raster and store it in a new list:

    ```
    means = []
    for n_array in nan_arrays:
        means.append(np.nanmean(n_array))
    print(means)
    ```

Run the cell and examine the mean value returned.

13. Finally, type the following code in a new cell to create a Matplotlib figure, pass the details about the chart to the figure, and then view the result:

```
# Create the plot
years = ['2009','2013','2018']
plt.xticks(range(len(means)), years)
plt.bar(years, means)
plt.xlabel('Year')
plt.ylabel('Mean NO')
plt.show()
```

Run the cell. You should see the following bar chart, which confirms that the mean pollution value has been decreasing over the 10-year study period:

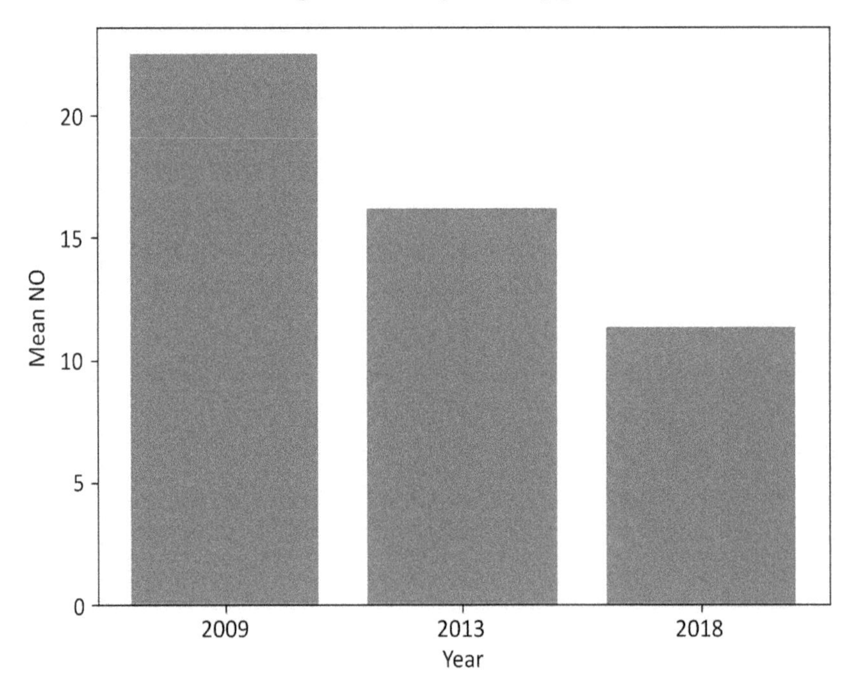

Figure 10.20: Bar chart showing decreasing mean pollution values

To get a more detailed view, you can create another chart that includes both bars and an estimation of the standard deviation.

14. With the means created above, you can skip recreating them. Instead, create a new list of the standard deviations of each year:

```
stdevs = []
for n_array in nan_arrays:
    stdevs.append(np.nanstd(n_array))
```

    Run the cell.

15. Now, use a new cell and create the chart using the `subplots` function. This function is a utility wrapper, which allows you to combine multiple plots together, and to add chart details such as labels, a title, grid ticks, and even bar color and opacity:

```
years = ['2009','2013','2018']
x_pos = np.arange(len(years))
fig, ax = plt.subplots()
ax.set_ylabel('Nitrous Oxide')
ax.set_xlabel('Year')
ax.set_xticks(x_pos)
ax.set_xticklabels(years)
ax.set_title('Monitoring of NO in NYC')
ax.yaxis.grid(True)
```

    Run the cell.

16. Next, you need to pass the mean data list and the standard deviation data list to the `bar` function, along with some parameters that describe the color of the bar and the error bar:

```
ax.bar(x_pos, means, yerr=stdevs, align='center',
alpha=0.5,  ecolor='black', capsize=10)
```

    Run the cell.

17. Finally, call the method to view the chart:

```
plt.show()
```

Run the cell. This is the result:

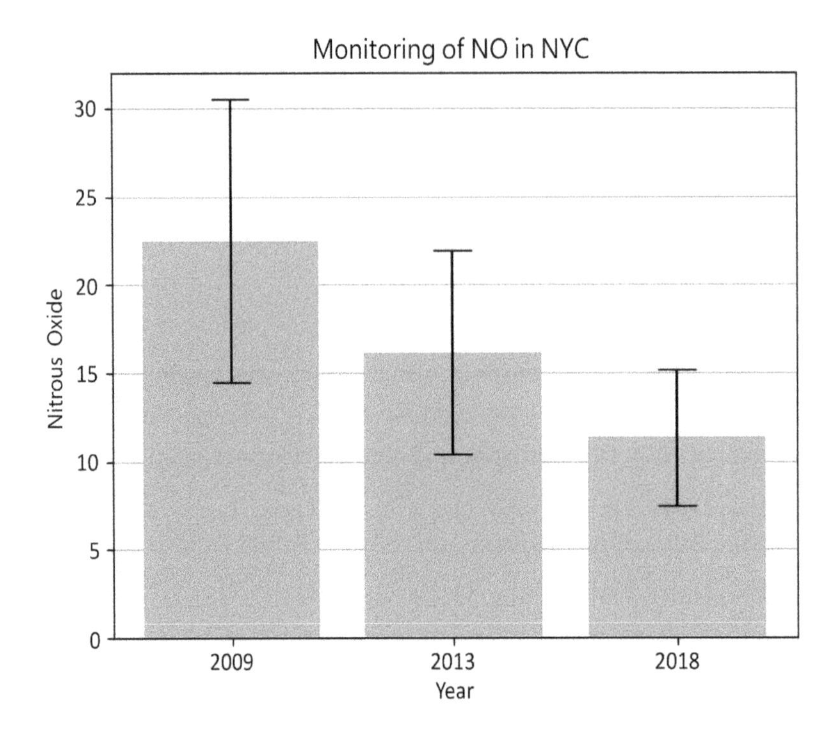

Figure 10.21: A more sophisticated bar chart with error bars

The output of your analysis proves that nitrous oxide has been decreasing in New York City over the decade of observation included in the raster set. You can explore the statistics of the other years included, as well as the other pollutants studied, by repeating the above analysis using those files. For instance, can you determine the rate of reduction of particulate matter (the PM300 files) over the 10-year period?

Using NumPy and Matplotlib, you can perform analysis and generate data visualizations quickly. There are other data visualization modules for Python that work well with Notebooks, including Seaborn, but Matplotlib is included with the installation of ArcGIS Pro, making it an easy choice for charting.

Summary

Using NumPy to process rasters (or vector data) can offer a unique way to create custom functions or complete custom tools. The ability to process *n*-dimensional arrays quickly makes NumPy a powerful tool for fast mathematical and statistical operations.

In this chapter, we reviewed many different functions NumPy has, including viewing and changing the properties of arrays and the mathematical operations that can be performed on arrays. Queries on arrays and converting rasters into arrays and back again were also covered. We explained the concatenation of arrays, and wrapped up by demonstrating how to generate charts from statistics using Matplotlib in an end-to-end exercise.

*

Up to now, you have learned how to use ArcPy, ArcGIS API for Python, Pandas, Spatially Enabled DataFrames, and NumPy to automate much of your analysis, data management, and map production. The next three chapters will be different, as they will be **case studies**. In each chapter, you will see how to apply what the previous chapters have taught you to real-world problems you may come across:

- *Chapter 11* will show you how to create Notebooks to manage ArcGIS Online administrator tasks from within ArcGIS Pro. This will highlight ways to make it more efficient for you to work on project tasks while still administering your ArcGIS Online account.
- In *Chapter 12*, you will learn how to set up a map layout for creating a map series. Then, you will create a map series to explore the environmental justice issues associated with the removal of a bus route.
- *Chapter 13* will show you an end-to-end crop yield prediction process, from gathering and processing data, to creating a random forest classifier, to uploading a layer to ArcGIS Online for use in a custom web map. We will go beyond Python at the very end and introduce you to the ArcGIS API for JavaScript, which is used in web mapping.

When you are finished with all three case studies, you will have broadened your knowledge of how to use Python in ArcGIS Pro, and seen how to solve and automate real-world problems. In addition, you will have Notebooks that you can modify to fit the problems you find in your job every day.

Part IV

Case Studies

11

Case Study: ArcGIS Online Administration and Data Management

Administrating your ArcGIS Online account and managing all of the users and data is an important yet difficult aspect for GIS professionals. You can make this process easier by creating Notebooks to help you manage users, create reports on credit usage, reassign items to a new user, and download attachments. These tools will allow you to be more efficient in switching between your project tasks and administering your organization's ArcGIS Online account.

This chapter will cover a few case studies with Notebooks that you can create to assist you in administering your ArcGIS Online account from within ArcGIS Pro. The benefit of being able to manage your ArcGIS Online account from within ArcGIS Pro is that you can automate tasks and you do not have to switch to the ArcGIS Online platform; everything can be completed within ArcGIS Pro.

This chapter contains the following case studies:

- **Administering your ArcGIS Online account**: Adding users, managing licenses and ArcGIS Online credits, creating item usage reports, and reassigning user data
- **Downloading photo attachments** from ArcGIS Online feature layers

To complete the exercises in this chapter, please download and unzip the `Chapter11.zip` folder from the GitHub repository for this book: `https://github.com/PacktPublishing/Python-for-ArcGIS-Pro/tree/main/Chapter11`.

Case study: Administering your ArcGIS Online account

In this case study, you will explore what you can do if you are the administrator. With admin privileges you can add users, move data ownership from one user to another, delete users, and manage ArcGIS Online credits.

As a GIS administrator, you are responsible for ensuring your ArcGIS Online account is current and your users have access and credits. You are also responsible for removing users and transferring data when someone leaves your organization. This is just a portion of your job, though, as you have projects and tasks that need your attention. These projects take place in ArcGIS Pro; you create ArcPy scripts and models in ModelBuilder for analysis tasks, or figures for publication in documents. Having to leave the ArcGIS Pro environment to go to the ArcGIS Online environment and manage your users can take time away from your other tasks. You can create Notebooks to do these same tasks from ArcGIS Online and save time and clicks, since it is easy to update the variables in the Notebooks and run them.

In the first two parts of this case study, you will create a user and manage their licenses and credits. In the third part, you will create a report to identify all the items each user owns. In the final part, you will transfer ownership of data from a user that has left your organization to a new user within your organization.

You must have admin privileges to do any of the following. To add users, you must have the credentials available for the new user.

Creating users

Your company is growing quickly and you have new GIS analysts who will need to be added to your organization's ArcGIS Online. Creating new users is a common task for a GIS administrator. A common way to create a new user is within ArcGIS Online. Once you are signed into your ArcGIS Online account, complete the following steps:

1. Click **Organization** in the top banner.

Figure 11.1: ArcGIS Online banner

2. Click **Members** in the **Organization** banner.

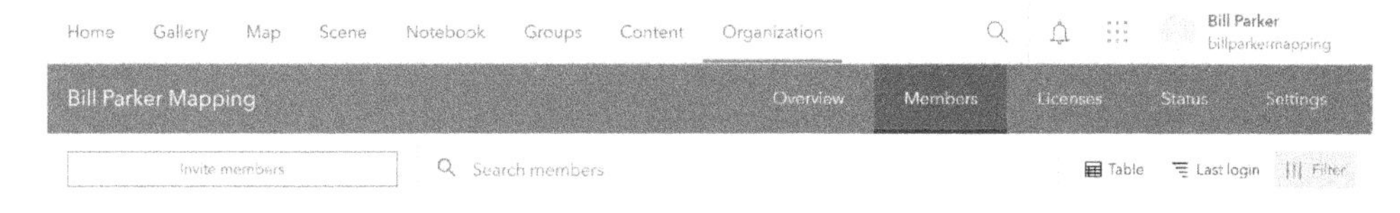

Figure 11.2: ArcGIS Online Organization banner

3. Click the **Invite members** button.
4. In the next window, you have choices to **Add members without sending invitations**, **Add members and notify them via email**, or **Invite members to join using an account of their choice.**

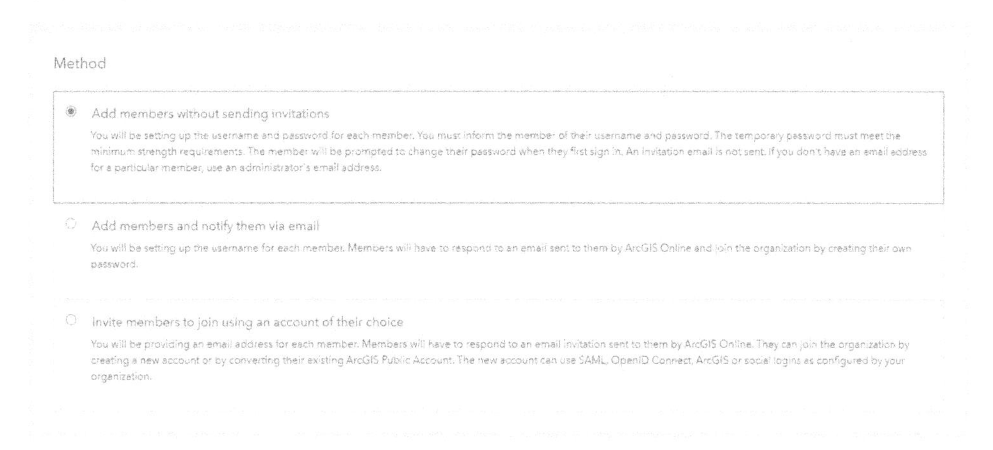

Figure 11.3: Options for adding a member

Select the first option, **Add members without sending invitations**, and click the **Next** button at the bottom of the screen.

5. On the next screen, you have to choose from **New member** or **New members from a file**.

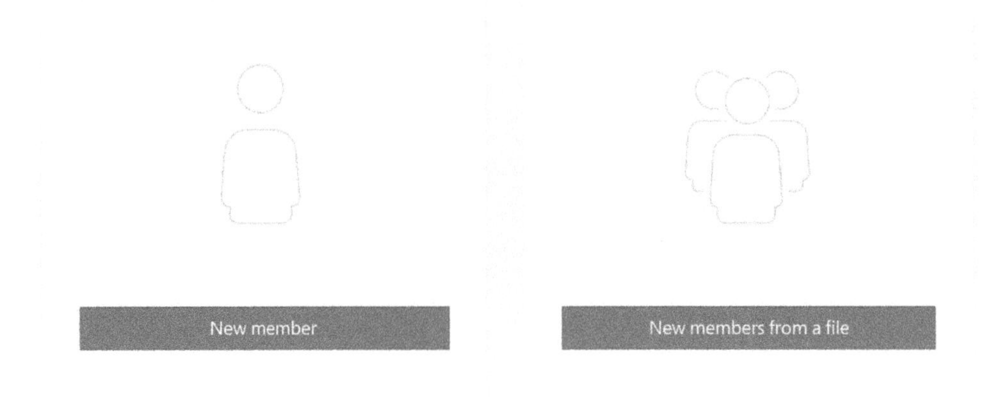

Figure 11.4: New member options

Select **New member**.

6. The next screen will appear, in which you need to fill in all the information for the new user. You will see later how creating a function in a Notebook makes adding all of this information a simple process.

Figure 11.5: New member information

Fill in all the information for the user and click **Next**.

7. The next screen will show the user type, role, and licenses the new user has.

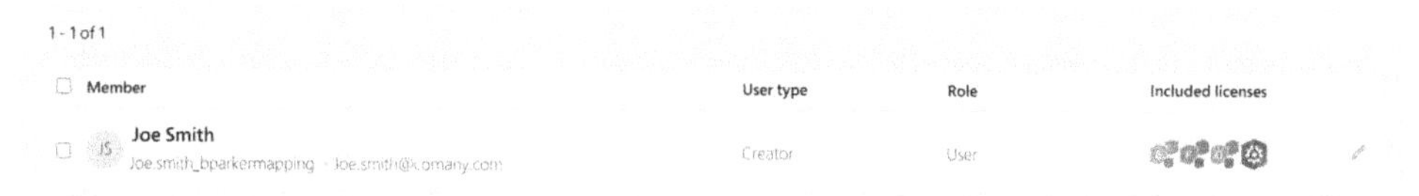

Figure 11.6: New member roles and licenses

After confirming the new member information, click **Next**.

8. On the next screen, you can add on licenses, assign the user to groups, set Esri access, allocate credits, and change user settings.

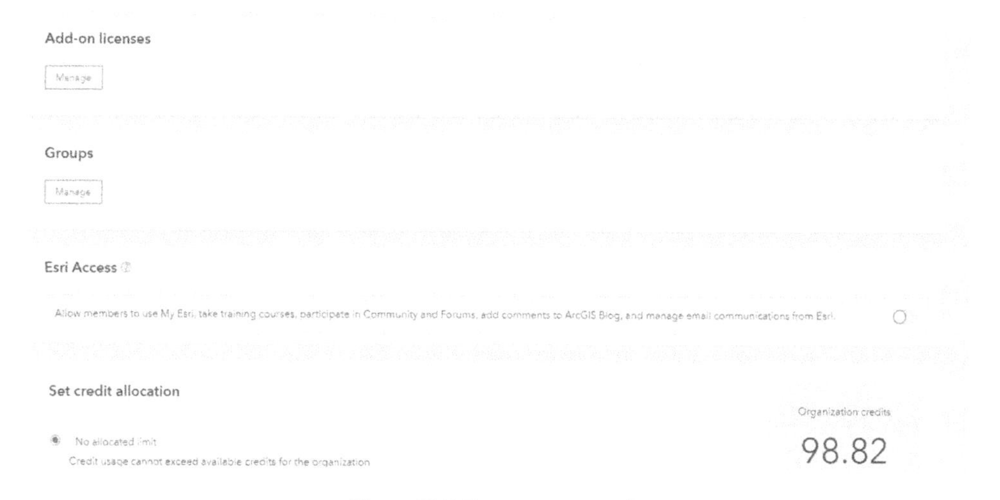

Figure 11.7: New user properties

Leave all as the defaults and click **Next**.

9. This is the final screen, where you can now add the user after reviewing that the information is correct:

Figure 11.8: New user add screen

If you want to add this user, click **Add members**; if not, click **Cancel.**

While you could do this for each new team member, the entire nine-step process can be performed in a single Notebook that can be used for all new team members you need to add. You simply open up the Notebook in ArcGIS Pro, type in the new user information, and run the Notebook to add them and give them access to any existing groups.

To create new users in ArcGIS Online, you use the `create()` operation, which is part of the `UserManager` class. We introduced the `UserManager` class in *Chapter 3* and you used it to search for users. The `create()` operation has the following five mandatory arguments:

- `username`: A 6–24-character-long string that must be unique in your organization.
- `password`: A password string that must be 8 characters long. When creating an ArcGIS Online account this can be left blank, and the user will receive an email that will allow them to set their password.
- `firstname`: A string that specifies the first name of the user.
- `lastname`: A string that specifies the last name of the user.
- `email`: A string that specifies the email of the user.

The `create()` operation also has eight optional arguments: `description`, `thumbnail`, `role`, `provider`, `idp_username`, `level`, `credits`, and `groups`.

You will use the `role` and `provider` arguments the most when creating new users. Visit this webpage: `https://developers.arcgis.com/python/api-reference/arcgis.gis.toc.html#user` and this webpage: `https://doc.arcgis.com/en/arcgis-online/reference/roles.htm` to read more about them.

Let's get started with creating the Notebook.

1. Right-click on the `Chapter11` folder and select **New** > **Notebook**. Rename the Notebook to `AdministeringYourOrg`.
2. You will log in to your organization's GIS through the user you are currently logged in to ArcGIS Pro with. In the next cell, type in the following:

   ```
   from arcgis.gis import GIS
   gis = GIS("home")
   ```

   Run the cell.

3. You are going to create a function to create a new user. This will allow you to create multiple users at a time easily by calling the function and passing the new user information to it. This function will act the same way as the information you would input in *Figure 11.5*, but you don't have to go through multiple pages in ArcGIS Online to get to it. The function is called `new_user`, and takes arguments for `user`, `password`, `firstName`, `lastName`, and `email`, as they are the most common ones you will use. In the next cell, type in the following:

   ```
   def new_user(user,password,firstName,lastName,email):
       new_user1 = gis.users.create(
           username = user,
           password = password,
           firstname = firstName,
           lastname = lastName,
           email = email,
       )
   ```

   Run the cell.

4. Now that you have created the function, call it and pass through to each argument the information to create the new user. This is where you can input all of the information you would have done in *Figure 11.5*. In the next cell, type in the following:

```
new_user("Jane.Smith.Company", "ch@ng3MeA$AP", "Jane", "Smith",
"Jane.Smith@company.com)
```

   Run the cell.

5. You can verify the user was created by searching for their username. Type in the following:

```
user = gis.users.search(query="username:Jane.Smith.Company")[0]
user
```

   Run the cell. You should see output similar to below:

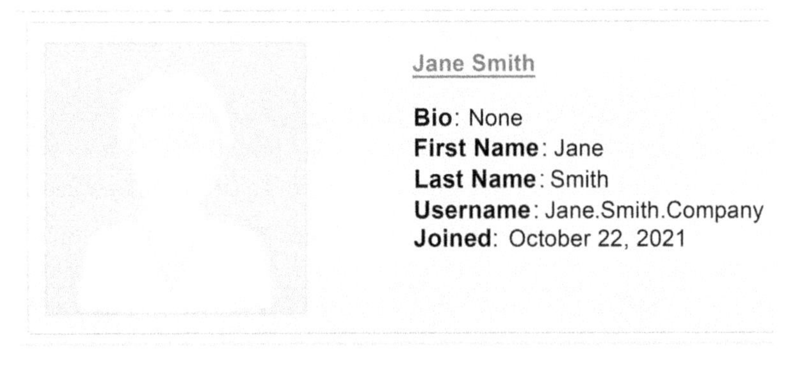

Figure 11.9: New user

In this example, only Jane Smith has joined your company. If you had more than one new hire, you could add each of them by calling the function for each user and passing the specific arguments for them.

Assigning licenses and credits

When you are adding users to ArcGIS Online in ArcGIS Pro, you can also manage user licenses and credits. This will allow you to more efficiently administer your team's ArcGIS Online account while continuing with your project work.

You are first going to see how to check the apps in your organization, and then you will view the licenses in your organization and see how to assign them.

1. Continue working in the `AdministeringYourOrg` Notebook from the previous exercise. First, you will find all the apps licensed in your organization. You will create a variable to hold the list of apps, and then a loop to print out the apps, each on a single line. Type in the following:

```
license = gis.admin.license.all()
for l in license:
    print(l)
```

2. Run the cell. You will see the apps your organization has licensed. The **Out** cell will look similar to the following:

```
<ArcGIS Insights License at https://www.arcgis.com/sharing/rest/>
<ArcGIS Pro License at https://www.arcgis.com/sharing/rest/>
<ArcGIS GeoPlanner License at https://www.arcgis.com/sharing/rest/>
<ArcGIS Business Analyst Web and Mobile Apps License at https://www.
arcgis.com/sharing/rest/>
```

3. In the next cell, you will use the `get()` method to get the license to your ArcGIS Pro app, so you can see what individual license you have. Type in the following:

```
proLic = gis.admin.license.get("ArcGIS Pro")
```

   Run the cell.

4. Now that you have the Pro app license saved in a variable, you can see what extensions you have licensed and if any are assigned. You are going to use the `report` object to get a table showing you what licenses you have, how many are assigned, and how many are remaining. Type in the following:

```
proLic.report
```

Run the cell. You should see a table similar to the following, just with the licenses in your organization:

```
In [5]:  1  proLic.report
```

Out[5]:

	Entitlement	Total	Assigned	Remaining
0	3DAnalystN	1	0	1
1	dataInteropN	1	0	1
2	dataReviewerN	1	0	1
3	geostatAnalystN	1	0	1
4	imageAnalystN	1	0	1
5	locateXTN	1	0	1
6	networkAnalystN	1	0	1
7	publisherN	1	0	1
8	spatialAnalystN	1	1	0
9	workflowMgrN	1	0	1

Figure 11.10: ArcGIS Pro licenses in your organization

5. Now that you know what licenses you have available, you can assign one. To do this, use the `assign()` method, specifying the `username` and the extension as the `entitlements` argument. The entitlements available to your organization are in the **Out** cell you just printed. Type the following to assign yourself a Spatial Analyst license:

```
proLic.assign(username = gis.users.me.username,
entitlements="spatialAnalystN")
```

You can also type in the username of the user you want to assign an entitlement to, surrounded by single or double quotes. In the code above, Bill would type in `'billparkermapping'` in place of `gis.users.me.username`.

Run the cell. If assigned, you should see `True` in the **Out** cell. You can run just this cell of the Notebook when you need to assign someone a license, instead of having to change over to ArcGIS Online and go into their profile.

6. You can check what extension a user has assigned with the `user_entitlement()` method, passing in the username as the argument. Check what extensions you have assigned to you by typing in the following, putting in your username between the quotes:

   ```
   proLic.user_entitlement("{yourUserName}")
   ```

   Run the cell. The results are returned in a dictionary with the following key/value pairs: `username`, `lastlogin`, `disconnected`, `entitlements`. The value of the entitlements is a list of your entitlements currently checked out. The **Out** cell will look similar to the following:

   ```
   {'username': 'billparkermapping', 'lastLogin': -1, 'disconnected':
   False, 'entitlements': ['spatialAnalystN']}
   ```

7. To revoke a license, you use the `revoke()` method on the license object. The arguments are the same as for `assign()`, a username and the extensions to be revoked. You can use the * wildcard to revoke all licenses. Revoke all the licenses assigned to you by typing in the following:

   ```
   proLic.revoke(username = gis.users.me.username, entitlements="*")
   ```

   Run the cell. If the license was revoked, you should see `True` in the **Out** cell. Just like when you need to assign a license, you can revoke one by running this cell. There is no need to switch to ArcGIS Online. You can open the Notebook, run the cell, and get back to your project.

8. To verify you have no license, use the `user_entitlement` method again:

   ```
   proLic.user_entitlement("{yourUserName}")
   ```

   Run the cell. The result in the **Out** cell should be `{}`, which is an empty dictionary.

9. You can use the `credits` property to view the number of credits available in your organization. Type in the following:

   ```
   gis.admin.credits.credits
   ```

   Run the cell. The **Out** cell will display the number of credits in your organization.

10. You can manage the credits available to each user by using credit budgeting. Turn on credit budgeting by using the `enable()` method on the `credits` property:

    ```
    gis.admin.credits.enable()
    ```

Run the cell. Your result in the **Out** cell should be `True`.

11. Now you can allocate credits to a user by using the `allocate()` method. The `allocate()` method takes the username and number of credits as its arguments. Allocate 10 credits to yourself by typing in the following:

```
gis.admin.credits.allocate(username=gis.users.me.username, credits=10)
```

Run the cell. Your result in the **Out** cell should be `True`.

You can also remove credits from a user by using the `deallocate()` method, which removes all the credits from a user. If you want to do this, type in the following: `gis.admin.credits.deallocate(username='{username}')`.

12. When credit budgeting is enabled, you can check the available credits for each user, as they get an `assignedCredits` and `availableCredits` property. Check the credits you have available by typing in the following:

```
gis.users.me.availableCredits
```

Run the cell. The result will be the number of credits available to you.

If you are using a single user account, you cannot assign credits to yourself, as you have access to all the credits.

You have created a time-saving Notebook that will allow you complete your administrator tasks with little interruption from your project tasks. You should store this Notebook in a folder that you have easy access to; open it in ArcGIS Pro and run the necessary cells, with minimal changes, to make updates to users quickly. This will reduce the amount of time you have to spend switching between platforms.

Creating reports for item usage

Data storage can be a major consumer of credits in your organization's ArcGIS Online account. As an administrator, you have the ability to view credit usage by user and run different reports. These can be accessed in the **Status** tab of the **Organization** tab.

Within the **Dashboard** tab, you can view a breakdown of how credits are being used in your organization over different time periods. The **Reports** tab is where any reports you run are stored.

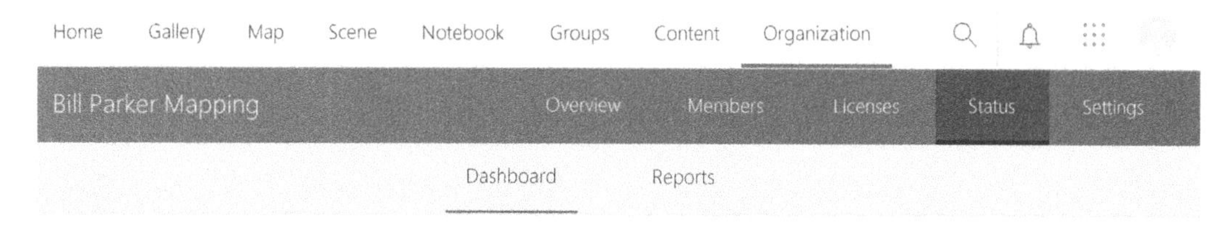

Figure 11.11: Status tab in the Organization tab

The **Credits** tab will show you the number of credits used over a given time period for storage, analytics, subscriber content, or published content. Many of the different charts and elements can be clicked on to get more details about the credit usage. By clicking on the cloud icon with the down arrow as shown in *Figure 11.12*, you can download a report of the credit usage over this time period:

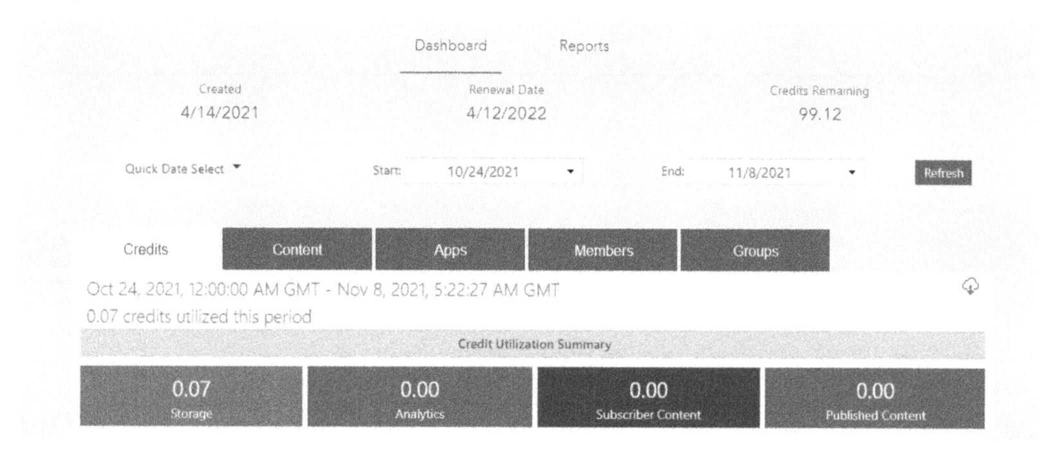

Figure 11.12: Credits tab

The **Content** tab will show you details about the different items you have in your ArcGIS Online account. An **item** is anything you have stored; it can be a CSV, shapefile, geodatabase, feature layer, raster layer, or anything else that ArcGIS Online can store.

Figure 11.13: Content tab

The **Apps** tab will show details about the apps you have created and stored:

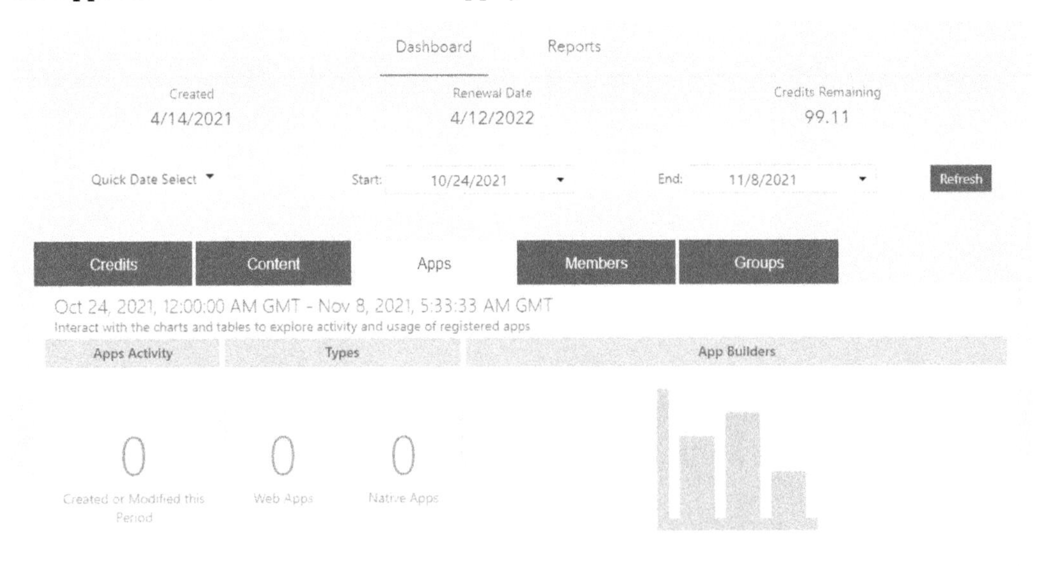

Figure 11.14: Apps tab

The **Members** tab will show you details about the members in your organization. You are only able to view a single member at a time:

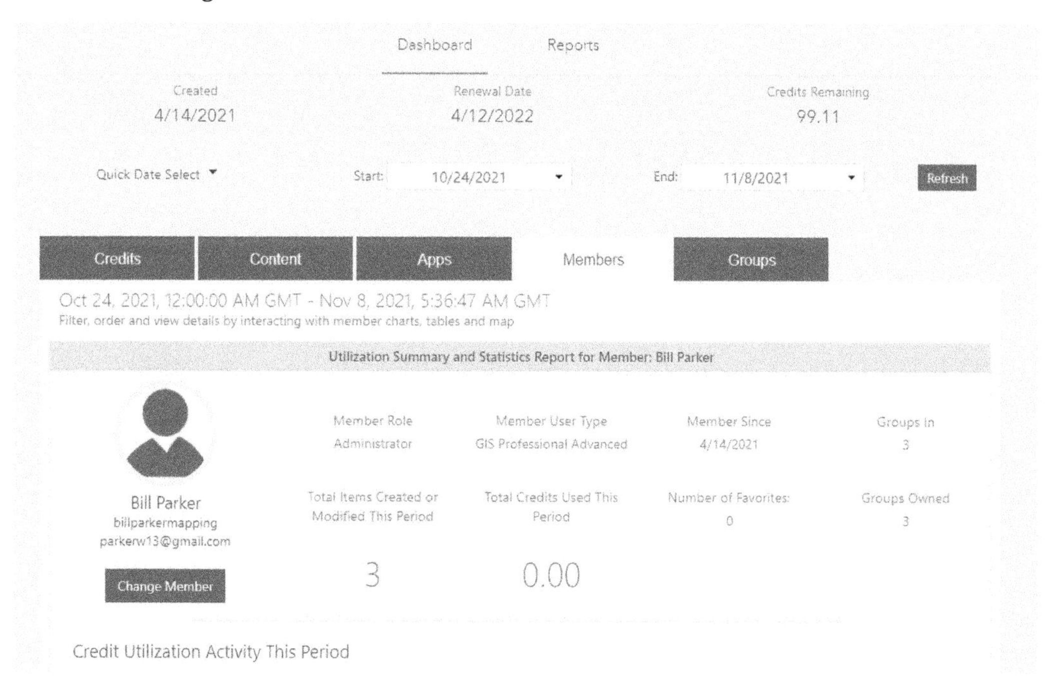

Figure 11.15: Members tab

The **Groups** tab will show details about the groups in your organization.

Figure 11.16: Groups tab

Within every tab, many of the different charts and elements can be clicked on if you want to see more details. Each tab will give you a lot of information about how credits are being used and which members are using credits for different analyses.

A big part of credits usage is often storage, and while the **Credits** tab can show you how many credits you are using for storage along with the size of the items, it can't tell you if the items have been viewed often or recently, which may indicate whether or not they can be removed. You have to go into each item's properties for that information.

Instead, you can leave ArcGIS Online behind and create a Notebook to efficiently identify all the items owned by each user and information about those items' usage. This can help your team to identify old items that can be removed from ArcGIS Online. Doing this will save your organization money, as you won't be storing items that you aren't using anymore.

In the following example, you will create a Notebook to export a CSV all of the items in all the folders owned by a user, along with the usage and last time accessed for each item.

This CSV will allow you to show each user what items they have that are not being used and could be deleted:

1. Right-click on the `Chapter11` folder and select **New** > **Notebook**. Rename the Notebook to `UserCreditsReport`.
2. You are going to import the `time`, `csv`, and `os` modules for use in the Notebook. The `time` module will allow you to convert the datetime value of an item to a month, date, year format. The `csv` module will allow you to export your results to a CSV. The `os` module will allow you to create a full path for the new CSV you are creating in the folder you are storing all the CSVs in. Along with those, you will import the GIS module and log in to your organization's GIS through the user you are currently logged in to ArcGIS Pro with.

   In the first cell, type in the following:

   ```
   import time, csv, os
   from arcgis.gis import GIS
   gis = GIS("home")
   ```

3. In the next cell, you are going to create a function to create a new CSV. This is the same function as you wrote in *Chapter 4* to export bus stop locations; please refer to the *Accessing the geometry of a feature class* section in that chapter for a full explanation of the function. Type in the following:

   ```
   def createCSV(data, csvName, mode = 'w'):
       with open(csvName, mode, newline = '') as csvfile:
           csvwriter = csv.writer(csvfile)
           csvwriter.writerow(data)
   ```

4. In the next cell, you will create a variable for the path of the CSVs you will be creating, as well as a list of the headers for the CSV. The path will be set to the `Chapter11` folder, and you will be creating a CSV for each user based on their username. By creating a CSV for each user, you will make it easy for your users to find all of their data. The headers will be for the item ID, item name, item size in megabytes, the date the item was last modified, the number of views the item has, and the last view date of the item. In the new cell, type in the following:

   ```
   csvPath = r'C:\PythonBook\Chapter11'
   csvHeaders = ['itemId', 'itemName', 'itemSize_MB', 'lastModified',
       'NumberOfViews', 'LastViewDate']
   ```

5. In the next cell, you will create a function that takes a list of items that a user owns and loops through all of them. You will create this list of items from items in the user's root folders and subfolders. For each item, you will find the ID, name, size, last modified date, number of views, and last view date, and write these to a line in the CSV. This information will allow the user to see what items they have that aren't being used and take up a lot of space. By adding the item number, you can later write another Notebook that will take all the item numbers that have been identified for deletion and delete them. **The next 6 steps will create this function and will all be written in the same cell.**

   You will start by declaring the function that takes a list of items as its parameter. In the function, you will loop through each item and extract information about it. You will find the item ID, the item title, the item size, the last modified date, the item type, and the number of views. The item size is reported in bytes, so you will multiply it by 0.000001 to convert it to megabytes. The last modified date returns the time from the last epoch in milliseconds, so you will divide it by 1000 and use the `time` module to convert it to a date and time.

   Refer to *Chapter 3* for more information on how to use the `time` module.

   You will also add `print` statements to print out all of the information you are extracting from the item. In a new cell, type in the following:

```
def userStorage(items):
    for item in items:
        itemId = item.id
        print(itemId)
        itemName = item.title
        print(itemName)
        itemSize = round(item.size*0.000001,2)
        print(itemSize)
        lastMod = time.localtime(int(item.modified)/1000)
        lastModDate = '{0}/{1}/{2}'.format(
            lastMod[1],lastMod[2],lastMod[0]
        )
        print(lastModDate)
        itemType = item.type
```

```
        print(itemType)
        numbViews = item.numViews
        print(numbViews)
```

6. Continuing in the same cell, you will create a variable to hold the usage of the item. The `usage` property has the following two optional arguments:

   - `date_range`: The time period used to query the item for the number of views or downloads. It has the following values: `24H` for past 24 hours, `7D` for past 7 days (default), `14D` for past 14 days, `30D` for past 30 days, `60D` for past 60 days, `6M` for past 6 months, and `1Y` for past year.
   - `as_df`: A Boolean; when set to `True` (default), a pandas DataFrame is returned, and when set to `False` a data dictionary is returned.

   You will set the `date_range` argument to `6M` and `as_df` to `False` to return a data dictionary with the usage data for the item. Each data dictionary returned will look like the code below when there are no views over the selected time period:

```
{'startTime': 1635897600000, 'endTime': 1636588800000, 'period':
'1d', 'data': []}
```

   In the same cell as above, type in the following:

```
        itemUsage = item.usage('6M',False)
```

   For longer time periods, like 6 months or 1 year, `usage` actually returns a list of data dictionaries that looks like the code below:

```
[{'startTime': 1620777600000, 'endTime': 1626048000000, 'period':
'1d', 'data': []}, {'startTime': 1626048000000, 'endTime':
1631232000000, 'period': '1d', 'data': [{'etype': 'svcusg', 'name':
'6c7d00d2be7844f49a75b2780992299d', 'num': [['1626048000000', '0'],
['1626134400000', '0'], ['1626220800000', '0'], ['1626307200000',
'0'], ['1626393600000', '0'], ['1626480000000', '0'],
['1626566400000', '0'], ['1626652800000', '0'], ['1626739200000',
'0'], ['1626825600000', '0'], ['1626912000000', '0'],
['1626998400000', '0'], ['1627084800000', '0'], ['1627171200000',
'0'], ['1627257600000', '0'], ['1627344000000', '0'],
['1627430400000', '0'], ['1627516800000', '0'], ['1627603200000',
'0'], ['1627689600000', '0'], ['1627776000000', '0'],
['1627862400000', '0'], ['1627948800000', '0'], ['1628035200000',
'0'], ['1628121600000', '0'], ['1628208000000', '0'],
```

```
['1628294400000', '0'], ['1628380800000', '0'], ['1628467200000',
'0'], ['1628553600000', '0'], ['1628640000000', '0'],
['1628726400000', '1'], ['1628812800000', '0'], ['1628899200000',
'0'], ['1628985600000', '0'], ['1629072000000', '0'],
['1629158400000', '0'], ['1629244800000', '0'], ['1629331200000',
'0'], ['1629417600000', '0'], ['1629504000000', '0'],
['1629590400000', '0'], ['1629676800000', '0'], ['1629763200000',
'0'], ['1629849600000', '0'], ['1629936000000', '0'],
['1630022400000', '0'], ['1630108800000', '0'], ['1630195200000',
'0'], ['1630281600000', '0'], ['1630368000000', '0'],
['1630454400000', '0'], ['1630540800000', '0'], ['1630627200000',
'0'], ['1630713600000', '0'], ['1630800000000', '0'],
['1630886400000', '0'], ['1630972800000', '0'], ['1631059200000',
'0'], ['1631145600000', '0']]}]}, {'startTime': 1631232000000,
'endTime': 1636416000000, 'period': '1d', 'data': []}, {'startTime':
1636416000000, 'endTime': 1636675200000, 'period': '1d', 'data':
[]}]
```

It is from this list of data dictionaries that you will need to extract the last viewed data of the item. This information is within the `'data'` key of each data dictionary. That `'data'` key has a value that is a list containing a single data dictionary. Within each data dictionary in that list, there is a key called `'num'` that is a list of pairs. The pairs consist of the date in milliseconds from the epoch and the number of views on that date. You can see in the above code that the item was last viewed `1` time on `162872640000`.

This structure and information can be used to loop through all of the data dictionaries and compare the dates that have views to find the most recent view date. You need to start by creating a variable for the view date set to `0` and a loop to loop through each usage data dictionary. In the same cell as above, type in the following:

```
viewDate = 0
for usage in itemUsage:
```

7. Within the loop, you will first use a conditional to determine if the `data` key has any values in its list. If it does, you will create a variable to store the list of date and view pairs in the `'num'` key of the data dictionary. In the same cell as above, type in the following:

```
    if len(usage['data']) > 0:
        listNumViews = usage['data'][0]['num']
```

8. You now have a list that contains lists of pairs consisting of a date and number of views. You will loop through this list and use a conditional to determine if there was a view that day. If there was a view, the date field will be converted from the string it is stored as to an integer, and stored in a variable. This value will be compared to the `viewDate` variable that was originally set at `0`; if it is larger than the current `viewDate`, the `viewDate` will be set to the new view date. This comparison will be made for each view, with the final `viewDate` variable being the largest number, representing the most recent view date. In the same cell as above, type in the following:

```
for views in listNumViews:
    if int(views[1]) > 0:
        print(views)
        newViewDate = int(views[0])
        if newViewDate > viewDate:
            viewDate=newViewDate
```

9. Now that you have the last view date for the item, you can convert it to a month/day/year format using the `time` module to make it easier for each user to see when it was last accessed. You will print the last view date to help track your progress. In the same cell as above, type in the following:

```
lastView = time.localtime(viewDate/1000)
lastViewDate = '{0}/{1}/{2}'.format(
    lastView[1],lastView[2],lastView[0]
)
print(lastViewDate)
```

10. You are now ready to create the list of all the data about the item you are working on and then write that list to the CSV using the `createCSV` function. In the same cell as above, type in the following:

```
    itemList = [itemId, itemName, itemSize, lastModDate,
numbViews, lastViewDate]
    createCSV(itemList, csvFull, mode = 'a')
```

11. In a *new* cell, you will create a list of all the users in your organization and then loop through it. First, you will create a variable for the user's full name and username, and, within the loop, print the full name so you can track which user the function is running.

Then, you will use the `items` property of the `user` object to get a list of all the items the user owns in their root folder, setting the `max_items` argument to a value of more than the total items in your organization. After this, you will create a full path for the CSV for the user with the username, replacing any dots with underscores, and use the `createCSV` function to create the CSV and write the headers.

You must set the `max_items` argument, or you will only get the first 10 items returned (as the default is set to 10).

You will then use the `userStorage` function from the above cell, passing to it the list of items. This will give you all the items in the user's root folder, but not any items in any other folders. To access those items, you must create a variable to hold a list of all the user's folders using the `folders` property, and then loop through each folder and get a list of all the items in the folder using the `items` property again. In the new cell, type in the following:

```
users = gis.users.search()
for user in users:
    print(user)
    userFullName = user.fullName
    userName = user.username
    print(userFullName)
    items = user.items(max_items=10000)
        csvFull = os.path.join(csvPath,'{0}_data.csv'
            .format(userName.replace('.','_')))
        createCSV(csvHeaders, csvFull, mode = 'w')
        userStorage(items)
    print(len(items))
    folders = user.folders
        for folder in folders:
            folder_items = user.items(
                folder=folder, max_items=10000)
            print(len(folder_items))
        userStorage(folder_items)
```

How do you find the total number of items in your organization?

Unfortunately, there is no easy way using the ArcGIS API for Python to find the total number of items in your organization. The quickest way is to go to the Content tab in the Organization dashboard, as seen above in *Figure 11.13*, and scroll down to the bottom. There, you will see the total items in your organization.

If you want to run the code for a single user only, you can update `gis.user.search()` using the search syntax from *Chapter 3*. To search for just Bill's username within his organization, the first row would look like this:

```
users = gis.users.search(query='username:billparkerma
pping')
```

12. You are now ready to run the Notebook; run all the cells. This will create a CSV for each user in your organization containing all of the data they own and information about the size and usage of that data:

	A	B	C	D	E	F
1	itemId	itemName	itemSize_MB	lastModified	NumberOfViews	LastViewDate
2	6c7d00d2be7844f49a75b2780992299d	Farmers Markets	0.01	8/11/2021	1	8/11/2021
3	674b4c2835984d2da04ec22b57a8e116	AdditionalAlamedaFarmersMarket_Test2	0.25	10/27/2021	1	10/27/2021
4	109ea42b13e74e758d92ab65e388e8ed	AdditionalAlamedaFarmersMarket_Test2	0.03	10/27/2021	2	11/9/2021

Figure 11.17: User credit reports CSV

You should also see the data from the `print` statements in the **Out** cell, looking similar to the following:

```
<User username:billparkermapping>
Bill Parker
6c7d00d2be7844f49a75b2780992299d
Farmers Markets
674b4c2835984d2da04ec22b57a8e116
AdditionalAlamedaFarmersMarket_Test2
...
15478346101f43548714e12b69278f17
Alameda County Farmers Market_stakeholder
```

You now have a CSV that tells you all of the items a user owns, along with how large the items are and when they were last viewed. This can be distributed to users to help them identify any data that can be removed from ArcGIS Online. Removing any out-of-date items will help you ensure your users are always using the correct data and will save you credits. You can run this Notebook for all users in your organization, which is more efficient than creating reports for each user from within ArcGIS Online.

Reassigning user data

You will inevitably have users leave your organization, and you will need to delete their user account when they do. Before doing that, you will have to move all of the content they own to a new user, as ArcGIS Online will not let you delete a user that owns items and groups. You can reassign all data and groups from one user to another when calling the `delete` method on a user, but that will just move all of the data into the target owner's root folder. This means the target owner has to move all of that data to folders. Instead, you can reassign the data to an existing folder in the target owner's account or create a new folder for the data in the target owner's account.

In the following case study, your company has lost one of its GIS analysts and you need create two different Notebooks to transfer their items to a new user. The first will reassign items to a new user and create a new folder for those items in the new user's account. The second will reassign items to a new user and place the items into an existing folder in the new user's account.

Transferring data to a different user and creating a new folder

The first example you will work through is transferring data from one user to another and creating a folder of the same name in the new user's account to hold all the data. This will keep the same data organization for the new user, and not clutter their content folder with new data:

1. Right-click on the `Chapter11` folder and select **New** > **Notebook**. Rename the Notebook to `TransferOwnershipCreateFolder`.
2. You are going to log in to your organization's GIS through the user you are currently logged in to ArcGIS Pro with. In the first cell, type in the following:

    ```
    from arcgis.gis import *
    from IPython.display import display
    gis = GIS("home")
    ```

    Run the cell.

3. In the next cell, you will create a variable to hold the username for the old user, a variable to hold the username for the new user, and a variable for the folder name in the old user's account that will be created in the new user's account. The code here has placeholders for the old and new users; replace them with users in your organization. Type in the following:

```
oldUserName = "John.Doe.Company"
newUserName = "Jane.Smith.Company"
folderName = "Folder From Old User"
```

   Run the cell.

4. In the next cell, you will use the variables above to create user objects for the old user and new user. In addition, you will get all of the folders in the old user's account and store them in a variable. Type in the following:

```
oldUser = gis.users.get(oldUserName)
newUser = gis.users.get(newUserName)
folders = oldUser.folders
```

   You have now done all the setup you need to write the code to transfer data from the old user to the new user into a newly created folder with the same name. **The code will be written in one cell, as it is all part of a single for loop. It will be broken up over the next six steps.** It is in this cell where you are really gaining efficiency. To do this within ArcGIS Online, you would have to select each item to transfer its ownership:

5. You will begin by looping though the `folders` variable you created above that holds all of the folders in the old user's account. The `folders` variable is a list of data dictionaries, with a data dictionary for each folder.

   Recall from *Chapter 3* that the data dictionary for a folder contains the `"username"`, `"id"`, and `"title"` keys.

   You will use this to print the title of each folder as you iterate through the folders. Type in the following:

```
for folder in folders:
    print(folder["title"])
```

6. You will use a conditional statement to test the folder title against the title in your `folderName` variable. All of the work to move the data will be done in this conditional and there will be no `else` statement at the end. This is because you are only concerned with a single folder in the list of folders; all the rest can just be skipped over. After the `if` statement, you will create a list of the current folders the new user has, and an empty list that will be used to store the folder names. You will then loop through the list of new user folders and append the title of each folder to the list. You are doing this so you can check later that the new user doesn't already have a folder with the same name. Type in the following in the same cell as above:

```
if folder["title"] == folderName:
    newUserFoldersDict = newUser.folders
    newUserFolders = []
    for f in newUserFoldersDict:
        newUserFolders.append(str(f["title"]))
```

7. Now you have a list of all the new user's folders, you will use a conditional to check if the new user has a folder of the same name. If the folder does not exist, a new folder with that name will be created. If it does exist, the script will print out that the folder already exists and call the `break` keyword, which will stop the loop and end the cell. Type the following in the same cell as above:

```
    if folderName not in newUserFolders:
        gis.content.create_folder(folderName,newUserName)
    else:
        print("{0} folder already exists, you need to use the
              TransferExistingFolder Notebook"
              .format(folderName))
        break
```

Why check if a folder exists?

It is good practice to always check if a folder or directory exists before creating it within the script. Even though you may have looked at the old user's content structure, you could have missed it, and if you did, trying to create a folder that already exists will cause the Notebook to fail. This check keeps it from failing and tells you what your error is.

8. In the same cell, you are going to create empty lists to store different types of ArcGIS Online data. You will create a list to hold all of the **service definition files, replica geodatabases**, and all other items. You need to do this because the replica geodatabases need to be reassigned first and the service definition files last. If you try to reassign a replica geodatabase after the layer it is replicating is reassigned, the script will be unable to find the replica geodatabase, as it was reassigned with the layer it was replicating. The opposite is true of service definition files. They need to be reassigned *after* all other maps and layers; if they are reassigned before a layer they are associated with, the layer will not be found, as it was reassigned with the service definition file.

Replica geodatabases are SQLite geodatabases that ArcGIS Online creates for files that are made available for offline use. You will not see them when looking at your data within ArcGIS Online. You will only see them when listing all the items owned by a user within a folder.

Type in the following in the same cell as above:

```
serviceDef = []
otherItems = []
replgdb = []
```

Why is the above splitting of type necessary? Why doesn't ArcGIS Online know that it moved files together?

When a layer is reassigned and the replica geodatabase is reassigned with it, it is unclear why ArcGIS Online doesn't recognize that both have had their ownership transferred. The above is a workaround to ensure that files are reassigned in the proper order, and was only found through trial and error. If you get an error that an item was not found when it is called to be reassigned, it most likely means that the item's ownership was already reassigned as part of another item that was reassigned. This means you can just run the cell again and that item that was not found when it was to be reassigned will not be in the list of items, as it was already reassigned.

9. Now that you have created the empty lists, you will loop through all the items in the folder and add an item to each list depending on its type. In addition to this, you will delete any **map area** items. A map area is a downloadable area for a map that was created in ArcGIS Online. Ownership of map areas cannot be transferred, and a map that already has map areas created for it cannot have additional areas created by anyone other than the owner of the map.

For more information about offline maps, see the documentation here: `https://doc.arcgis.com/en/arcgis-online/manage-data/take-maps-offline.htm#`.

Because of these rules around map areas, you will have to delete them and allow the new owner to create their own. If you were to do this manually in ArcGIS Online, you would have to go into each map to delete the map areas. Instead, your code will simply identify any map areas in the folder and delete them. Something that would have taken separate steps in ArcGIS Online is now part of your entire process in the Notebook.

Type in the following in the same cell as above:

```
folderItems = oldUser.items(folder=folderName)
for item in folderItems:
    print(item.name)
    print(item.type)
    if item.type == "Map Area":
        print("deleting map area")
        item.delete()
    elif item.type == "Service Definition":
        serviceDef.append(item)
    elif item.type == "SQLite Geodatabase":
        replgdb.append(item)
    else:
        otherItems.append(item)
```

10. You have now deleted all the map areas and separated out the different item types so you can transfer them in the correct order to remove potential errors. To move the items, you will loop through each list of items and move all of the items in that list. Because of the ordering issues discussed in *step 9* above, you will first move the replica geodatabases, then the other items, and finally the service definitions.

You will add `print` statements to track which list you are in and which items are being moved. In the same cell as above, type in the following:

```
print("---Moving replicag gdbs (needed for offline work)---")
for item in replgdb:
    print("Moving {0}".format(item["title"]))
    item.reassign_to(newUserName, target_folder=folderName)
print("---Moving all other non service definitions---")
for item in otherItems:
    print("Moving {0}".format(item["title"]))
    item.reassign_to(newUserName, target_folder=folderName)
print("---Moving service definitions---")
for item in serviceDef:
    if item["title"] not in otherItems["title"]:
        print("Moving {0}".format(item["title"]))
        item.reassign_to(
            newUserName,
            target_folder=folderName
        )
```

11. Your Notebook is now ready to run. Run all the cells. When completed, you will see that the new user has a new folder created and all of the items have been transferred to that folder.

You now have a Notebook that will create a folder and reassign the data from the old user to the new user, placing each item in the newly created folder. It allows you to transfer ownership of projects to different team members and easily transfer the ownership of all the items for that project. This process works well for projects that have one team member who owns all of the data and is working on the project by themselves.

In larger projects, you may have multiple team members assigned to them. In those situations, you may have a standard folder structure that your team is supposed to follow. This would mean that each team member will already have a folder with the same name for the project to store their data. In this case, you would not need to create a new folder for the new user but find the existing folder that already contains their data for the project. In the next section, you will modify the above Notebook to reassign ownership of items from the old user to the new user, in the scenario where the new user has an *existing folder*.

Transferring data to a different user with an existing folder

In the above example, the new user has never worked on this project before and does not have a folder set up to contain any items for it. However, in this example, the new user was already working on the project and has a folder with the same name as the old user. The above Notebook will not work because a user cannot have two folders with the same name. Instead of creating a new folder for the new user, you will just find their existing folder and transfer ownership of the items to that new user, placing the items in the existing folder:

1. Right-click on `TransferOwnershipCreateFolder` and select **Copy**.
2. Right-click on the folder with `TransferOwnershipCreateFolder` and select **Paste**. The result will be a new file created called `TransferOwnershipCreateFolder_1`.
3. Right-click on the `TransferOwnershipCreateFolder_1` file and select **Rename**. Type in `TransferOwnershipExistingFolder`.
4. The first cell will remain the same as it is, importing in the modules you need and creating a connection to your ArcGIS Online account.
5. In the second cell, you will add a line at the bottom of the cell for a variable to hold the folder name in the new user's account. Type in the following on a new line in the second cell:

```
newFolderName = "Existing New User Folder"
```

6. The third cell will remain the same as it is, creating user objects for the old and new users, and a list of all the folders in the old user's account.
7. The fourth cell will require some changes that will be covered in this step and the next step. First, you will delete the lines that create a list of folders within the new user's account, check if the folder exists, and create a new folder with the same name as the folder from the old user. Delete the following lines from the fourth cell:

```
        newUserFoldersDict = newUser.folders
        newUserFolders = []
        for f in newUserFoldersDict:
            newUserFolders.append(str(f["title"]))
        if folderName not in newUserFolders:
           gis.content.create_folder(folderName,newUserName)
        else:
            print("{0} folder already exists, you need to use the
                  TransferExistingFolder Notebook"
                  .format(folderName))
            break
```

8. You now just have to change the `target_folder` argument for each item being reassigned to the new folder. Since you have to call the `reassign_to` property for each of the three lists, you will have to change the `target_folder` three times. Update each of the lines with `item.reassign_to` in the fourth cell, so the code looks like the following:

```
for item in replgdb:
    print("Moving {0}".format(item["title"]))
    item.reassign_to(newUserName, target_folder=newFolderName)
print("---Moving all other non-service definitions---")
for item in otherItems:
    print("Moving {0}".format(item["title"]))
    item.reassign_to(newUserName, target_folder=newFolderName)
print("---Moving service definitions---")
for item in serviceDef:
    if item["title"] not in otherItems["title"]:
        print("Moving {0}".format(item["title"]))
        item.reassign_to(
            newUserName,
            target_folder=newFolderName
        )
```

9. Your Notebook is now ready to run. Run all the cells. When completed, you will see that the new user has been reassigned all the data from the old user, and it has been placed in the new user's existing folder for the project.

You now have two Notebooks that will reassign all of a user's items in their folder to a new user, in either a new folder or an existing folder for the user. Both Notebooks will save you time when transferring projects between users in your organization, as running them is much quicker than transferring item ownership within ArcGIS Online.

You can alternatively reassign all of a user's items when you delete them; you are only able to reassign them to the root folder. To do this, use the following code: `newUser.delete(reassign_to ="{userNameToAssignTo}")`.

In the next section, you will create a Notebook and a script tool from the Notebook to assist with downloading attachments from data collected in the field.

Case study: Downloading and renaming attachments

Field data collection is an important way users interact with ArcGIS Online. Through the use of apps like Field Maps and Survey123, field staff can collect data. That data can be stored as a feature layer or feature layer collection on ArcGIS Online. On many occasions, it is necessary for field staff to take pictures of the data they are collecting.

Reviewing those pictures on ArcGIS Online or in the app the data was collected in is useful, but sometimes the pictures need to be downloaded from your account. While this can be done by exporting the feature layer to a geodatabase and then running a tool in ArcGIS to extract the photos, the photos do not have names associated with the features. It would be useful for the photos to have names of attributes from the feature layer.

In the case study for this section, you have been collecting survey data from Survey123 on farmers' markets in Oakland and Berkeley, which checks on different produce at the stands across Oakland. You are looking to track the availability of produce at the markets throughout the season.

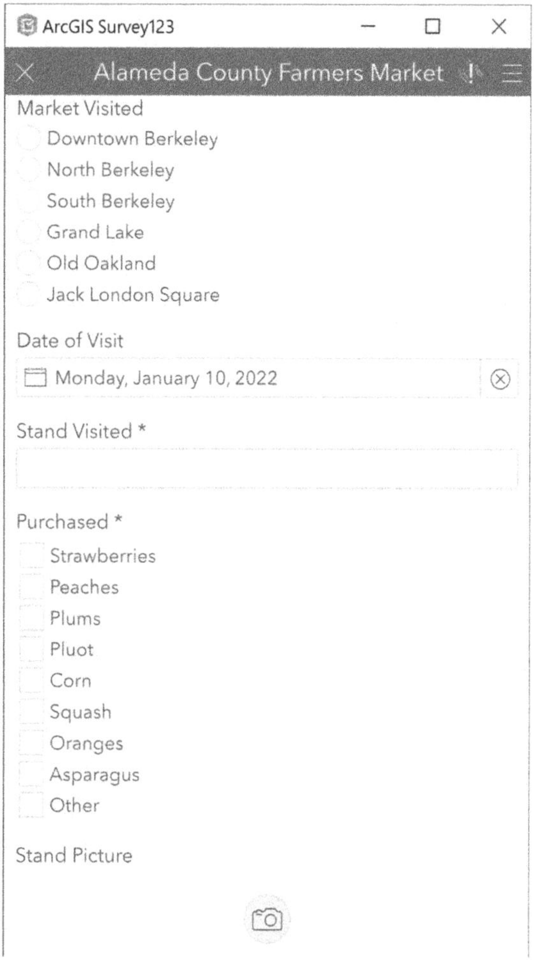

ArcGIS Survey123

Alameda County Farmers Market

Market Visited

- Downtown Berkeley
- North Berkeley
- South Berkeley
- Grand Lake
- Old Oakland
- Jack London Square

Date of Visit

Monday, January 10, 2022

Stand Visited *

Purchased *

- Strawberries
- Peaches
- Plums
- Pluot
- Corn
- Squash
- Oranges
- Asparagus
- Other

Stand Picture

Figure 11.18: The farmers' market survey

You have been tasked with taking pictures of the stands and produce to help show how the different stands are set up. The team will be producing a written report in which the photos will be associated with each stand.

Figure 11.19: A photo from Survey123

This data from Survey123 is stored as a **feature layer** in ArcGIS Online, with the photos as **attachments**. This means you need to extract the photos from ArcGIS Online and download them. Esri has a technical support document that walks you through the process to download a ZIP file with a geodatabase that contains the photos (`https://support.esri.com/en/technical-article/000012232`).

However, the problem still remains that the photos are in a geodatabase; they will need to be individually extracted in order to be associated with the market and stand, by clicking on them and saving them with a name.

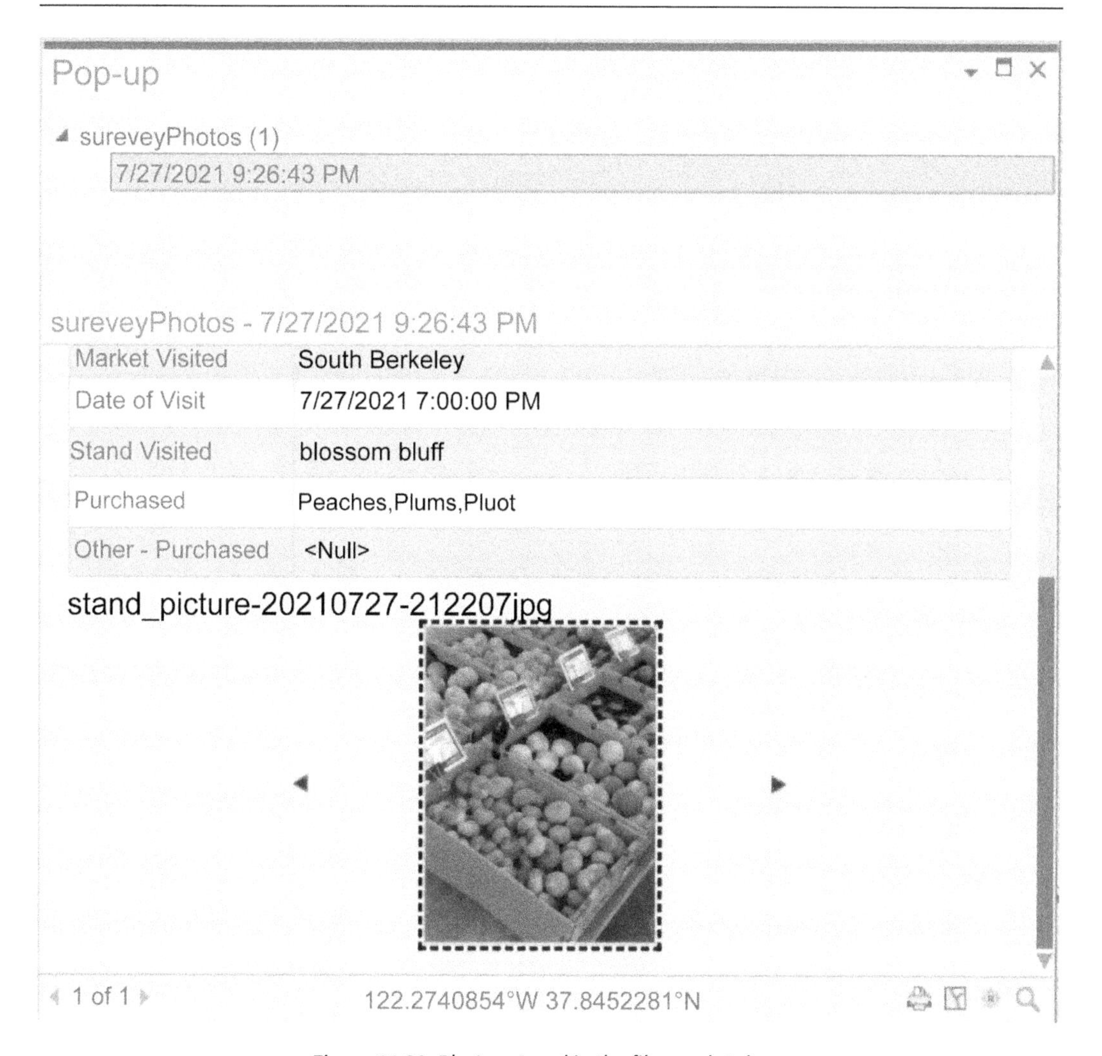

Figure 11.20: Photos stored in the file geodatabase

There is an Esri technical document about how to batch export photos by creating a script tool in ArcMap (`https://support.esri.com/en/technical-article/000011912`). However, that script tool is written for ArcMap, so you would have to update it to work in ArcGIS Pro. It also takes the name of the picture as it is stored in the file geodatabase and downloads it using that name. As you can see in the figure above, the name is not very useful in associating that photo with a certain market and stand. If you have many pictures, it will take a lot of time to sort through them and rename them appropriately.

It would be much more useful to download the photos directly from ArcGIS Online and name them using the attributes within the feature layer. You can then easily find all the photos associated with a farmers' market and stand, as that information will be part of the name of the photo.

You will now create a Notebook to do just this.

1. Right-click on the `Chapter11` folder and select **New** > **Notebook**. Rename the Notebook to `DownloadPhotos`.
2. You are going to log in anonymously, as you will be looking for public data available outside your organization. You will also import the `os` module, which you will need to create folder names and rename the pictures. Type in the following:

   ```
   from arcgis.gis import GIS
   import os
   gis = GIS()
   ```

   Run the cell.
3. In the next cell, you are going to create a variable for the location of the folder to store your folder of downloaded photos in. Type in the following:

   ```
   folderLoc = r"C:\PythonBook\Chapter11"
   ```

   Run the cell.
4. In the next cell, you will search for the feature layer. In this case, it is a feature layer collection that is created from the Survey123 survey, but it would work with any feature layer. If you had survey crews using ArcGIS Collector or ArcGIS Field Maps, this would work the same way. You would just need to find the correct layer using the title and owner.

   To find the layer, you need to query for both the title and the owner by putting an & between the title and owner queries. Type in the following:

   ```
   fmSurveySearch = gis.content.search(
       query='title:Alameda County Farmers Market &
           owner:billparkermapping',
       item_type="Feature Layer"
   )
   print(fmSurveySearch)
   print(len(fmSurveySearch))
   ```

Run the cell. The result will be a list containing the list of feature layers and its length, showing that there is just one layer in the list:

```
[<Item title:"Alameda County Farmers Market_fieldworker"
type:Feature Layer Collection owner:billparkermapping>]
1
```

5. Now that you have a list of results, you can extract the first and only item from the list using the list index and display the feature layer. Type in the following:

```
fmSurvey = fmSurveySearch[0]
display(fmSurvey)
```

Run the cell. The result will be the details of the feature layer:

Alameda County Farmers Market_fieldworker

Feature Layer Collection by billparkermapping
Last Modified: August 09, 2021
0 comments, 10 views

Figure 11:21: Feature layer collection with survey data

6. In the next cell, you can check how many layers are in the feature layer collection by looping through each layer and printing out the results. Type in the following:

```
for lyr in fmSurvey.layers:
    display(lyr)
```

Run the cell. The result should be a single layer within the feature layer collection:

```
<FeatureLayer url:"https://services3.arcgis.com/HReqYJDJNUe3sQwB/
arcgis/rest/services/survey123_85b524e1efac48a6bf3d96a8bfb07022_
fieldworker/FeatureServer/0">
```

7. Since there is just the one, you can again use the list index to select that layer. Type in the following:

```
fmSurveyLyr = fmSurvey.layers[0]
fmSurveyLyr
```

Run the cell. The result is the same as above, a single layer within the feature layer collection:

```
<FeatureLayer url:"https://services3.arcgis.com/HReqYJDJNUe3sQwB/
arcgis/rest/services/survey123_85b524e1efac48a6bf3d96a8bfb07022_
fieldworker/FeatureServer/0">
```

In this survey, there is just one layer in the feature layer collection. If, for example, your survey team is completing biological surveys, they may be collecting data in different layers all within one feature layer collection. It is important to know which layer within the feature layer collection contains the photo attachments you want to download. If you are unsure, you can navigate to the feature layer collection on ArcGIS Online and view the list of layers. The top layer is layer 0 and the layers after that are 1, 2, and so on. In the following *Figure 11.22*, **survey** is layer 0, **Summer21RouteShape** is layer 1, and **AlamedaContraCostaCounty_RaceHispanic_BlockGroup** is layer 2.

Figure 11.22: Layers in a feature layer collection

8. Now that you have the layer with the photo attachments, you can create a folder to hold those photos based on the name of the layer. You use the `properties.name` property of the layer to get the name of the layer.

Making the folder in your script allows you to use the name of the layer for the folder. It is another step that the Notebook is doing, so you don't have to do it manually. This helps you keep your entire process contained in one Notebook.

Then, you use the `os.mkdir` function to create a folder after using the `os.path.exists` function to make sure it doesn't exist yet. Type in the following:

```
lyrName = fmSurveyLyr.properties.name
PhotoPath = os.path.join(folderLoc,lyrName+"_Photos")
```

```
if not os.path.exists(PhotoPath):
    os.makedirs(PhotoPath)
```

Run the cell.

9. Next, you want to check and make sure that the layer has attachments. That is part of the `properties` property. If you are using field survey data where there are many layers within a feature layer collection, this step will help to ensure you have selected the correct layer. Type in the following:

```
fmSurveyLyr.properties.hasAttachments
```

Run the cell. The **Out** cell will return the following:

```
True
```

10. To extract the photos, you need to query the feature layer. First, you should check the `capabilities` property to make sure the layer can be queried. Type in the following:

```
fmSurveyLyr.properties.capabilities
```

Run the cell. The **Out** cell will return the following:

```
Create, Query, Editing, Sync
```

This shows that the layer can indeed be queried.

11. Now you can use the `fields` property of the layer to find the field names available for naming the photos:

```
for field in fmSurveyLyr.properties.fields:
    print(field["name"])
```

Run the cell. The **Out** cell will return all of the field names to you:

```
objectid
globalid
CreationDate
Creator
EditDate
Editor
market_visited
date_of_visit
stand_visited
```

```
purchased
purchased_other
```

These are the fields that you can use to pull attributes from for naming your photos. You should work with your survey team on their specific survey to identify which fields they will find the most useful in naming photos. For example, a team doing an archaeological survey may want the site ID in the photo name so they can find all the photos for a particular site.

You need to make sure that the field you choose is fully populated and does not have any null values. It is important to work with your survey teams before they go out to identify the potential fields for naming photos and, if you are using Survey123 or Field Maps, set up those fields so they cannot be left blank.

12. You plan to use the `market_visited` and `stand_visited` fields. In the next cell, create two variables to hold the field names for use later:

    ```
    nameField1 = "market_visited"
    nameField2 = "stand_visited"
    ```

    Run the cell. Since there are no `print` statements, nothing will be returned.

13. In the next cell, you will create a list of the object IDs. You will iterate over this list later and use the values to query each feature in the layer. To get the list of object IDs, you use the `query` function and pass in the argument `return_ids_only` set to `True`. This will give you a dictionary of just object IDs. From this you can create a list, as one of the values of the dictionary is a list of the object IDs. Type in the following:

    ```
    objIds = fmSurveyLyr.query(return_ids_only = True)
    print(objIds)
    listObjIds = objIds["objectIds"]
    print(listObjIds)
    ```

    Run the cell. The first print statement will return the dictionary from the `query` function. The second print statement will return the list extracted from the dictionary:

    ```
    {'objectIdFieldName': 'objectid', 'objectIds': [1, 2, 3, 4]}
    [1, 2, 3, 4]
    ```

14. This next cell is where you are going to download and rename all of the photos. **The code for this will be written in one cell, as it is all part of a single for loop. It will be broken up over the next two steps.**

    You will loop through the list of object IDs from above. Within the loop, you will get the list of attachments for each feature in the layer and print it to see what the list looks like and what values you have access to. You will then create a query for the object ID of the layer and create a layer query, using that layer query to extract the values of the two fields you are going to use to rename the photos. You will run a `replace()` on those values to replace any spaces with underscores, and print out the names to track your progress. Type in the following:

    ```
    for objID in listObjIds:
        objAtt = lyr.attachments.get_list(oid=objID)
        print(objAtt)
        sql = "OBJECTID = {}".format(objID)
        lyrQuery = fmSurveyLyr.query(where = sql, out_fields="*")
        lyrQueryFeatures = lyrQuery.features
        name1 = lyrQueryFeatures[0].attributes["{0}".
    format(nameField1)].replace(" ","_")
        print(name1)
        name2 = lyrQueryFeatures[0].attributes["{0}".
    format(nameField2)].replace(" ","_")
        print(name2)
    ```

15. In the same cell as above, you will create another `for` loop to loop through the attachments list. The attachments list contains the data dictionary for each attachment. You can access the attachment name and ID and set each to a variable. Next, you will download a photo by passing arguments for the object ID of the layer, the attachment ID of the photo, and the path for the folder to save it to, to the `download` function. This will download the photo with the attachment name, which does not contain the useful information collected in the survey. You will then create a new name for the photo using the name of the layer, the values from the attributes, and the attachment ID. To rename the photo, you use the `os.rename` function, passing the downloaded photo name and the new name you just created. In the same cell as above, type in the following:

    ```
        k = 0
        while k < (len(objAtt)):
    ```

```
        attachmentName = objAtt[k]["name"]
        print(attachmentName)
        attachmentID = objAtt[k]["id"]
        print(attachmentID)
        pic = lyr.attachments.download(
            id=objID,
            attachment_id=attachmentID, save_path=PhotoPath
        )
        newName = os.path.join(PhotoPath,lyrName+"_"+str(name1)+"_"
+str(name2)+"_"+str(attachmentID)+".jpg")
        os.rename(pic[0],newName)
        k+=1
```

Run the cell. You will see the data dictionary for the object attachment, which are the photos, along with the names of the two values to name the photos with, the attachment name, and the attachment ID for each photo. The first object ID will look like the code below:

```
[{'id': 1, 'globalId': '22a7b523-03a3-423a-9a92-9984fd093fb5',
'parentGlobalId': '143d832f-faf4-4fbe-9943-a4459ec081ee', 'name':
'stand_picture-20210727-212156.jpg', 'contentType': 'image/jpeg',
'size': 804201, 'keywords': 'stand_picture', 'exifInfo': None},
{'id': 2, 'globalId': '0580dc2a-d45e-41af-910a-f4b840efea8c',
'parentGlobalId': '143d832f-faf4-4fbe-9943-a4459ec081ee', 'name':
'stand_picture-20210727-212207.jpg', 'contentType': 'image/jpeg',
'size': 973045, 'keywords': 'stand_picture', 'exifInfo': None}]
South_Berkeley
blossom_bluff
stand_picture-20210727-212156.jpg
1
stand_picture-20210727-212207.jpg
2
```

You have now downloaded all of the photos from your survey and renamed them based on attributes. This makes reviewing the photos offline easier, as they now have recognizable names based on the data collected.

This process was just for a single layer within a feature layer collection. It could be modified to go through all the layers in a feature layer collection. It could also be modified to download only certain photos from a feature layer based on an attribute or spatial query.

All of this allows you to extract photo attachments from your survey without having to download an entire geodatabase of data, and then rename the photos manually.

As always, you have the option of converting this Notebook to a script tool so you can deploy it to your entire team, no matter their level of Python knowledge. Creating script tools to help your team administer their data collection without your assistance will allow you to focus more on managing your projects and create a more self-sufficient team.

We covered the Notebook-to-script tool conversion process in *Chapter 6, ArcToolbox Script Tools,* so we will not proceed with it here. It is a good idea when creating a script tool to create a help document to go along with the script tool. In this document, you should explain the different parameters and where to find them. Once the help document is created, you can deploy the script tool to your team, so your GIS staff and field staff can self-serve downloading photos and renaming them from field visits.

You can find a copy of the final script tool along with an accompanying help document in the GitHub repository for this chapter, at `https://github.com/PacktPublishing/Python-for-ArcGIS-Pro/tree/main/Chapter11`.

Summary

In this chapter, you have seen multiple case studies that show how you can streamline administrative tasks. You looked at creating users and usage reports, assigning licenses and credits, and reassigning user data, as well as downloading attachments. These case studies highlight how you can write Notebooks in ArcGIS Pro to handle tasks in ArcGIS Online that take multiple clicks through multiple pages. This allows you to easily switch from your project work with ArcGIS Pro to your admin tasks without having to change platforms. All of the Notebooks can be kept in an administrator toolbox in the **Favorites** tab of your **Catalog** pane to give you easy access to them.

In the next chapter, we will look at a case study on creating an advanced map automation to view the impact of suspended bus lines on minority groups.

12

Case Study: Advanced Map Automation

You may be familiar with **data-driven pages** in ArcGIS Desktop and using them to export multiple pages to PDF from one layout view to make a map book. In ArcGIS Pro, data-driven pages have been replaced with **map series**, which also use a single layout to export multiple pages to PDF. This chapter's case study will see you using ArcPy to take a layout and create, add, and style layers to the layout, along with extracting data from those layers and adding that data as text to the layout. Then, you will export a map to PDF before moving to the next selected block group and completing the process again. Finally, you will merge all the PDFs into a single map book.

In *Chapter 7*, you learned how to automate map production using ArcPy. In this chapter's case study, you will take what you have learned and extend that to create a set of maps displaying minority populations along a discontinued bus route. You will be highlighting block groups with a minority population greater than that of a **reference community**. A reference community is a larger community that is compared to smaller areas, like block groups or tracts. This is a practice you will see in Environmental Justice sections of Environmental Impact Reports/Statements to determine if minority populations are being disproportionately impacted. The skills you learn in this case study can be applied to creating custom wetland delineation maps, habitat maps, parcel maps, or any other maps where the data remains the same but the view changes along a project corridor.

This chapter will cover:

- Setting up a layout for use in map automation

- Creating and adding data to a map
- Querying layers within the map to improve display
- Changing the map view in the map frame
- Updating the map title for each page
- Exporting each page to PDF
- Combining all the pages into one PDF

To complete the exercises in this chapter, please download and unzip the `Chapter12.zip` folder in the GitHub repository for this book: `https://github.com/PacktPublishing/Python-for-ArcGIS-Pro/tree/main/Chapter12`.

Case study introduction

In this case study, you will use ArcPy to create a **custom map series** displaying Census Block Group data along AC Transit Transbay bus routes.

In the summer of 2020, due to COVID-19, many Transbay bus routes were suspended, leaving limited service from Alameda and Contra Costa County to San Francisco. You are working with a group on a preliminary Environmental Justice study to see if one of the lines that was suspended had a disproportionate impact on minority communities. You have been asked to produce a series of maps that highlight the block groups along the route of one bus line that have a higher-percentage minority population than that of block groups along the routes that were not suspended.

In addition to highlighting those block groups, the population of each race group in the selected block group needs to be displayed as a dot density map and a table included on the figure that contains the percentage of each race group.

This requires you to create a custom map series in an ArcGIS Notebook in ArcPy, as the map series function in ArcGIS Pro cannot add and style layers or extract attribute data for display on the map in a table.

For more information on creating a map series by using the ArcGIS Pro tool for map series, see the documentation here: `https://pro.arcgis.com/en/pro-app/latest/help/layouts/map-series.htm`.

Setting up a layout for map automation

To create your custom map series, you need to first create a layout that contains all the elements you will need. In most cases, your layout will include the following elements, with many of the elements defined before you start writing Python code:

- A map frame for your main map
- A legend with any static data properly styled
- A scale bar
- A north arrow
- A text box for the title
- A map frame for an inset map

Optionally, you may also include the following elements:

- Text boxes for additional information
- Additional map frames as needed

You need to make sure that all of the settings that you cannot change using ArcPy are already set correctly for your map. Often, this is an iterative process of changing your map extent and exporting sample pages to ensure all the settings are correct. In this case study, we have already done that and created a layout for you in the `Chapter12.aprx` file. You will inspect all of the elements of the **Map**, **Layout**, and **Inset** to view the settings that have been chosen for this map series.

1. Open up ArcGIS Pro.
2. Click **Open another Project**, navigate to where you unzipped the `Chapter12.zip` folder, and select `Chapter12.aprx` to load the `Chapter12` project.
3. Click on the map to view the data already loaded in the map. You should see a World Street Map set as the basemap, along with the following feature classes:
   - `AC_Transit_AdditionalTransbay`: A feature class containing the Transbay bus routes that were discontinued in 2020.
   - `Summer21RouteShape`: A feature class containing all the AC Transit bus routes in Summer 2021. It has a definition query on it so it only displays Transbay routes. To view this, right-click on the layer and select **Properties**. In the **Layer Properties** dialog box, click on **Definition Query** to see the query. It reads:

     ```
     Pub_RTE In ('W', 'V', 'U', 'P', 'O', 'NX', 'NL', 'LA', 'L',
     'J', 'G', 'F', 'DB1', 'DB')
     ```

- `AlamedaContraCostaCounty_RaceHispanic_BlockGroup`: This is the *American Community Survey* (ACS) 5-year data for 2019 at the block group level for race counts. It was created using the script tool from *Chapter 6*, with block group shapefiles and census CSVs for Alameda and Contra Costa County. The script tool has been included in the Toolbox in the `Chapter12` project, and the census CSV is in the `censusCSV` folder in the `Chapter12.zip` folder. The feature class has already been styled using dot density symbology. It is set to 1 dot = 1 person, which will look much better when the scale is set to a block group.

4. Click on **Layout** to look at the default starting layout:

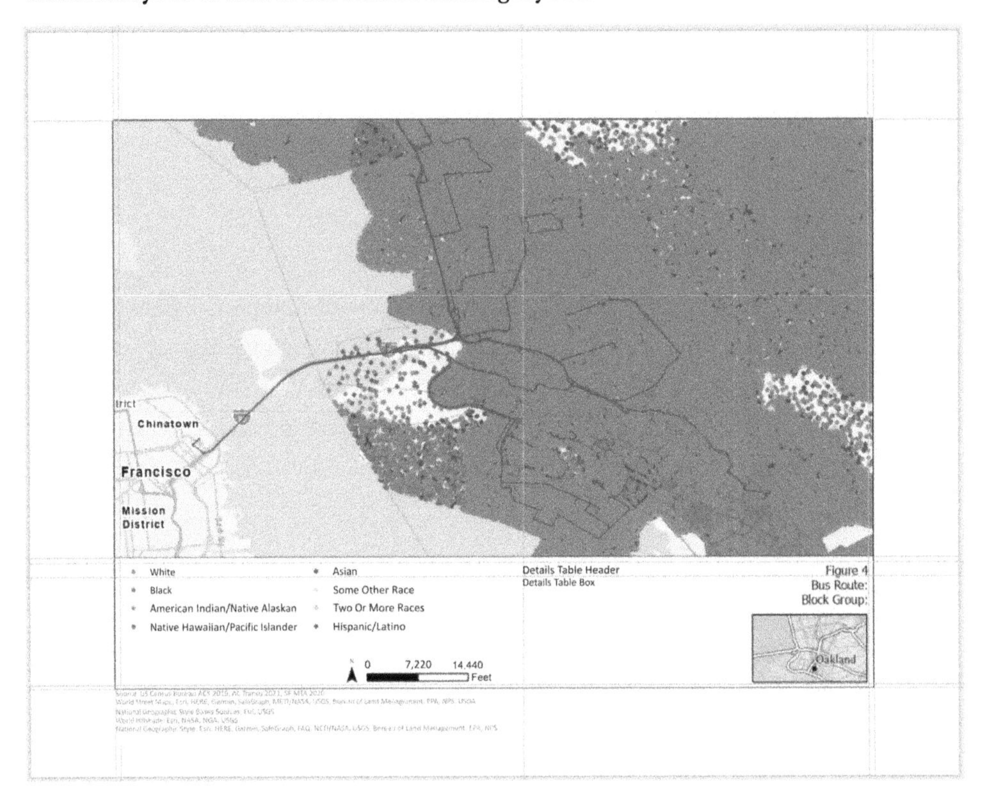

Figure 12.1: Map layout

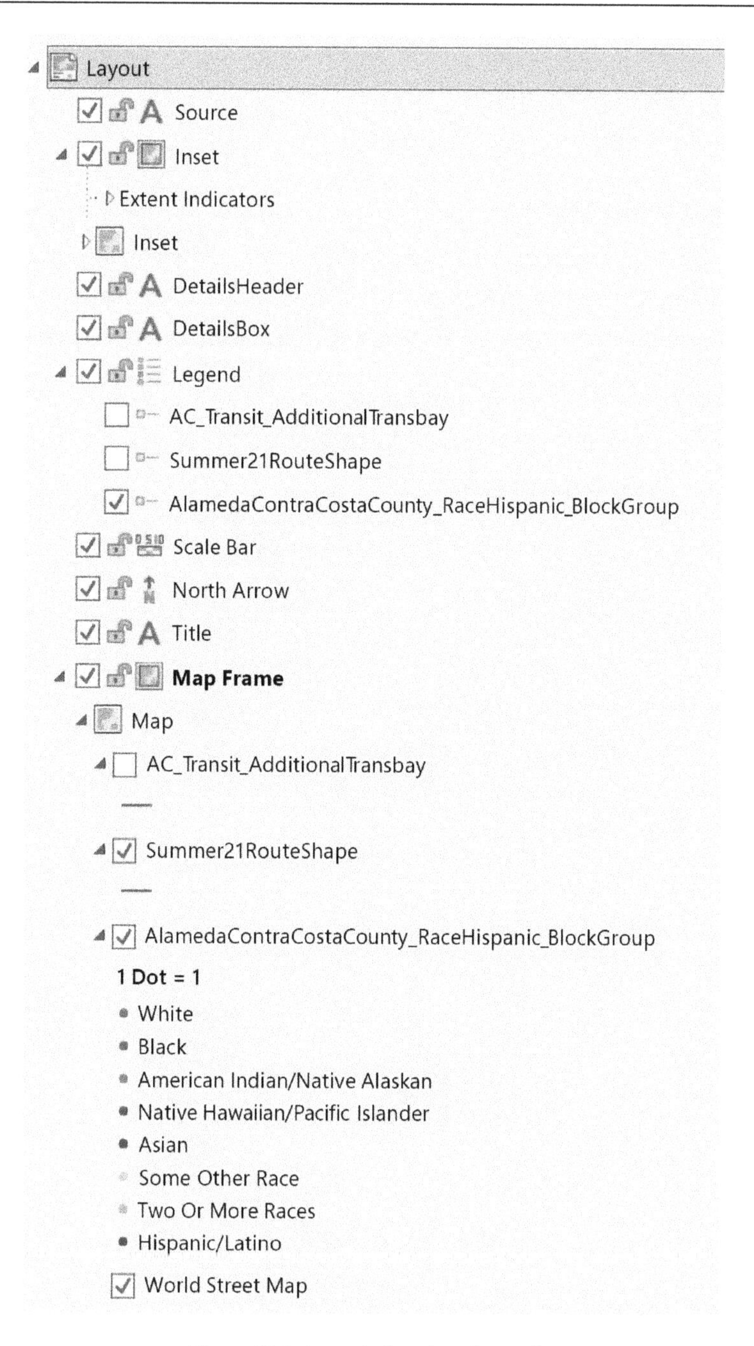

Figure 12.2: Layout view drawing order

We will go through the layout elements in the drawing order to observe the settings of each element, noting which settings cannot be modified with Python and must be set manually before starting your automation.

Source text element

The top element in the layout drawing order is the text element named **Source**. This contains the source information for your map. Below, you will explore the different settings available for a text element, highlighting those that can be set using ArcPy and those that cannot.

1. Click on **Source** to bring up the element pane.
2. Within the **Text** tab, you will find the **General** and **Text** tabs.

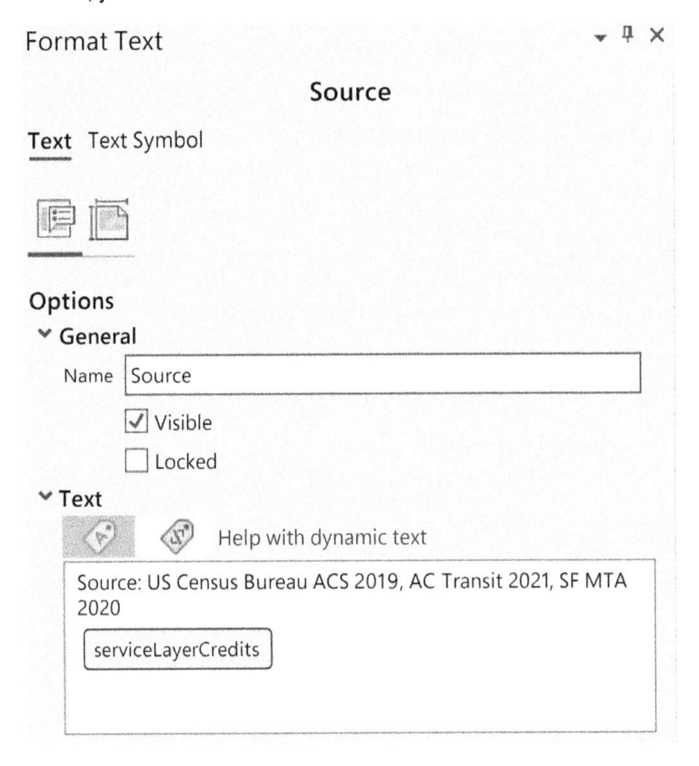

Figure 12.3: Source element Text tab options

You can see the name of the element and the text written in the text box.

3. The **Text** box shows the text printed in the text box. It contains the source for the census data and the bus route data. It also contains the **serviceLayerCredits** button for the dynamic text for the service layers. Click on the button to see how that is written:

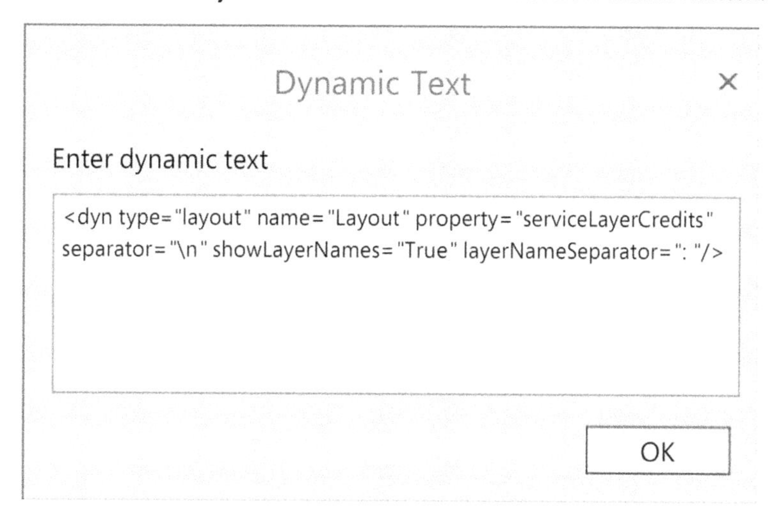

Figure 12.4: Dynamic text in serviceLayerCredits button

This service layer credit text has been modified from the standard text inserted when you click **Dynamic Text > Service Layer Credit**. The following code has been added:

```
separator="\n" showLayerNames="True" layerNameSeparator=":"/>
```

The separator tag is set to `"\n"`, which will create a new line after each service layer. The `showLayerNames` tag is set to `True` to show the basemap layer names. The `layerNameSeparator` is set to `": "` to place a colon and a space between the layer name and the credits. This all makes the source for the maps easier to read. When you are finished viewing this, click **OK** to close the box.

4. Now view the **Placement** tab to see the **Size** and **Position** tabs:

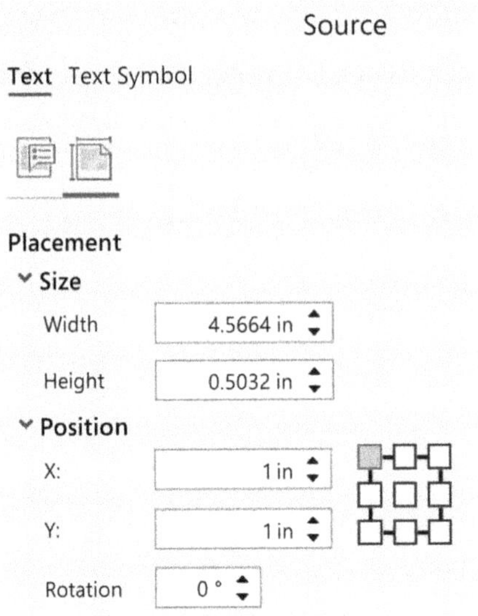

Figure 12.5: Text Placement tab

This shows the width and height of the text box, its position, and the **anchor point**. This is what the X and Y locations are based on when moving elements in ArcPy. Verify that the anchor location is the top left.

While the width, height, and positions can be changed using ArcPy, the anchor location cannot and must be set manually here. Always check the anchor point location. If you do not have it set to the anchor point you expect, your element will be in the wrong place. You cannot change the anchor point in ArcPy. It must be set in the element pane.

5. Now look at the **Text Symbol** tab:

Figure 12.6: Text Symbol General tab

> Only font size under the **Appearance** setting and angle under the **Rotation** Setting, are available to be modified in ArcPy. Everything else must be set manually and must work for all views in your automation. All of the elements in the **Formatting** tab (the icon containing the "A") must be set prior to any automation, as they cannot be accessed via ArcPy.

For all of the text properties you can modify in a text element, you are limited to what you can call in ArcPy. It is important to test different settings out on different views to understand how they work. In this template, I have modified the source text element to wrap the dynamic text to a new line for each source, set the font size to 6 pts, the font color to gray, and the anchor point to the top left. If you want to modify any of these, you must set them in the layout as they cannot be changed in the Notebook.

Inset map frame

The next layout is the **Inset** map frame, which is the map frame layout element for the inset map that shows the extent of your study area. Below, you will explore the different settings available for a map frame element.

1. Click on **Inset** to bring up the element pane to format the **Map Frame**:

Map Frame

Options
General
Name Inset
Visible
Locked
Map Frame
Map Inset
Accessibility

Figure 12.7: Inset Map Frame element

There are more options for a map frame element than for a text frame. You have a drop-down that is set to **Map Frame** and the **Options** tab as your starting view. This is where you can change the name of the layout element and the map associated with that map frame. You can adjust the map displayed in a map frame and the name of the map frame using ArcPy. You will be leaving the name and map as **Inset**.

2. Click on the dropdown to the right of the top **Map Frame**. You will see options for **Map Frame**, **Background**, **Border**, and **Shadow**:

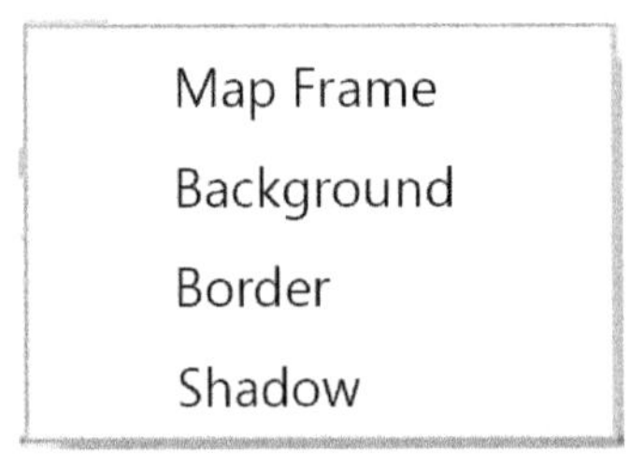

Figure 12.8: Map Frame dropdown choices

You can modify the background, border, and shadow effects for the **Map Frame** from these tabs. They must be set here, as none of the properties can be controlled with ArcPy.

3. Now look at the **Display Options** tab:

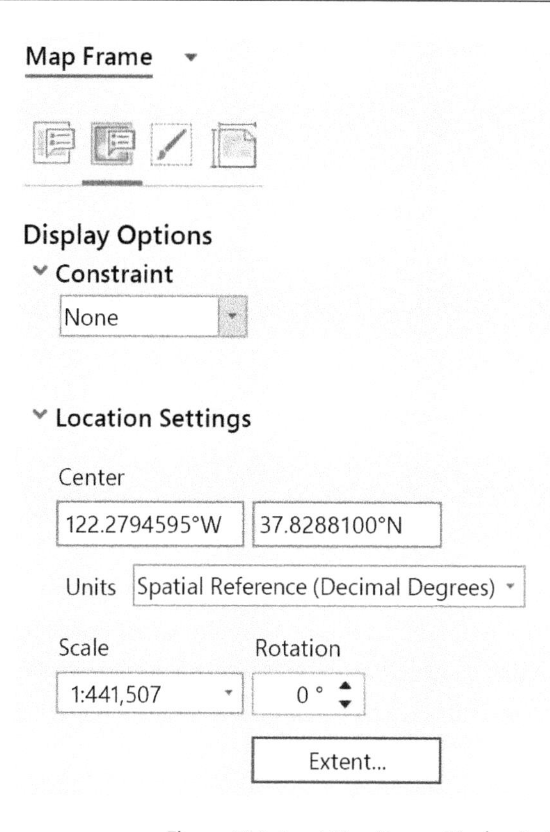

Figure 12.9: Inset Map Frame Display Options

The **Constraint** option cannot be accessed with ArcPy and must be set here for your map series. The extent and scale can both be changed through the ArcPy Camera object. You will use this later in this chapter to change the scale and extent of your maps.

4. Look at the **Display** tab next. You can make changes to the display of the border, background, and shadow here by setting their colors, offset, and rounding. None of these settings can be controlled using ArcPy and must be set manually.
5. Now look at the **Placement** tab. These are the same placement settings as from the Source text element. The width, height, and position elements can be modified through ArcPy, but not the anchor point. Verify that the anchor point is set to the bottom left.
6. The last setting to check on the Inset map frame is the Extent Indicator. It has already been placed in the map frame and set to a cyan color.

Click on **Extent of Map Frame** in the **Contents** tab:

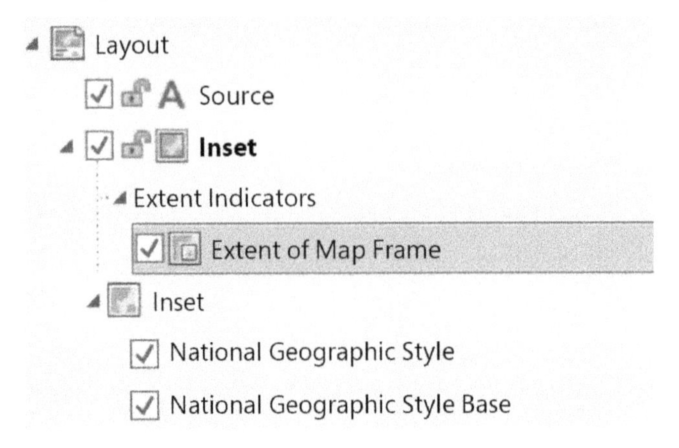

Figure 12.10: Extent Indicator in the Contents tab

This will bring up the **Extent of Map Frame** elements pane. In this tab, you can change the name of the element, the source of the extent, the color and shape of the extent indicator, the setting for a leader line if needed, and the setting for turning an extent indicator into a point on collapsing. You will leave all of these as their current settings.

None of the extent indicator settings can be accessed in ArcPy and must be set here manually before creating your automation.

You have seen all the settings for a map frame and how they are set up for an inset map for map automation. You must set any of these that cannot be accessed through ArcPy before creating your automation. In many cases, you will need to test different views to ensure that your settings have the look you expect them to.

DetailsHeader and DetailsBox text elements

DetailsHeader and DetailsBox are both text elements. **DetailsHeader** is used as a header for a box of text information you will be printing out for each page. **DetailsBox** holds the different information for each page, which will change with what is being viewed on each page. Since it is a text box like the Source element, it has the same settings. You should check the anchor point to make sure it is in the correct location.

1. Click on **DetailsHeader** to bring up the element pane to format the **DetailsHeader** text box.
2. Click on the **Placement** tab under the **Text** tab to show the placement settings. Check to make sure that the anchor point is the top-left corner.

Do the same for the DetailsBox element. You can further explore the settings for both elements in your own time. Like with all text elements, you are limited in what you can control in ArcPy and you will need to set some of the different properties of your text element before creating your automation.

You will be extracting data from a map layer and writing to this text box in your script, and it has been set to take up to nine lines of data. If more lines were needed, you would need to modify the settings here to allow for that. You would then verify those settings on a single page before continuing on to create your automation.

Legend element

The Legend element has many settings that can be changed through the Element tab. Many of the settings cannot be changed by using ArcPy; the settings that can are accessed through the **LegendElement** and **LegendItem** classes. Much of what you can change in ArcPy involves adding layers, moving them, and changing layer names and group names. Because of this, it is important to think about what legend elements will be displayed on your map and how you want them displayed.

You will explore the different settings that have been set for this legend that allow for layers to be added to the map and the legend through your automation script.

1. Click on **Legend** to bring up the element pane to format the legend element.

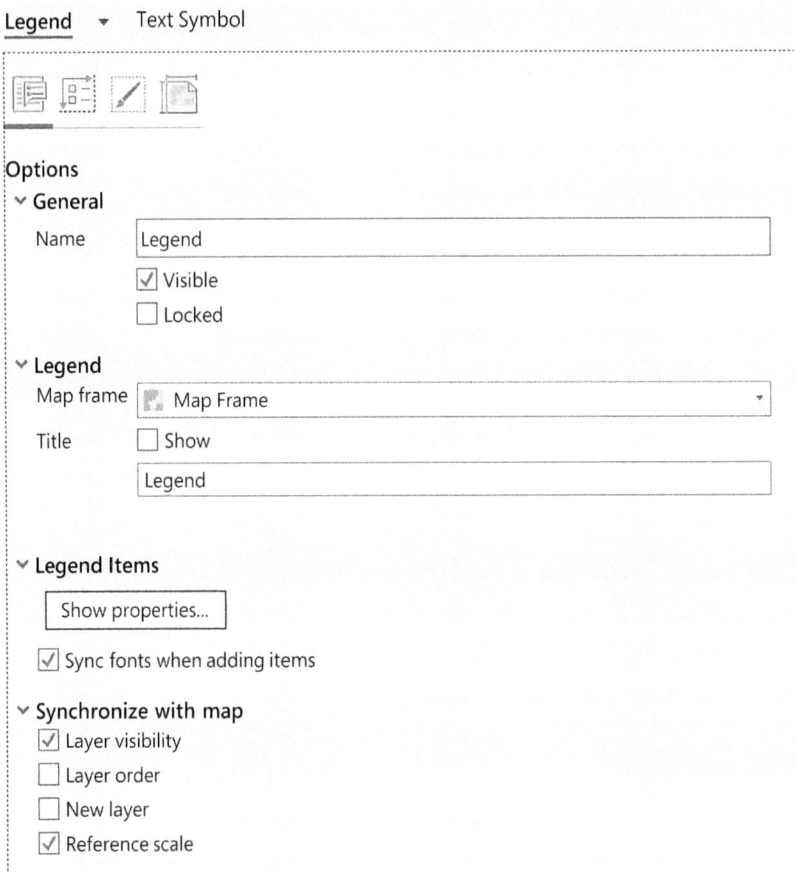

Legend ▾ Text Symbol

Options

General

Name: Legend

☑ Visible

☐ Locked

Legend

Map frame: Map Frame

Title: ☐ Show

Legend

Legend Items

Show properties...

☑ Sync fonts when adding items

Synchronize with map

☑ Layer visibility

☐ Layer order

☐ New layer

☑ Reference scale

Figure 12.11: Legend element Options tab

The above view is of the **Options** tab within the **Legend** element tab. This shows the name of the legend element, the map frame for the legend, a button to access the properties of all the legend items, and settings for synchronizing with the map. The synchronize settings can all be controlled through ArcPy with the following properties:

- `syncLayerVisibility`: A Boolean that corresponds to the **Layer visibility** checkbox. When `True` or checked, a layer will automatically appear in the legend if it is visible on the map.
- `syncLayerOrder`: A Boolean that corresponds to the **Layer order** checkbox. When `True` or checked, the layer order in the legend is the same as the map.

- `syncNewLayer`: A Boolean that corresponds to the **New layer** checkbox. When `True` or checked, a layer will be automatically added to the legend when added to the map.
- `syncReferenceScale`: A Boolean that corresponds to the **Reference scale** checkbox. When `True` or checked, the symbols in the legend will match the scale of the symbols in the map.

You will set these here before creating your automation as they will be the same throughout. Set **Layer visibility** and **Reference scale** to checked and **Layer order** and **New layer** to unchecked. This will ensure that the only layers on the legend are layers you have added. You will be adding layers to the legend in the order you want as you create them in your automation.

2. Click on the dropdown next to the word **Legend** to see what the options are:

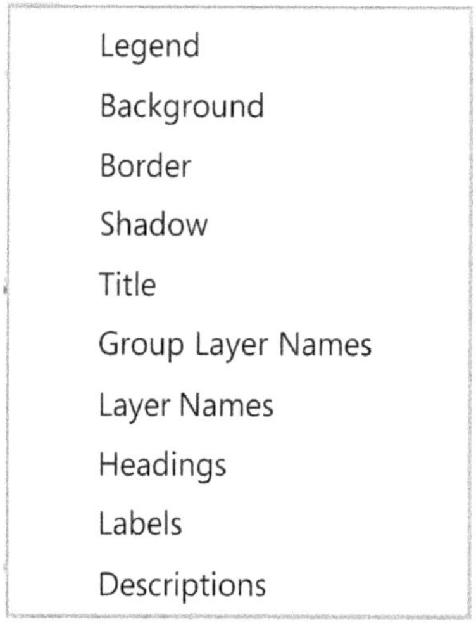

Legend
Background
Border
Shadow
Title
Group Layer Names
Layer Names
Headings
Labels
Descriptions

Figure 12.12: Legend dropdown options

Legend is the option you are in right now. **Background**, **Border**, and **Shadow** give you access to those display settings for the legend box. They cannot be set using ArcPy and must be set manually here before your automation. **Title**, **Group Layer Names**, **Layer Names**, **Headings**, **Labels**, and **Description** will allow you to change all of the text formatting options for all of the legend elements. These options are the same as the text formatting options you have for a text box. When you set them here, they are universal for all legend elements.

None of these settings can be accessed with ArcPy. You need to set them either here or for the individual legend elements.

3. Within the **Legend Item** tab, click the **Show Properties (...)** button. This will select all the legend items and show the different settings you can set for *all* of them.

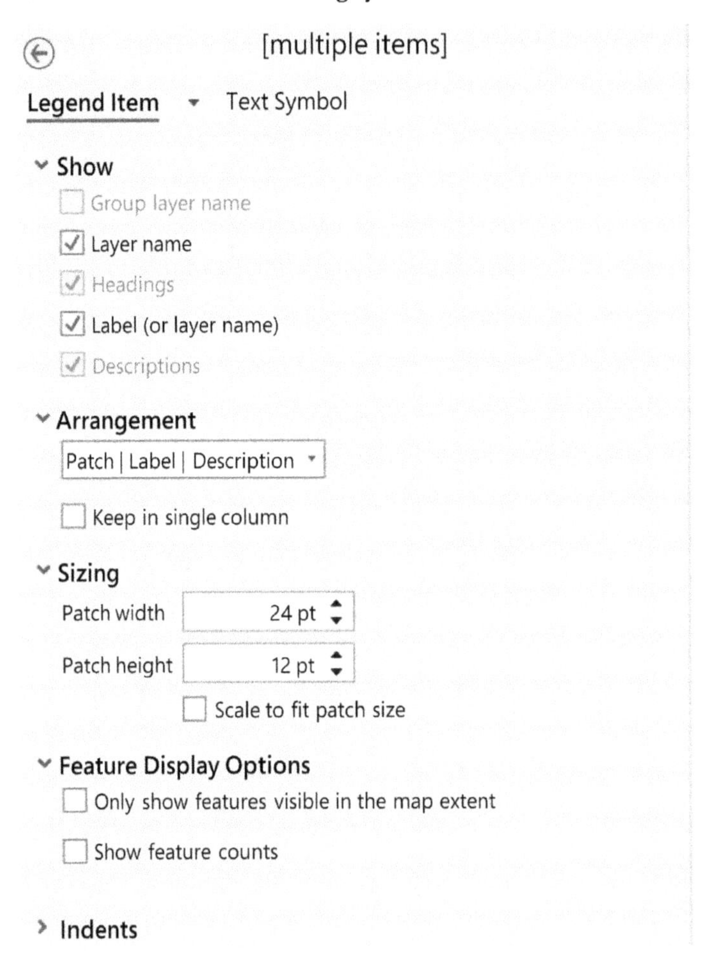

Figure 12.13: Legend Item multiple selection

If something is grayed out, it means the different items selected have different settings selected for that option, and you will have to change them individually. The **Text Symbol** tab will bring up the tabs for formatting the text. When you are done viewing this, click the back arrow to get back to the **Legend Options** tab.

The **Show Properties (...)** button is useful for creating universal settings for all of your legend items. It can save you time in creating a legend that is ready for automation if you want all of your patch sizes, arrangements, and fonts to be the same for each element. Otherwise, you have to create the same settings for each element.

4. Click the **Legend Arrangement Options** tab next to the **Options** tab.

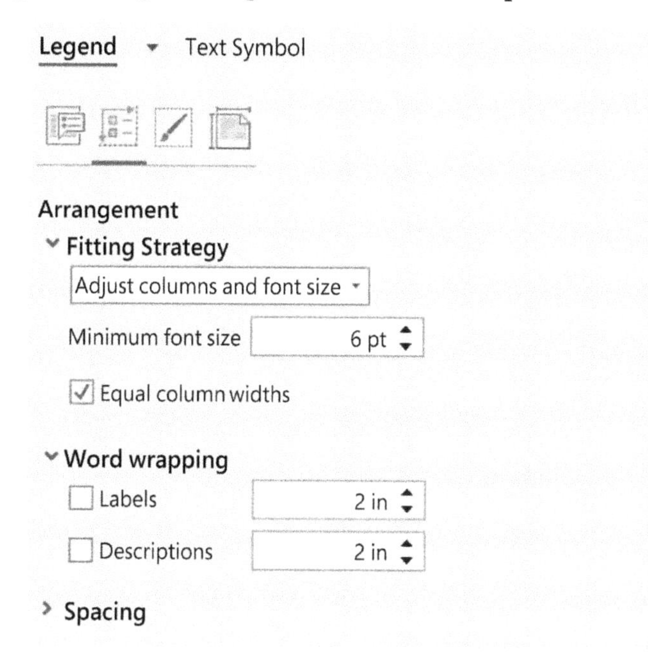

Figure 12.14: Legend Arrangement tab

The only setting here that can be controlled using ArcPy is the **Fitting Strategy**; refer to *Chapter 7* for the different `fittingStrategy` methods available. **Adjust columns and font size** has been selected for this map. This Fitting Strategy should ensure that when you add items to your map and legend, the legend will display all of them.

5. Look at the **Display** tab. It has the same border, background, and shadow settings that cannot be modified using ArcPy.
6. View the **Placement** tab and make sure the anchor point is set to the top left.

You have checked all of the legend elements and set them for your map. Now it is time to look at the individual legend elements and set them up so when you automate your map creation, the legend items are displayed correctly.

Legend Item elements

You have three legend items in your legend right now:

- AC_Transit_Additional_Transbay
- Summer21_RouteShape
- AlamedaContraCostaCounty_RaceHispanic_BlockGroup

Only AlamedaContraCostaCounty_RaceHispanic_BlockGroup should be checked to make it visible.

From here, you are going to view the legend item elements for AlamedaContraCostaCounty_RaceHispanic_BlockGroup and then save a legend item setting as a default.

Saving a default will mean that whenever a legend item is added it will have those settings.

1. Click the **AlamedaContraCostaCounty_RaceHispanic_BlockGroup** legend item to view the **Legend Item** elements pane.

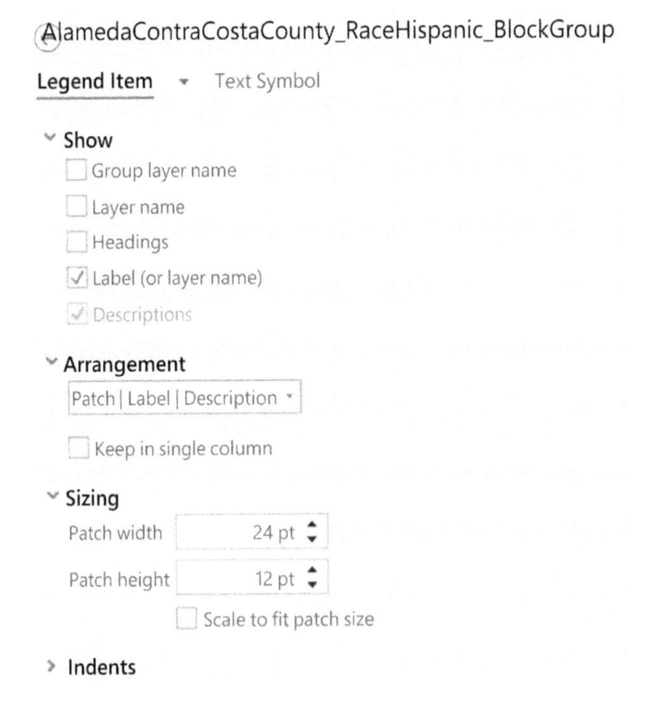

Figure 12.15: Legend Item element tab

You will set the display settings for the different elements of this **Legend Item** here.

2. Click the dropdown next to **Legend Item** to view the dropdown options.

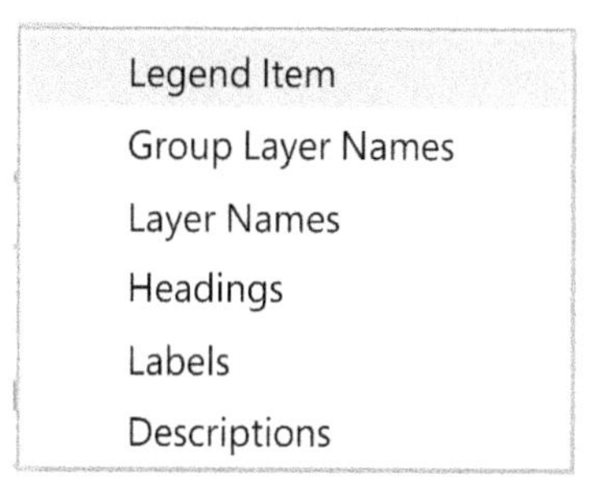

Figure 12.16: Legend Item dropdown

Legend Item is the top level where you select what legend items will be displayed. **Group Layer Names**, **Layer Names**, **Headings**, **Labels**, and **Descriptions** all give you access to the text formatting settings. You can select all the same text formatting settings as with the text element here. None of these settings can be accessed with ArcPy and must be set here.

3. Click on **Legend Item** to stay with the legend item settings. The legend item for `AlamedaContraCostaCounty_RaceHispanic_BlockGroup` is set to only display the **Label** (or **Layer Name**). The legend right now is showing only the label for each race and the dot symbol for each. You want the layers you add to also only display the label or layer name when they are added. To do that, you will save this legend item as the default.
4. Right-click on **AlamedaContraCostaCounty_RaceHispanic_BlockGroup** and select **Save As Default**. Now every layer that is added to the legend will have the same setting as the `AlamedaContraCostaCounty_RaceHispanic_BlockGroup` item.

You can update and make changes to the `AlamedaContraConstaCounty_RaceHispanic_BlockGroup` legend item now and they will not impact the default; that default is saved based on the settings at the point of saving.

5. Click the checkbox next to **Layer name** in the legend item element tab. This will turn on the layer name for the `AlamedaContraCostaCounty_RaceHispanic_BlockGroup` item. You should use the layer name for display on the `AlamedaContraCostaCounty_RaceHispanic_BlockGroup` item, as you can access and change that in ArcPy.
6. Check the box next to **Headings** on the `AlamedaContraCostaCounty_RaceHispanic_BlockGroup` item tab.

This will display the heading for a dot density that shows the ratio for what each dot equals. Your legend should now look like this:

AlamedaContraCostaCounty_RaceHispanic_BlockGroup
1 Dot = 1 Person
White
Black
American Indian/Native Alaskan
Native Hawaiian/Pacific Islander
Asian
Some Other Race
Two Or More Races
Hispanic/Latino

Figure 12.17: Legend with updated settings

You have set your defaults for your legend and set a default setting for legend items that will be added. You can now add items to your legend and rename the layer names and labels through ArcPy. It is important to set as many of these settings as possible when creating your layout template, as you cannot change many of them through ArcPy.

Make sure that the text formatting for your legend layer names, headings, labels, and description are all set before working on your automation script, as you cannot change those in the script.

Scale bar and north arrow elements

The scale bar and north arrow settings that are accessible through ArcPy are both `MAPSURROUND_ELEMENTS`. You can only change the height, width, X and Y position, map frame the element is associated with, name, and visibility. You will set the scale bar and north arrow details within ArcGIS Pro.

Scale bar

You will check the scale bar element to see the settings for the layout. You will also set the anchor point from within the scale bar element.

1. Click on **Scale Bar** to bring up the element pane, where you will format the scale bar element. It should look like this:

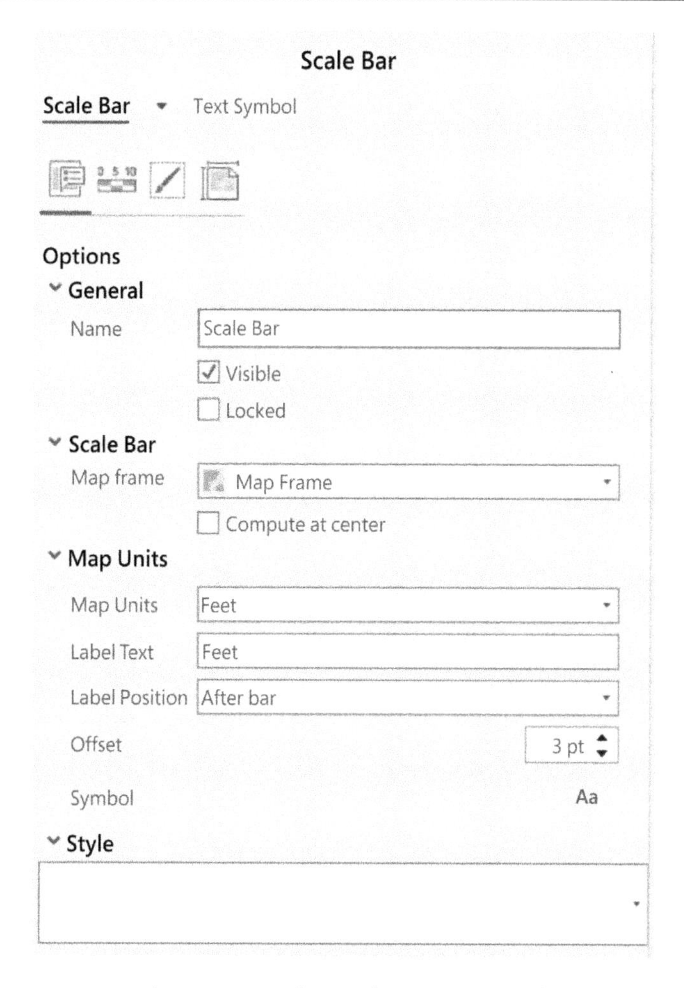

Figure 12.18: Scale Bar element Options tab

2. The **Options** tab shows you the **General**, **Scale Bar**, **Map Units**, and **Style** settings. None of these settings can be modified with ArcPy and need to be set here.

3. The **Properties** tab should look like the figure below. This is where you have many options available to you for styling your scale bar:

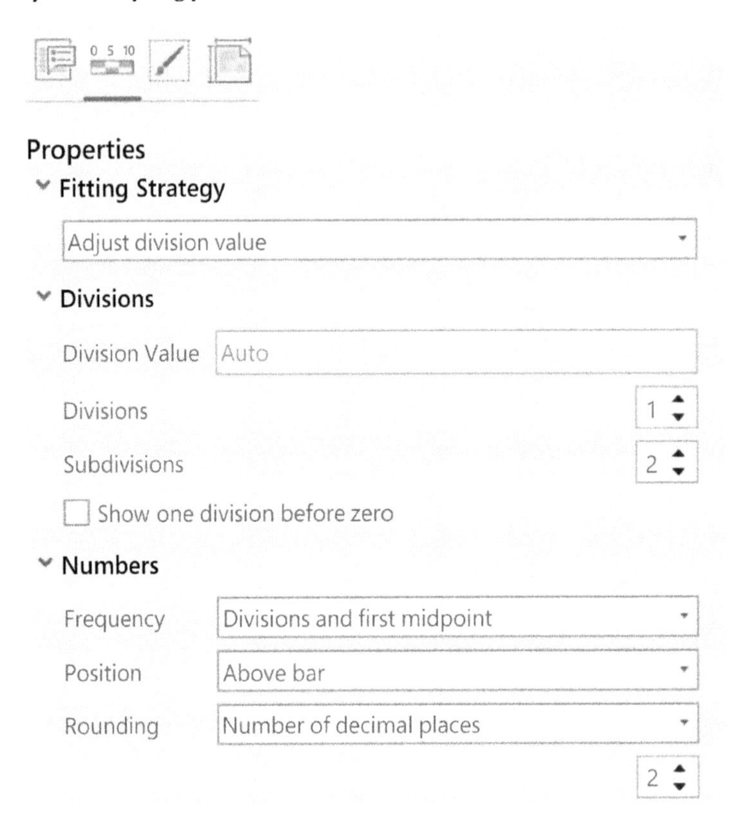

Figure 12.19: Scale Bar element Properties tab

None of these settings can be modified with ArcPy and must be set here. These settings play an important part in the look of your scale bar.

Note that the **Fitting Strategy** is set to **Adjust division value.** This will ensure that the scale bar stays the same size and the division values are what will change as your scale changes. This is a good setting to use when creating map automations where the scale may change a great deal. It allows you to make sure that your scale bar will always be the same size no matter what scale it is showing.

4. The other two tabs are the **Display** and **Placement** tabs, which are similar to the display and placement tabs in the other layout elements. The **Display** tab will set the border, background, and shadow display settings. The **Placement** tab will show you the size, position, and anchor points. Click the **Placement** tab and make sure the anchor point is the bottom left.

A lot of details in the scale bar can be set in the element pane. In addition to what was discussed, you can click the drop-down arrow next to **Scale Bar** to access the following graphic elements and their settings: **Background, Border, Shadow, Numbers, Unit, Symbol 1, Symbol 2, Division Marks, Subdivision Marks,** and **Midpoint Marks.** You can also click on the **Text Symbol** tab to access the same text formatting options available to all text elements.

The background, border, and shadow graphic elements can be set from either the dropdown next to **Scale Bar** or in the **Display** tab.

These settings allow for a sizeable number of cartographic options for creating rich maps. However, none of them can be accessed and changed in map automation. Because of that, it is important that you set them before you begin your automation and make sure that they will work with all the different scales and map views your automation will create.

North arrow

The north arrow is similar to the scale bar in that most of its settings can only be set in the **Element** tab in ArcGIS Pro. You will check the settings on the north arrow to see what has been set before creating your automation.

1. Click on **North Arrow** to bring up the element pane to format the north arrow element.

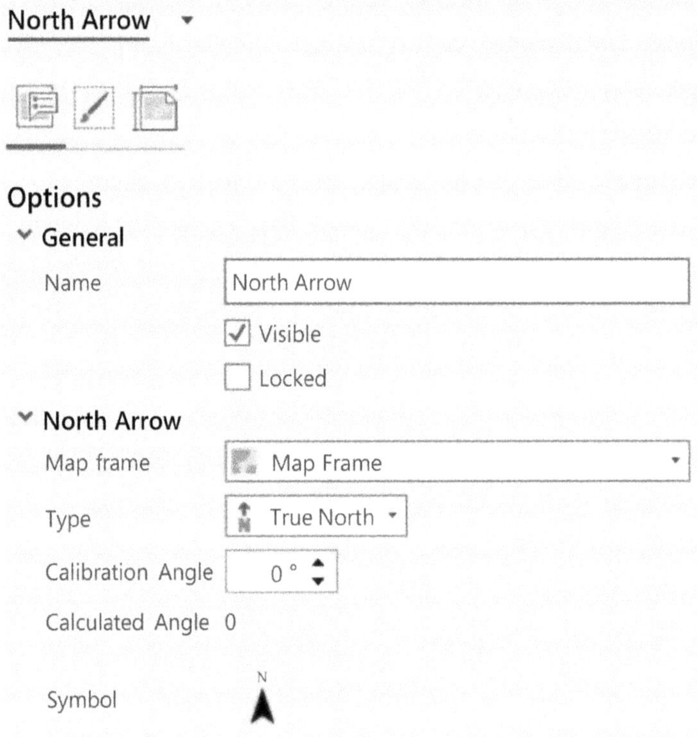

Figure 12.20: North Arrow

The first tab is the **Options** tab and shows you the name of the element, its visibility, the map frame it is associated with, the type of north arrow, the calibration angle, and the symbol. Of these, only the **Name** and **Visible** settings can be changed with ArcPy in your map automation.

2. The north arrow also has the same **Display** and **Placement** tabs as the other layout elements. Click in the **Placement** tab to see the height, width, and position settings. These can all be changed in your automation. Check to make sure the anchor point is the bottom middle.

In addition to those settings, you can also access more details by clicking the drop-down arrow next to **North Arrow**. This will give you access to display settings available in **Point Symbol**, **Background**, **Border**, and **Shadow**. All of these can add rich details to your map. Like with the scale bar, though, they must be set here and cannot be modified in your automation. So, it is important that whatever settings you choose are acceptable for all maps you will export in your automation.

Title text element

The **Title** text element is a text element that is being used for the map title. Since it is a text box like the Source element, it has the same settings. You should check the anchor point to make sure it is in the correct location.

1. Click on **Title** to bring up the element pane to format the Title text box.
2. Click on the **Placement** tab under the **Text** tab to show the placement settings. Check to make sure that the anchor point is the top-right corner.

You can further explore the settings for the Title text element in your own time. Like with all text elements, you are limited in what you can control in ArcPy. It is important to set any specific text fonts and formatting now, as they cannot be changed in the map automation process.

Map Frame element

The **Map Frame** element is a map frame element like the Inset map frame, so it will have the same settings available; refer to that section above for more details. As this is the main map frame, you will be keeping all the settings as they are currently set.

You have now set up a layout that you can add new layers to and then iterate through different views to create a series of maps. Often, when setting up your layout, you will need to experiment with the settings to make sure they work for the data you are adding and changing through the automation. This part of map automation is not any different from creating layouts and following the principles of cartography. The only difference is you are looking to ensure your settings work not just on the template map but also on the tens, hundreds, or even thousands of pages you will be creating and exporting.

Always pay attention to what you can and cannot change through ArcPy. Spend time creating a template that sets all of the elements that cannot be changed before starting your automation.

Creating and adding data to your map

You now have a layout that is ready for creating your map automation. It contains:

- The existing Transbay bus routes
- The Transbay bus routes that were suspended
- The race data by block group displayed using dot density symbology

The legend is set up to add in the new layers you will be creating using ArcPy in an ArcGIS Pro Notebook.

The final product will be a map series for a suspended bus line that highlights each block group within a 0.5-mile study area of the route with a minority population larger than that of a reference community. The reference community is the minority percentage of all the block groups within 0.5 miles of the routes that were not suspended. This will allow you to understand how many block groups in the suspended line have a higher percentage of minority population than the population along the bus routes that remained.

Each block group with a minority percentage higher than the reference community will be highlighted by being the only one on the page displaying the dot density map. Any surrounding block groups in the bus route's study area will be symbolized by whether they have a minority population above or below the reference community. Each page will utilize the **DetailsBox** element to display the percentage of each race group in the block group. From this, you can identify areas of potential disproportionate impact on minorities by the suspension of this bus line. This will help your team identify the impact on minority communities if the bus line were to be resumed.

1. Open the **Catalog** pane if it is not already open and select the **Project** tab.
2. Within the `Chapter12` project folder, right-click and select **New** > **Notebook**.
3. Right-click on **New Notebook.ipyb** and select **Rename.** Rename the Notebook to `CreateMapSeriesForOneBusLine`.
4. In the first cell, you import the modules you need and set an environment setting. You will import the os module to help you work with filenames and paths. You will also need the `glob` and `PyPDF2` modules, which will be used together to combine the individual PDFs in a later step. The `glob` module is used to find files in a workspace based on a file pattern. The `PyPDF2` module will be used to merge the PDFs into a single PDF. You will also set the `overwriteOutput` environment setting to `True`, so that any output that has the same name as an existing file will overwrite that existing file. This is useful when working through and testing map automation scripts. Type in the following:

   ```
   import os, glob
   from PyPDF2 import PdfFileMerger, PdfFileReader
   arcpy.env.overwriteOutput = True
   ```

   Run the cell.

5. In the next cell, you will define all of the variables that you will need throughout the script. There will be a lot of variables, as you will be creating feature layers and new layers. The first two variables are the project geodatabase and project folder. They may be different from what you see below, depending on where you downloaded the data to. In addition, you need to define the feature classes that you are starting with. This is also the place where you will create your variables for your map and layout elements. You will also create a variable to hold the bus route name from the suspended bus route dataset. Type in the following:

```
# Project gdb and folder locations
projectGDB = r"C:\PythonBook\Chapter12\Chapter12.gdb"
projectFolder = "\\".join(projectGDB.split("\\")[:-1])

# Bus and census data that already exists in the project gdb
busLines = os.path.join(projectGDB,"Summer21RouteShape")
censusPoly = os.path.join(
    projectGDB,"AlamedaContraCostaCounty_RaceHispanic_BlockGroup"
)
newBusLines = os.path.join(
    projectGDB,"AC_Transit_AdditionalTransbay"
)

# Summary stat table to be created for the reference community
calculations
sumStatTable = os.path.join(projectGDB,"SummStat_RaceHispanic_
StudyArea")

# Feature layers created for selections
studyAreaFL = "SelectedCensus_FL"
busLinesFL = "SelectedBusLines_FL"
newBusCBGs = "NewBusCBGsRefComm"

# Suspended bus route and data created for suspended bus route
busRoute = "C"
newBusLinesSel = os.path.join(projectGDB,"NewTransbayLine_{0}".
format(busRoute))
cbgStudyArea = os.path.join(projectGDB,"CBG_StudyArea_Bus_{0}".
format(busRoute))
```

```
# Project and map and layout object for the map, inset map, and map
frames for map and inset map
project = arcpy.mp.ArcGISProject("CURRENT")
m = project.listMaps("Map")[0]
inset = project.listMaps("Inset")[0]
layout = project.listLayouts("Layout")[0]
mf = layout.listElements("mapframe_element","Map Frame")[0]
mfInset = layout.listElements("mapframe_element","Inset")[0]

# Table header, table box, title text elements, and legend element
tableHeader = layout.listElements("TEXT_ELEMENT","DetailsHeader")[0]
tableBox = layout.listElements("TEXT_ELEMENT","DetailsBox")[0]
title = layout.listElements("TEXT_ELEMENT","Title")[0]
legend = layout.listElements("LEGEND_ELEMENT","Legend")[0]
```

Run the cell.

6. In the next cell, you will create a summary statistics table of the total population for each race in the block groups that are within 0.5 miles of a bus route. This summary statistics table is how you will find the minority percentage for your reference community. The entire process will be wrapped in a `with` statement that will change the environment settings so any output feature layers or feature classes will not be added to the map. This will prevent you from having to remove layers you only created for analysis later. Within the `with` code block, you will create a feature layer containing the existing Transbay bus lines and a feature layer containing the race block group. You will then use the **Select By Location** option to select the block groups in the race block group feature layer that are within 0.5 miles of the existing Transbay bus route feature layer. Type in the following:

```
with arcpy.EnvManager(addOutputsToMap=False):
    arcpy.management.MakeFeatureLayer(censusPoly,studyAreaFL)
    arcpy.management.MakeFeatureLayer(
        busLines,busLinesFL,
        "PUB_RTE IN ('W', 'V', 'U', 'P', 'O', 'NX', 'NL', 'LA', 'L',
                     'J', 'G', 'F', 'DB1', 'DB')"
    )
    arcpy.management.SelectLayerByLocation(studyAreaFL,"INTERSECT",
                                           busLinesFL,"0.5 Miles")
```

7. *Continuing in the same cell*, read in the fields to be used in the summary statistic calculation from the `AlamedaContraCostaCounty_RaceHispanic_BlockGroup` feature class.

This is done by looping through the fields in the feature class and adding a field to a `summStat` list only if the field is an integer. In addition, you need to add the summary statistic to be applied to that field. It will be `"SUM"`, as you want to sum up the totals of each race group for all the selected block groups. The summary statistic tool will then take that list of summary statistic fields as a parameter. Within the loop, you will add a print statement to ensure you are getting the correct fields. Type in the following:

```
summStats = []
for field in arcpy.ListFields(studyAreaFL):
    if field.type == "Integer":
        print("Added {0} field name to summary stat list"
              .format(field.name))
        summStats.append([field.name,"SUM"])
arcpy.analysis.Statistics(studyAreaFL,sumStatTable,summStats)
```

Run the cell. You will see the following output for each field that is added to your `summStats` list, confirming the creation of the summary statistics table:

```
Added total_pop field name to summary stat list
Added white field name to summary stat list
Added black field name to summary stat list
. . .
Added total_minority field name to summary stat list
```

How do you know what field types you have in your attribute field?

Right-click on the `AlamedaContraCostaCounty_RaceHispanic_BlockGroup` feature class in the **Contents** pane and select **Attribute Table.** Within the attribute table, right-click one of the fields and select **Fields.** The **Data Type** column is the type. You can see that for the `AlamedaContraCostaCounty_RaceHispanic_BlockGroup` feature class, the only Long values are the population counts.

8. In the next cell, you will add a field for the minority reference community, calculate the percentage minority in that reference community, and assign it to a variable to be used for finding and symbolizing census block groups. To do this, you will use the field name and the `.find()` function to find the **total population** and **minority** fields.

The `variable.find()` function looks for the exact string value provided within the `()` in the `variable`. If it finds it, it returns the position in the string where it starts. If it doesn't find anything, it returns -1.

Once you have those field names, you will use them to calculate the percentage minority of all the block groups within 0.5 miles of existing Transbay lines. You will use the `CalculateField` tool to add the percent minority you just calculated to the minority reference community field you added to the summary statistics table. Next, you need to extract that minority percentage from the summary statistics table and store it in a variable for later use in comparing each block group's minority percentage. To do that, you create a search cursor on the summary statistic table and the `MinorityRefCom_Prct` column and use `row = next(cursor)` to extract the first row and stop the cursor. You can do this here because the summary statistic table is just one row. But the row is a list of values, so you need to take the first value as that is the minority percentage of the reference community.

Refer to *Chapter 4, The Data Access Module and Cursors,* for a refresher on the search cursor.

Finally, a print statement at the end will verify that you have extracted the reference community percentage. Type in the following:

```
arcpy.management.AddField(sumStatTable,"MinorityRefComm_
Prct","FLOAT")

for f in arcpy.ListFields(sumStatTable):
    print(f.name)
    if f.name.find("minority") != -1:
        print("Minority find value is {0} for field {1}"
              .format(str(f.name.find("minority")),f.name))
        numerField = f.name
    if f.name.find("pop") != -1:
        print("Pop find value is {0} for field {1}"
              .format(str(f.name.find("pop")),f.name))
        denomField = f.name
arcpy.management.CalculateField(sumStatTable,"MinorityRefComm_
Prct","(!{0}!/!{1}!)*100".format(numerField,denomField))
```

```
with arcpy.da.SearchCursor(sumStatTable,["MinorityRefComm_Prct"]) as
cursor:
    row = next(cursor)
    refCommPrct = row[0]

print("Reference Community Number is {0}".format(str(refCommPrct)))
```

Run the cell. You will see the following output:

```
Reference Community Number is 65.56812286376953
```

9. In the next cell, you will create a new feature class for the selected suspended bus route, make a feature layer of the block groups within 0.5 miles of that route, and remove any GEOID that has a population of 0. You now have a feature layer that contains all of the block groups with a population greater than 0 that are within 0.5 miles of the selected suspended bus route. This feature layer is your study area for the suspended bus route. You will take this study area feature class and use a search cursor to create a list of all the GEOIDs, and a list of just the GEOIDs that have a percent minority population greater than the reference community.

   Then, you will convert the list of all the block groups to a tuple, creating a list surrounded by parentheses that can be inserted directly into a SQL `in` query. This SQL statement will be used to create a feature class of the census block groups study area. You will include some print statements to track your process and review the results.

   All of this will again be wrapped in a `with` statement, to set the environment settings so outputs are not added to the map. You will add and style the layers you need later. Type in the following:

```
with arcpy.EnvManager(addOutputsToMap=False):
    arcpy.analysis.Select(
        newBusLines, newBusLinesSel,
        "route_s_nm = '{0}'".format(busRoute)
    )
    arcpy.management.MakeFeatureLayer(censusPoly,newBusCBGs)
    arcpy.management.SelectLayerByLocation(
        newBusCBGs,"INTERSECT", newBusLinesSel, "0.5 Miles"
    )
    arcpy.management.SelectLayerByAttribute(
```

```
        newBusCBGs,"REMOVE_FROM_SELECTION","total_pop = 0"
    )
    minorityGEOIDs = []
    allGEOIDs = []
    with arcpy.da.SearchCursor(
        newBusCBGs,["GEOID","percent_minority"]) as cursor:
        for row in cursor:
            allGEOIDs.append(row[0])
            if row[1] >= refCommPrct:
                minorityGEOIDs.append(row[0])
                print("Added {0} to minority GEOID list"
                      .format(row[0]))

    cbgStudyAreaTuple = tuple(allGEOIDs)
    cgbStudyAreaSQL = "GEOID in {0}".format(cbgStudyAreaTuple)
    print(cgbStudyAreaSQL)
    arcpy.analysis.Select(censusPoly,cbgStudyArea,cgbStudyAreaSQL)
```

Run the cell. You will see the following output (truncated here):

```
Added 060014036001 to minority GEOID list
Added 060014013001 to minority GEOID list
...
Added 060014251023 to minority GEOID list
GEOID in ('060014036001', '060014012003', ..., '060014007004')
```

10. In the next cell, you are going to add in the census block group study area you created above and style it. You will add the study area block group to the map, create a layer object for its layer in the map, and create a symbology object from that layer object. You will also update the renderer to a graduated colors renderer with a classification field of `percent_minority`, a break count of 2, and a break value of the reference community variable. This will create the study area layer of block groups surrounding your highlighted minority block group. It will help put the surrounding area of the highlighted block group into context and allow readers to see if there are large areas of connected block groups with high minority populations.

For more information on using break counts and break values with the graduated colors renderer, refer to *Chapter 7*.

Once those are set, you will loop through the class breaks in the renderer and set the label, symbol color, and outline color for the block groups with a minority population below the reference community, and for those with a minority population above the reference community. All block groups will be set with a symbol color that is 70% transparent with an outline that is 50% transparent. The block groups above the reference community will be red, and those below will be green.

The last step is setting the new layer symbology equal to the new symbology object. Type in the following:

```
m.addDataFromPath(cbgStudyArea)
cbgStudyAreaLyr = m.listLayers("CBG_StudyArea_Bus_{0}".
format(busRoute))[0]
cbgStudyAreaLyrSym = cbgStudyAreaLyr.symbology
cbgStudyAreaLyrSym.updateRenderer('GraduatedColorsRenderer')
cbgStudyAreaLyrSym.renderer.classificationField = "percent_minority"
cbgStudyAreaLyrSym.renderer.breakCount = 2
breakValue = refCommPrct
firstVal = 0
for brk in cbgStudyAreaLyrSym.renderer.classBreaks:
    brk.upperBound = breakValue
    if firstVal == 0:
        brk.label = "Minority Population > Reference Community"
        brk.symbol.color = {'RGB' : [255, 0, 0, 30]}
        brk.symbol.outlineColor = {'RGB' : [255, 0, 0, 50]}
    else:
        brk.label = "Minority Population < Reference Community"
        brk.symbol.color = {'RGB' : [0, 255, 0, 30]}
        brk.symbol.outlineColor = {'RGB' : [0, 255, 0, 50]}
    breakValue = 100
    firstVal += 1
cbgStudyAreaLyr.symbology = cbgStudyAreaLyrSym
```

Run the cell.

There will be no output, but the layer `CBG_StudyArea_Bus_C` will be added to the map and symbolized with a red polygon for the **Minority Population > Reference Community** and a green polygon for **Minority Population < Reference Community**:

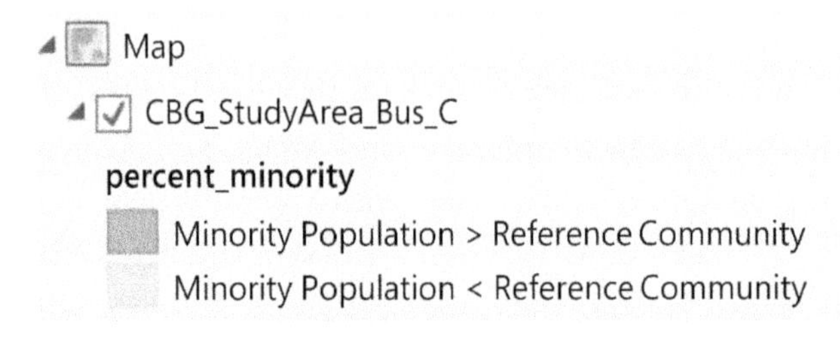

Figure 12.21: Study area census block groups in the map

11. In the next cell, you will add the selected suspended Transbay bus line you created in *step 9* and symbolize it. You will add the data to the map, create a layer object from it, and create a symbology object from the layer object. You will leave the Transbay bus line renderer as a simple renderer, only changing the line color to purple and increasing the line size to 1.5 points. You will also change the layer name to something that will look better in the legend. Type in the following:

```
m.addDataFromPath(newBusLinesSel)
newBusLyr = m.listLayers("NewTransbayLine_{0}".format(busRoute))[0]
newBusLyrSym = newBusLyr.symbology
newBusLyrSym.renderer.symbol.color = {"RGB" : [169, 0, 230, 100]}
newBusLyrSym.renderer.symbol.size = 1.5
newBusLyr.symbology = newBusLyrSym
newBusLyr.name = "Transbay Bus Route"
```

Run the cell. There will be no output, but the layer `NewTransbayLine_C` will be added to the map, the symbol will change to a purple 1.5-point line, and the name will be **Transbay Bus Route**:

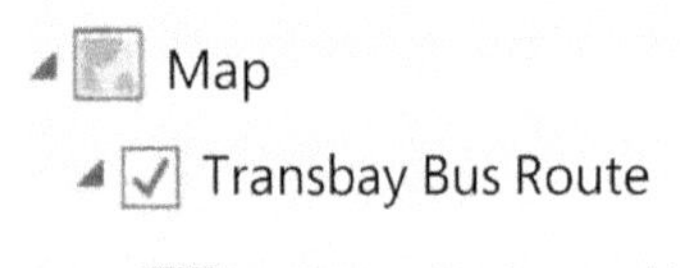

Figure 12.22: New Transbay Bus Route layer

12. In the next cell, you will add the selected suspended Transbay bus route and the block group study area to the inset map. This will help the reader place the highlighted block group and the surrounding study area along the full route. Before adding the layers to the inset map, you need to check if they are already there. If there are already layers in the inset with the same name, you need to remove them and add in these layers. You should make sure the layers being displayed in the inset are the current ones in the main map.

    To start, you use a **list comprehension** to build a list of layers in the inset map. You will then use conditionals to check if the Transbay bus route layer and census block group study areas are in the inset. If they are, you will delete that layer and then add in the new layer from the map. If they are not, you will add in the new layer. You will add print statements to track your progress and print out results. Type in the following:

    ```
    curInsetLayers = [l.name for l in inset.listLayers()]
    if cbgStudyAreaLyr.name in curInsetLayers:
        insetCBGLyr = inset.listLayers(cbgStudyAreaLyr.name)[0]
        inset.removeLayer(insetCBGLyr)
        print("Removing old CBG Study Area Layer to Inset")
        inset.addLayer(cbgStudyAreaLyr)
        print("Adding new CBG Study Area Layer to Inset")
    else:
        inset.addLayer(cbgStudyAreaLyr)
        print("Adding new CBG Study Area Layer to Inset")
    if newBusLyr.name in curInsetLayers:
        insetBusLyr = inset.listLayers(newBusLyr.name)[0]
        inset.removeLayer(insetBusLyr)
        print("Removing old Transbay Bus Route to Inset")
        inset.addLayer(newBusLyr)
        print("Adding new Transbay Bus Route to Inset")
    else:
        inset.addLayer(newBusLyr)
        print("Adding new Transbay Bus Route to Inset")
    ```

    Run the cell. If your inset does not have any old layers, you will see the following printed out:

    ```
    Adding new CBG Study Area Layer to Inset
    Adding new Transbay Bus Route to Inset
    ```

The study area layer and the Transbay bus routes layers have now been added to the inset.

A list comprehension is just a shorter way to create a list of values. Above, you wrote:

```
curInsetLayers = [l.name for l in inset.listLayers()]
```

That is the same as writing the following:

```
curInsetLayers = []
for l in inset.listLayers():
    curInsetLayers.append(l.name)
```

By using a list comprehension, you have condensed three lines of code into one.

13. In the next cell, you will get the extent of the study area, set the inset extent to that extent, and then increase the scale to add some buffer between the study area and the edge of the inset map. You will use the **Camera object**, which allows you to control the scale and extent on 2D maps and the camera position on 3D maps.

    You will first create a layer object for the study area layer. You will then get the extent of that layer. You will use the camera object to set the extent of the inset frame to the extent of that layer. Next, you will update the scale by using the camera object to set the scale to the current scale plus 2,000, and round it to a whole number. You will include print statements to track your progress and results. Type in the following:

```
insetStudyAreaLayer = inset.listLayers("CBG_StudyArea_Bus_{0}".
format(busRoute))[0]
insetExtent = mfInset.getLayerExtent(insetStudyAreaLayer,False,True)
print(insetExtent)
mfInset.camera.setExtent(insetExtent)
mfInset.camera.scale = round((mfInset.camera.scale + 2000),0)
print(mfInset.camera.scale)
```

Run the cell. You should see the following printed out:

```
-122.347072470221 37.8072665294239 -122.211846529491
37.8503534704394 NaN NaN NaN NaN
441507.0
```

The inset map now displays the selected bus route and its study area, at an extent and scale where you can see the entire study area.

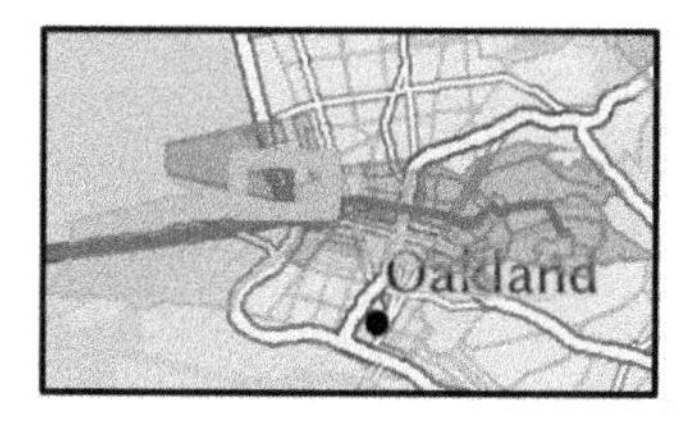

Figure 12.23: Inset map with study area layers

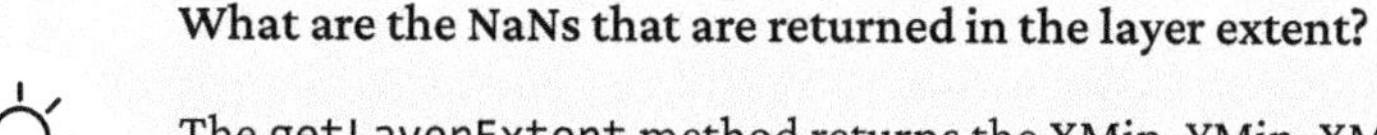

What are the NaNs that are returned in the layer extent?

The `getLayerExtent` method returns the XMin, YMin, XMin, YMin, ZMin, ZMax, MMin, and MMax extent values of the layer. If your layer is not Z- or M-enabled, those values will be NaN for Null.

14. In the next cell, you need to turn on the labels for the study area in the map. Labeling the surrounding study area block groups will allow readers to identify any of the surrounding block groups that also have a minority population percentage larger than the reference community and find its map in the map series. This can help people to see any patterns and groupings of minority communities that may have been disproportionately impacted by this bus line suspension. You are limited to changing the label's expression, visibility, SQL query, and name of the label class when using the `LabelClass`.

You will create a label class and set its expression. The label expression is what you would write in the expression window of the labeling pane when labeling feature classes manually.

Figure 12.24: Label expression

When labeling using a field, you wrap the field in quotation marks and add `$feature.` before the field name. You will set the label class to Visible and the layer property `showLabels` for the census block group study area to `True`, to display the labels. You will be labeling using the GEOID field.

This needs to be done after having added this layer to the inset map, or the labels will be displayed on the inset map, which is too small for labeling each block group in the study area.

Type in the following

```
cbgLayerLabels = cbgStudyAreaLyr.listLabelClasses()[0]
cbgLayerLabels.expression = "$feature.GEOID"
cbgLayerLabels.visible = True
cbgStudyAreaLyr.showLabels = True
```

Run the cell. You will not see any output, but if you select the Map you will see that the census block group study area polygons are now labeled:

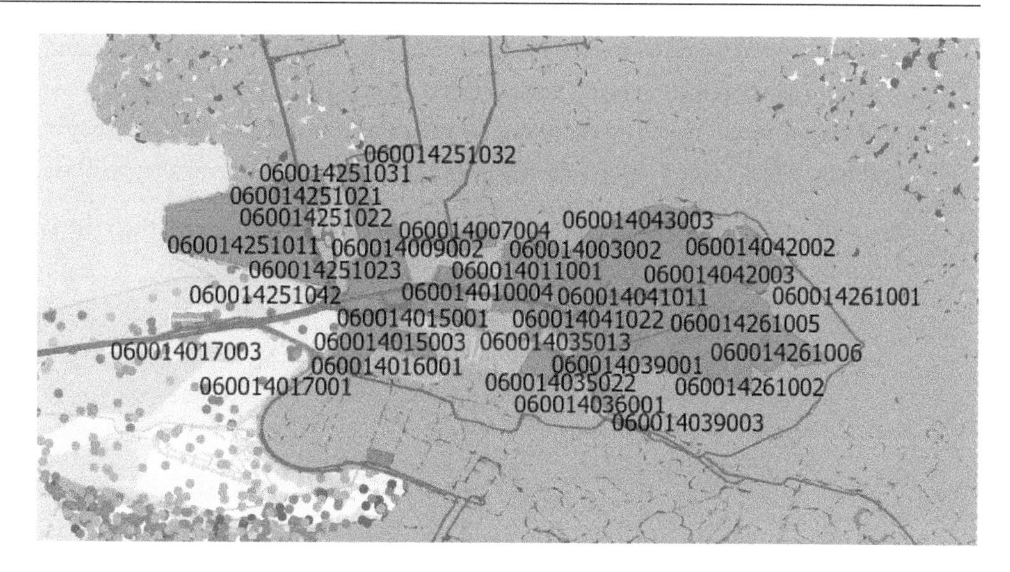

Figure 12.25: Study area block groups labeled

You now have all of your data created and added to the map. The next section will update the legend and work on setting up the data for the table box and table header.

Working with legend and text elements in the layout

Now that you have all of your layers on your map, you need to make sure those layers are properly called out in your legend. In this section, you will take layers you just added to the map and add them to your legend. They will be added using the default legend style you set above in the *Legend Item elements* section.

In addition, you also want to add some details about each highlighted block group. Your map would be much more useful to readers if you included a table that contained the percentage of each race group in the highlighted block group on your map. This will allow them not only to see the race groups in the dot density map but also to reference the percentage each occupies in the block group.

You will do this by creating a list and data dictionary that will be used to extract data from the attribute table of the `AlamedaContraCostaCounty_RaceHispanic_BlockGroup`. This data will be extracted in the next section to insert into the table box text element.

1. If you do not have the `CreateMapSeriesForOneBusLine` Notebook open, open it back up.
2. In the cell at the bottom after the last cell, you will add the selected Transbay bus route and the census block group study area to the legend.

You do not want to add a layer if it is already in the legend, so you will use the same process from above when adding layers to the inset map: creating a list comprehension of all the layer names in the legend and checking them against the names of the Transbay bus route and census block group study area layers. If they are not in the legend, they will be added in.

You will add the census block group study area first and put it on top, then the Transbay bus line, also putting it on top. Type in the following:

```
legendItemNames = [item.name for item in legend.items]
if cbgStudyAreaLyr.name not in legendItemNames:
    legend.addItem(cbgStudyAreaLyr,"TOP")
if newBusLyr.name not in legendItemNames:
    legend.addItem(newBusLyr,"TOP")
```

Run the cell. There will be no output results. Select the **Layout**, and observe that those layers have been added to the legend:

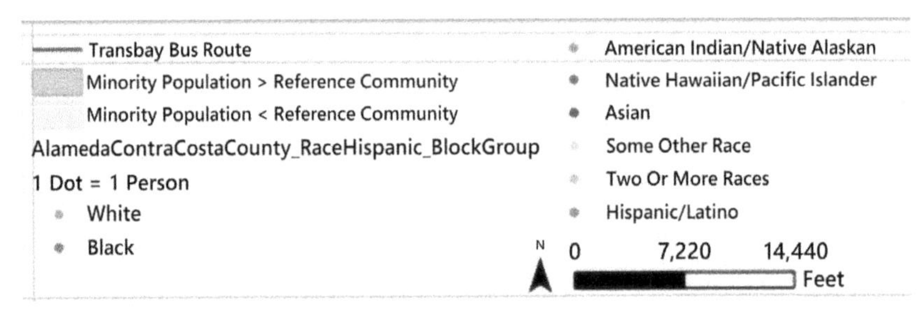

Figure 12.26: Legend with new layers

3. In the next cell, you are going to make sure that only the three layers you want to be displayed in the legend are being displayed. At this point, you should have only the two items you just added and the dot density layer of the race by block group, but it is still a good idea to make sure that nothing unexpected is being displayed.

   You will loop through the legend items to find item names matching those of the layers you want to display, setting their `visible` property to `True` if matching and `False` otherwise. Type in the following:

   ```
   for item in legend.items:
       if item.name in (
           cbgStudyAreaLyr.name,
           newBusLyr.name,
   ```

```
        "AlamedaContraCostaCounty_RaceHispanic_BlockGroup"
    ):
        print('{0} is displayed in the legend'.format(item.name))
        item.visible = True
    else:
        item.visible = False
```

Run the cell. You should see the following print statements:

```
Transbay Bus Route is displayed in the legend
CBG_StudyArea_Bus_C is displayed in the legend
AlamedaContraCostaCounty_RaceHispanic_BlockGroup is displayed in the
legend
```

4. Select the **Layout** and observe that the correct layers are displayed.
5. In the next cell, you will use the field type information in the `AlamedaContraCostaCounty_RaceHispanic_BlockGroup` to create a list of the fields with the percentage of each race group. Along with this, you will create a data dictionary containing the percentage of each race group field name as the key and the field alias as the value. You will start by creating an empty list and an empty data dictionary. You will then loop through the fields in `AlamedaContraCostaCounty_RaceHispanic_BlockGroup` and check for the field type `"Single"`.

   When you look at the attribute table for this layer, you will see that the only **Float** fields are all of the percentage fields, so only those fields will pass this test. You will add those field names to that list, also adding the field name as the key and the field alias as the value to the data dictionary. Outside the loop, you will set the text of the `tableHeader` text element to `"Percent Race/Hispanic"`. You will add a print statement to track your results. Type in the following:

```
prcField = []
prcDataDict = {}
for field in arcpy.ListFields(censusPoly):
    if field.type == "Single":
        print("Added {0} field name to prcField list"
              .format(field.name))
        prcDataDict[field.name] = field.aliasName
        prcField.append(field.name)
tableHeader.text = "Percent Race/Hispanic"
```

Run the cell. You will see the following results (truncated here):

```
Added prct_white field name to prcField list
Added prct_black field name to prcField list
...
Added percent_minority field name to prcField list
```

Select the **Layout**; the header for the text box will now be **Percent Race/ Hispanic.**

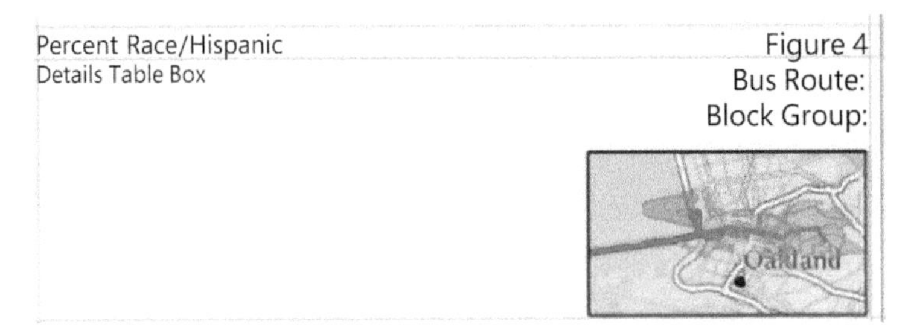

Figure 12.27: Updated details box header

You have now added the data to your legend that you will display in the final layouts you will export. You have also created a list and data dictionary that will be used in the next step to extract the percent race for each selected block group. In the next section, you will export the individual pages for the block groups that have a higher percentage minority than the reference community.

Changing the map view and exporting

At this point, you have added all the elements you need to your map and your legend.

You also have:

- A list of the GEOIDs within your study area of all the census block groups that have a minority percentage that is greater than the reference community
- A list and data dictionary that will be used to create the text box containing the specific race population percentages within the selected census block group

In this section, you will use all of the above to export to PDF a map for each of the census block groups with a minority percentage greater than the reference community. **All the code in this section is going to be written into two cells, and will contain many print statements so you can track the progress of your code.** The first cell is the largest, as it contains the code to loop through the block groups, make all the changes to the map, and export it.

The code for this cell will be split into multiple steps.

1. If you do not have the CreateMapSeriesForOneBusLine Notebook open, open it back up.
2. In a new cell at the bottom, after the last cell from the section above, you will start by creating a layer object for the AlamedaContraCostaCounty_RaceHispanic_BlockGroup layer and a counter that will be used for figure numbers. Type in the following:

```
cbgLayer = m.listLayers("AlamedaContraCostaCounty_RaceHispanic_
BlockGroup")[0]
i = 1
```

3. Continue on the next line and start a for loop through all the values in the minorityGEOIDs list you created above. You will create a definition query on the census block study area layer to remove the highlighted GEIOD from its display. You will also create a definition query on the AlamedaContraCostaCounty_RaceHispanic_BlockGroup to select just the highlighted GEOID. This will make it so your map only shows the dot density map for the selected block group and does not cover it with the red or green of the study area block groups. Finally, you will create an empty list for the text box that you will add data to as you extract it from the AlamedaContraCostaCounty_RaceHispanic_BlockGroup. Type in the following:

```
for geoid in minorityGEOIDs:
    print(i)
    print(geoid)
    cbgStudyAreaLyr.definitionQuery = "GEOID <> '{0}'".format(geoid)
    cbgLayer.definitionQuery = "GEOID = '{0}'".format(geoid)
    textBox = []
```

4. Now you are going to collect the different race group percentages for display for the selected block group and store them in a list. Continuing on the next line, create a with statement for a SearchCursor, which will search through the AlamedaContraCostaCounty_RaceHispanic_BlockGroup feature class. It will use the fields list created in the step above, and a where clause to limit the results to just the highlighted GEOID. Since your where clause is for a specific block group, the search cursor only contains one row and you can use the next() function to get that single row. Next, you set a variable to 0 as your counter and start a while loop to loop through each of the rows returned to the cursor.

Inside the `while` loop, you will write a conditional statement that formats the last row differently from all other rows. You will read in field names using the `cursor.fields[]` method to give you the field name associated with the row of data. That field name is used to look up the field's alias in the data dictionary `prcDataDict` you created in the previous cell. Using `.format()`, you will write to a temporary string variable the field name and the percent minority row value associated with it. This string variable will have the return (`\r`) and new line (`\n`) characters at the end for all rows except the last row, and will be appended to the text box list for each row. Finally, you will increment your counter by 1.

Type in the following:

```
with arcpy.da.SearchCursor(
        censusPoly,prcField,"GEOID = '{0}'".format(geoid)
    ) as cursor2:
        row = next(cursor2)
        j=0
        while j < len(row):
            print(cursor2.fields[j])
            print(row[j])
            if j != (len(row)-1):
                tempVal = "{0}: {1}\r\n".format(
                    prcDataDict[cursor2.fields[j]],round(row[j],2)
                )
                print(tempVal)
            else:
                print("the last value")
                tempVal = "{0}: {1}".format(
                    prcDataDict[cursor2.fields[j]],round(row[j],2)
                )
                print(tempVal)
            textBox.append(tempVal)
            j+=1
```

5. You can now take that list and iterate through it, adding each value to the text box. This will create your table of race percentages on your map. Continuing on the next line, you will use the `textBox` list to set the data in your table box. After clearing the text of the table box to make sure it is empty, you will loop through the data in the `textBox` variable.

Remember that this is a list of strings where each string is the percentage of the minority population in the highlighted block group. Back in the original `for` loop, with the same indentation as the `with` statement, you will add each string to the table box. After all the text has been added, you will ensure the table box height fills the space available and the Y position is in the correct location, as sometimes these may change when the text is changed. Type in the following:

```
tableBox.text = ""
for text in textBox:
    tableBox.text += text
    print(text)
tableBox.height = 1.23
tableBox.elementPositionY = 2.2813
```

How do you determine the correct location for the text box X and Y positions?

The X and Y locations for the text boxes are determined when you set your layout view. It is important to spend time carefully setting up your template and noting where the anchor point is and what the X and Y locations are. This can be used to reset any of your layout elements as elements are added or removed from them.

6. You need to change the title of your map to include the bus line and the block group, along with a figure number, and update the layer name in the legend to include the block group. The title will include the figure number that comes from your counter, the `i` variable set at the beginning, and the GEOID. You will also change the `AlamedaContraCostaCounty_RaceHispanic_BlockGroup` name to include the block group and be more descriptive for the legend. Continuing on the next line, type in the following:

```
title.text = "Figure 2.{0}\r\nBus Route: {1}\r\nBlock Group: {2}" \
    .format(str(i),busRoute,str(geoid))
print(title.text)
cbgLayer.name = "Population in Block Group {0}".format(geoid)
```

7. You have not actually moved the extent of the map to the selected block group yet. Now you will find the extent of your selected layer and use that to set your layout's extent. Continuing on the next line, you will first get the layer extent of the highlighted block group. Then you will use the Camera object to set the extent of the map frame to that extent. Since you want to see some of areas around the block groups, the scale needs changing. As this is an urban area with smaller block groups, you will check to see if the scale is less than 10,000 and, if so, set it to 10,000. If it is greater than 10,000, you will add 2,000 to the scale, round it to a whole number, and set the scale to that. Type in the following:

```
# If scale is less than 10000 set to 10000,
# otherwise add 2000 and round to a whole number
extent = mf.getLayerExtent(cbgLayer,False,True)
print(extent)
mf.camera.setExtent(extent)
print(mf.camera.scale)
if mf.camera.scale < 10000:
    mf.camera.scale = 10000
else:
    mf.camera.scale = round((mf.camera.scale + 2000),0)
print(mf.camera.scale)
```

How to pick the scale(s) on a map series

The choice of 1:10,000 came from manually selecting block groups in the map and seeing what they looked like at different scales. Even though the final creation of all the maps is an automated process, determining things like the different scales to use requires you to check what the options look like. In some cases, a single minimum scale will work. In others, you may want more than one scale option. It will depend on your data, and how your scale bar is set up. You need to make sure that your scale bar will still look appropriate based on all of your scales, as you cannot change the scale bar settings other than location and length.

8. You now have everything set for your map and it is time to export it to PDF. Continuing on the next line, you will start by creating a string of the figure number, using `if` statement to check if the figure number has only 1 digit. If it does, you will add a leading 0 to it.

You need to add the leading 0 to ensure that the PDFs are in the correct order when they are read in in the next step. When ordering a string, 10 comes before 1 through 9.

You will then call the `exportToPDF` method and name the figure with the figure number, bus route, and GEOID. The last step is to reset the census block layer name to `AlamedaContraCostaCounty_RaceHispanic_BlockGroup` and increment the counter by 1. Type in the following:

```
pdfFigNum = str(i)
if len(pdfFigNum) == 1:
    pdfFigNum = "0"+pdfFigNum
layout.exportToPDF(
    os.path.join(
        projectFolder,
        "Figure_2_{0}_BusRoute_{1}_GEOID_{2}.pdf"
        .format(pdfFigNum,busRoute,geoid)
    )
)
cbgLayer.name = "AlamedaContraCostaCounty_RaceHispanic_BlockGroup"
i+=1
```

You have finished with the cell and can now run it, using the print statements to track your progress as the script moves through the list and exports the different figures. You will also see the figures appear in the folder you are exporting them to.

9. *In the next cell*, you will be using the `glob` module to find all of the PDFs for a given bus route and the `Py2PDF` module to combine them into a single PDF. First, you will create a PDF merge object using the `PdfFileMerger()` function, then use the `glob.glob` method to search for all the PDFs for a bus line.

The `glob.glob` method takes one argument that, in most cases, is the path you want to search for, combined with what you are searching for surrounded by an asterisk (*) for a wildcard. The wildcard can be used to match anything before or after it.

Within the glob.glob method, you will use os.path.join() to set the path of the PDFs and the wildcard to find the PDFs for your bus route. Once you have the list of PDFs, you will loop through all of them using the PdfFileReader() function from the Py2PDF module to open, read, and append them to the PDF merge object. PdfFileReader() needs the path of the PDF and how to open them; 'rb' tells it to read them. Once all of the PDFs have been opened, read, and appended to the PDF merge object, you will write that to a new file that contains all the PDFs. Type in the following:

```
pdfMergeObj = PdfFileMerger()

pdfFiles = glob.glob(
        os.path.join(
            projectFolder,
            "*BusRoute_{0}*.pdf".format(busRoute)
        )
)
for pdf in pdfFiles:
    print(pdf)
    pdfMergeObj.append(PdfFileReader(pdf,"rb"))

pdfMergeObj.write(
        os.path.join(
            projectFolder,
            "Figure_2_BusRoute_{0}_MinorityRace_Greater_RefComm.pdf"
            .format(busRoute)
        )
)
```

Run the cell. Since you added a print statement, you can track the cell's progress. When it is finished, you will have a new PDF that contains all the pages of the highlighted census block groups.

10. The final step is resetting everything in the map and layout to the way it was when you started. You should do this so your map and inset template are ready for any future automations. You will remove the study area layer and the selected bus route layer. Type in the following:

```
m.removeLayer(newBusLyr)
m.removeLayer(cbgStudyAreaLyr)
```

Run the cell. There will be no output in the Notebook, but the two layers will be removed from the map.

You now have a single PDF that highlights all the block groups along the selected bus route that have a higher minority percentage than that of the population for existing bus routes. These maps can be used in community meetings or in an environmental justice section to show how removing this bus line impacted minority communities and the effect restarting it will have.

In addition, this Notebook can easily be modified to run additional bus lines by changing the `busRoute` variable in the second cell. With maps for those additional bus lines, you can compare which suspended bus lines had a larger impact on minority communities.

Summary

In this chapter, you have created a map book that highlights block groups along a suspended bus route with a high-percentage minority population. This information can be used to identify potential disproportionate impacts on minority communities that can be an environmental justice issue. It can also be used to give guidance on reopening bus routes by identifying the most impactful routes to reopen.

First, the layout template was created, with special attention paid to the settings that cannot be changed with ArcPy. Then you created a Notebook, in which you defined a reference community, selected a bus route, created a study area, and added it to the map. You selected block groups within your study area with a minority percentage greater than the reference community, added the study area and selected bus route to an inset map, ensured a readable legend, selected a block group and set the extent to it, and created a table on the map of the percentages of races in the block group. Finally, you exported the maps to PDF and merged them into one file. It is a long process, but still much more efficient than creating a new layout in ArcGIS Pro for each selected block group.

Now that you have this code, you can modify it for other projects where complex map books are required.

In the next and final chapter, you will see how to use ArcPy, ArcGIS API for Python, and ArcGIS Online to gather data from various sources and create a complete web mapping application for a crop yield case study.

13

Case Study: Predicting Crop Yields

In our final case study, we will explore the real-world problem of crop yields. To do this, we will demonstrate an **Extract, Transform, Load (ETL)** workflow that uses many of the Python methods explained in previous chapters – ArcPy, ArcGIS API for Python, Pandas, and scikit-learn – as well as some of the web tools that Python allows you to use. The ETL process combines worldwide agricultural data into a format that can be used to predict crop yields using machine learning and loads it into ArcGIS Online. The resulting combined dataset is geographically enabled and can be updated with the latest data at any time using code.

To top it all off, we will display the final combined data in a simple web app built with HTML, CSS, and JavaScript, to illustrate the kinds of tooling that Python makes possible.

The following topics are covered in this chapter:

- Introducing the problem, data, and study area
- Downloading the data using Requests, World Bank API, and ArcGIS Online
- Cleaning up and combining the data
- Fitting the random forest model
- Loading the result into ArcGIS Online
- Generating an HTML file using ArcGIS API for JavaScript

For the code used in this chapter, please download and unzip the `Chapter13.zip` folder in the GitHub repository for this book: `https://github.com/PacktPublishing/Python-for-ArcGIS-Pro/tree/main/Chapter13`.

Case study introduction

Despite the vast agricultural land expansion and technical advancements around the world, crop yields will need to increase exponentially to meet the needs of our growing global population. Crop monitoring and yield estimation will be crucial to ensuring food security, especially as climate change continues to intensify and our natural resources are depleted. Crop yield prediction can be time-consuming and complex, so creating a GIS-enabled data pipeline can improve the efficiency of the prediction process.

Researchers working on food- and agriculture-related topics want readily available data that they can download and study. They require a resource that can be used to extract agricultural data from various sources, clean it, and use it to predict crop yields for countries around the world. This will save them time and money, and result in more informed and timely decisions regarding aid for the countries involved.

You will create the ETL workflow to do this using Python in a Notebook, so that detailed explanations and visualizations can be provided throughout. Following the creation and testing of the workflow, the data will be displayed in a simple web application built with HTML, CSS, and JavaScript, so stakeholders and decision makers can easily and quickly understand the data gathered.

Data and study area

We will restrict the workflow and subsequent tools to the years 1960 through 2019 and only to the countries around the world that have data available from the **Food and Agriculture Organization (FAO)** and the **World Bank**. *Figure 13.1* below shows the countries around the globe that have data available:

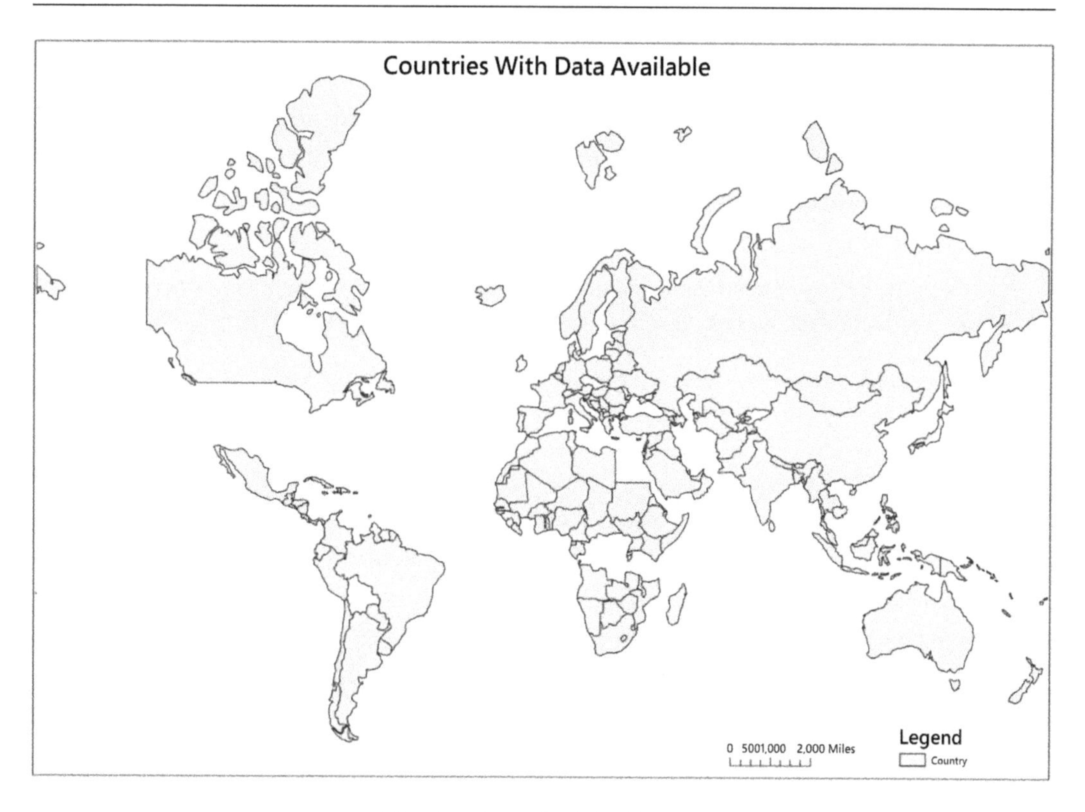

Figure 13.1: Data is available for these countries

There are seven primary datasets that we will utilize for this project: world country boundaries, population, rainfall, agricultural land, crop yields, pesticide and fertilizer use, and temperature change. All the datasets are either from the World Bank, Food and Agriculture Organization, or ArcGIS Online.

Crop data will be gathered from the FAO for some of the most important crops produced around the world. These include maize, cassava, rice, soybeans, wheat, potatoes, yams, and sorghum. Other data collected will include the country producing the crop (FAO), crop yield (FAO), agricultural land (World Bank), rainfall (World Bank), temperature change (FAO), pesticide and fertilizer use (FAO), and population (World Bank).

Data concepts

We will clean the data based on the principles and techniques described in *Tidy Data* (Wickham, 2014), and then feed it into a random forest machine learning model described in *Random Forests for Global and Regional Crop Yield Predictions* (Jeong, Resop, Mueller, et al., 2016).

Read the paper on data cleaning by Hadley Wickham here: `https://vita.had.co.nz/papers/tidy-data.pdf`

Read the random forests paper here: `https://dash.harvard.edu/bitstream/handle/1/27662263/4892571.pdf?sequence=1`

The *Tidy Data* paper discusses an important part of **data cleaning**, which is data tidying, or the structuring of datasets to facilitate analysis. Within data that is **tidy**, each variable forms a column, each observation forms a row, and each type of observational unit forms a table. Ordering variables and observations is key to making your data even easier to understand and analyze.

Wickham's assessment of the five most common issues with unorganized datasets and the broad solutions are listed below:

1. Column headers are values, not variable names.
    - *Solution*: **Melt** the data, by converting the data frame so that it contains one row for every observed value.
2. Multiple variables are stored in one column.
    - *Solution*: Melt the data and then split string values into their own columns.
3. Variables are stored in both rows and columns.
    - *Solution*: **Cast** or **unstack** the data, which is the opposite of melting the data.
4. Multiple types of observational units are stored in the same table. For example, this could be a data frame holding both weekly crime data as well as weather data.
    - *Solution*: **Normalize** the data, which we looked at how to do in *Chapter 8*.
5. A single observational unit is stored in multiple tables.
    - *Solution*: Combine the tables and add a new column for each table.

Once the data has been cleaned following the principles above, we will split it into a **training** and **testing set** to be used as input for a random forest machine learning model.

Random forests are a method used for regression and classification, which involve predicting the outcome of an experiment based on training data from previous experiments.

A random forest is a group of **decision trees**, which are, in the simplest form, flowcharts that show a clear pathway to a decision. The decision tree below demonstrates how we might start off with a crop (on the left) and follow the branches to determine the expected yield from the crop type and the amount of rainfall it receives:

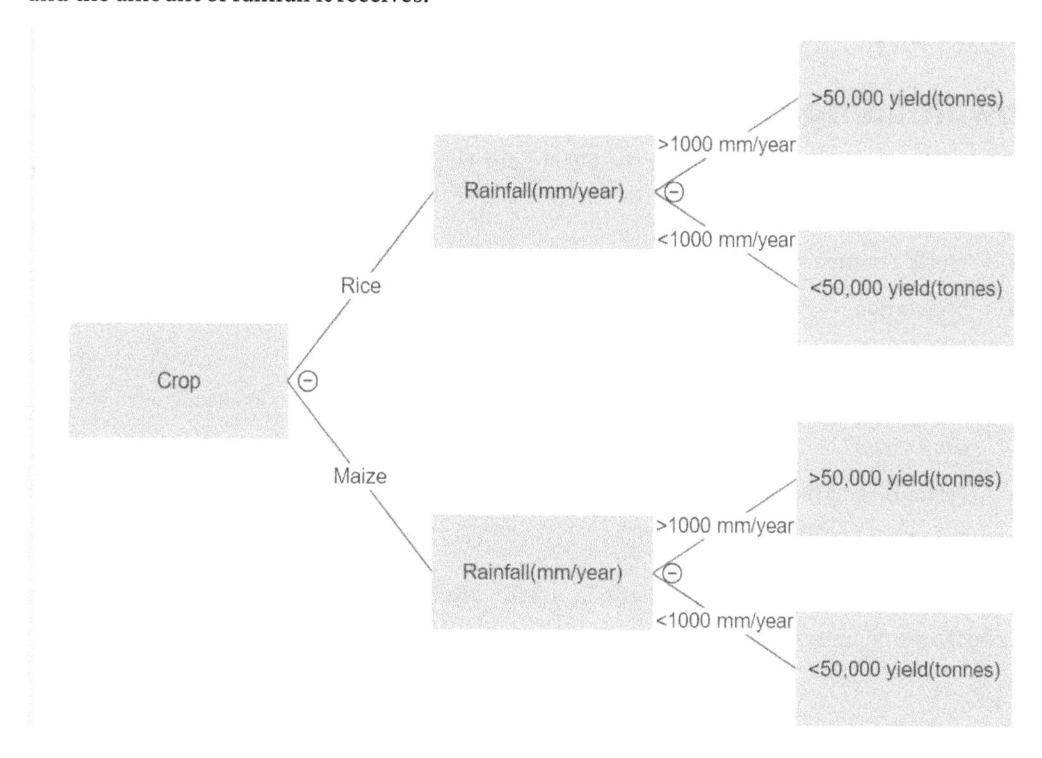

Figure 13.2: An example decision tree

The training data we gather will be used to create the group of decision trees by splitting the data into random subsets, generating the decision trees for each subset, and finally putting them together. Random forests are good when working with large datasets that might contain missing data. However, they are much less interpretable than an individual decision tree because they cannot be visualized as a singular sequence of decisions. It is best to use them when a significant amount of memory is available for storage because all of the information from the individual trees needs to be retained.

The data will need to be engineered to include the qualitative values in the model by changing them to quantitative values.

The data will also be split into a **features** section, containing all the columns except for the column we want to predict – crop yield – and the **target** section, containing only the crop yield column. This will allow us to understand the importance of each feature when it comes to predicting our target.

In terms of splitting the data, *Random Forests for Global and Regional Crop Yield Predictions* suggests a split of around 70% training and 30% testing or 80% training and 20% testing, stating that "including more data for model training is likely to improve the predictability of random forest regression."

Downloading datasets

In this section, we will start building the ETL in our Notebook by downloading the following datasets:

- World countries
- Population
- Rainfall
- Agricultural land
- Crop yields
- Pesticide and fertilizer use
- Temperature change

Let's get started. Create a new Notebook in ArcGIS Pro named `CropYieldETL` using the **New Notebook** option under the **Insert** tab:

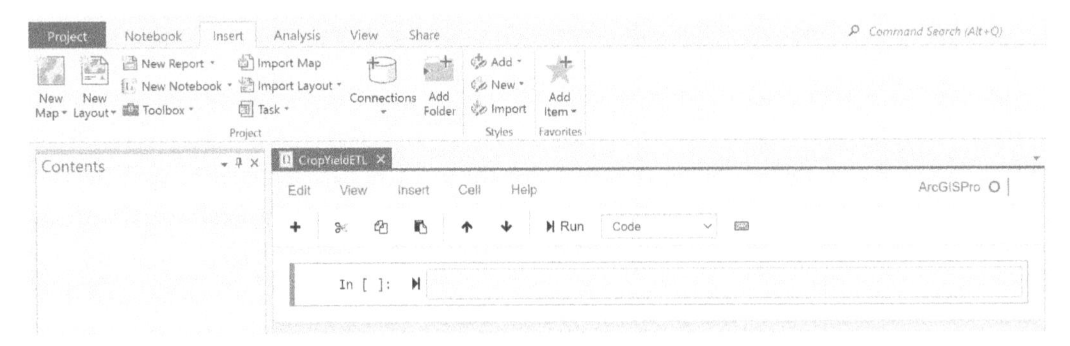

Figure 13.3: New empty Notebook named "CropYieldETL"

World countries

First, you will search for and download a country boundary dataset from ArcGIS Online using the ArcGIS API for Python.

1. In the first cell, you will import the API with the following code:

```
from arcgis.gis import GIS
```

   Run the cell.

2. Once imported, you will connect to two separate instances. The first one is anonymous and will be used for searching for datasets; the second takes your ArcGIS Pro login credentials for download purposes:

```
gis_search = GIS()
gis = GIS('pro')
```

   Run the cell.

3. Now you can search for the world boundary feature layer and return the first result:

```
items = gis_search.content.search(query='title:World
Countries(Generalized)', item_type='Feature Layer')
items[0]
```

   Run the cell. You should see the following output:

World Countries (Generalized)
World Countries (Generalized) provides optimized country boundaries with useful name and country code attribute fields. Current to November 2019.

Feature Layer Collection by esri_dm
Last Modified: October 12, 2021
0 comments, 37,034,987 views

Figure 13.4: World Countries (Generalized) thumbnail during an ArcGIS Online search

4. After confirming that the returned item is correct, it can be downloaded using the ArcGIS Online instance associated with your personal account. Access the item via the item's `id` and the `gis.content.get()` method, as shown:

```
world_get = gis.content.get(items[0].id)
```

   Run the cell.

5. In this cell, you will export the item to your ArcGIS Online account, specifying the title and format that you want. First, you will export the item as a shapefile named appropriately, and then download that shapefile to your current project folder. Type in the following code, which uses the built-in `zipfile` module to unzip the data and place it in your project folder:

```
from zipfile import ZipFile
world_export = world_get.export(title='world', export_
format='Shapefile')
world_path = world_export.download('world')
with ZipFile(world_path, 'r') as zipObj:
    zipObj.extractall()
```

Run the cell.

6. There is now a shapefile in the local project folder. A **Spatially Enabled DataFrame (SEDF)** can be created from the shapefile, and visualized using `sdf.spatial.plot()`:

```
import pandas as pd
sdf = pd.DataFrame.spatial.from_featureclass('World_Countries_
(Generalized).shp')
m = sdf.spatial.plot()
m
```

Run the cell. The output should look like this:

Figure 13.5: World Countries shapefile downloaded from ArcGIS Online

Population

There are many sources you could collect population from, but it is convenient to obtain the population for each country from 1960 to 2020 from the World Bank API. There is a Python module, `world_bank_data`, that makes it easier to explore the World Bank Indicators published by the World Bank.

To obtain this dataset, the `world_bank_data` Python module needs to be installed on a clone of your Python project environment, discussed in *Chapter 3, ArcGIS API for Python*:

7. Install and import the API with the following code:

   ```
   pip install world_bank_data
   import world_bank_data as wb
   ```

   Run the cell.

Once installed, there is a function, `wb.get_indicators()`, that shows all the datasets that are available. The population dataset is called **SP.POP.TOTL**. There is another function, `wb.get_series()`, which puts the dataset into a Pandas Series. As we saw in *Chapter 8*, a Series is a one-dimensional array that holds data. To continue to process the data and eventually join this dataset to others, we will turn this Series into a Pandas DataFrame, using the `to_frame` function.

8. In the next cell, you will read the population dataset into your Notebook as a Series and turn it into a DataFrame using `wb.get_series().to_frame()`. `id_or_value` is set to `'id'` to ensure that the full names of each country are not returned, and `simplify_index` is set to `True` so that only the country ID and year are returned in the index:

   ```
   population = wb.get_series('SP.POP.TOTL', id_or_value='id',
                 simplify_index=True).to_frame()
   ```

   Run the cell.

This data frame will be used to collect the rest of the datasets from the World Bank. The last step for this data is to change the name of the column from `"SP.POP.TOTL"` to `"population"` to represent all 16,226 rows of data. Population is considered the count of all residents within a country, whether they are a citizen or not, and all the values are mid-year estimates. All other series will be added to this data frame.

9. In the next cell, you will change the name of the column using the Pandas rename function and display the top five rows using the head function:

```
df_wb = population.rename(columns={'SP.POP.TOTL':'population'})
df_wb.head()
```

Run the cell. The output should look like this:

		population
Country	**Year**	
AFE	**1960**	130836765.0
	1961	134159786.0
	1962	137614644.0
	1963	141202036.0
	1964	144920186.0

Figure 13.6: Population data

Rainfall

We will also collect a rainfall dataset, called **AG.LND.PRCP.MM**, from the World Bank API using the same technique as above to return the data as a series. We will place it directly in the previously created data frame above, df_wb, in a new column titled "rainfall(mm/year)".

10. Read the dataset as a series and place the result directly in a new column in the df_wb data frame:

```
df_wb['rainfall(mm/year)'] = wb.get_series('AG.LND.PRCP.MM',
                                 id_or_value='id', simplify_index=True)
```

Run the cell.

There are a total of 16,226 rows in the AG.LND.PRCP.MM dataset, representing the total millimeters of rain per year from 1960 to 2020 for every country. This data is measured as the long-term average depth over space and time of annual precipitation in the country.

Agricultural land

We will collect agricultural land data, **AG.LND.AGRI.K2**, from the World Bank API using the same technique as above and add it to the df_wb data frame. Agricultural land is measured in sq. km and there are 16,226 rows of data from 1960 to 2020 for every country that the World Bank keeps data about.

11. Read the dataset as a series and place the result directly in a new column in the `df_wb` data frame:

    ```
    df_wb['agland(sq/km)'] = wb.get_series('AG.LND.AGRI.K2',
                                           id_or_value='id', simplify_index=True)
    ```

    Run the cell.

Agricultural land is considered land that is either arable, under permanent crops, or under permanent pastures. **Arable land** is any land that is under temporary crops, temporary meadows for mowing or pastures, land under market or kitchen gardens, and land that is temporarily fallow. **Land under permanent crops** is any land that is cultivated with crops that occupy the land for long periods of time and do not need to be replanted after harvest. Lastly, **permanent pasture** is land used for 5 or more years for forage, including natural and cultivated crops.

Crop yields

Crop yield data was obtained from the FAO. The FAO does not have an active API, where the data can be easily accessed using a Python module like the World Bank. This data is gathered by using a URL that points to the web page containing the data. The URL points to a JSON object containing a breakdown of all the information.

12. You will first create a variable holding the FAO datasets URL. In the next cell, type in the following:

    ```
    fao_url = 'http://fenixservices.fao.org/faostat/static/
    bulkdownloads/datasets_E.json'
    ```

    Run the cell.

This data is accessible in Python using the **Requests** module, which sends HTTP requests and receives responses from a specified URL. Several requests can be made using the module, including `get`, `post`, `put`, `patch`, or `head` requests. These also serve as the module's methods. Once a response object is returned, several methods can be used to explore the returned object. This instance calls for a `get` request to return a response object containing all the information on the FAO URL. Since the returned object was written in JSON format, the `json()` response method is used to further explore the information.

If the returned object was not written in JSON, the next best option is using the `content()` or `text()` response methods for exploration.

13. You will import the `requests` module, send a get request, and create a variable to hold the JSON object.

    Type the following in the next cell:

    ```
    import requests
    response = requests.get(fao_url)
    data = response.json()
    ```

    Run the cell. Once the JSON object is obtained, all the datasets within the FAO URL can be explored by indexing.

14. In the next cell, print the contents of the JSON file:

    ```
    for x in range(len(data['Datasets']['Dataset'])):
        print(f"{x}. {data['Datasets']['Dataset'][x]['DatasetName']}")
    ```

    Run the cell and examine the output, which should look like the following:

    ```
    0. Discontinued archives and data series: ASTI-Expenditures
    1. Discontinued archives and data series: ASTI-Researchers
    2. Food Balance: Commodity Balances (non-food)
    3. Investment: Country Investment Statistics Profile
    4. Prices: Consumer Price Indices
    5. Macro-Economic Indicators: Capital Stock
    6. Investment: Development Flows to Agriculture
    7. Land, Inputs and Sustainability: Fertilizers indicators
    8. Climate Change: Emissions intensities
    9. Land, Inputs and Sustainability: Livestock Patterns
    10. Land, Inputs and Sustainability: Land use indicators
    11. Climate Change: Emissions shares
    12. Land, Inputs and Sustainability: Livestock Manure
    13. Land, Inputs and Sustainability: Pesticides indicators
    14. Land, Inputs and Sustainability: Soil nutrient budget
    15. Climate Change: Temperature change
    16. Discontinued archives and data series: Food Aid Shipments (WFP)
    17. Food Balance: Food Balances (2014-)
    18. Food Balance: Food Balances (-2013, old methodology and population)
    ```

    Figure 13.7: Contents of the JSON file from the FAO URL

    Each dataset above contains a URL that points to a ZIP file containing all the data. To obtain the correct URL for crop yields, the returned list above should be searched to find the corresponding keys to supply to the dictionary. In this case, the crop yield corresponds to the number **47** above.

15. In this step, you will obtain the URL that holds the crop yield data. Type in the following:

    ```
    crop_yield = data['Datasets']['Dataset'][47]['FileLocation']
    crop_yield
    ```

Run the cell and check the URL. Once the dataset needed is found, the URL is put into a variable, `crop_yield`. The Requests module is used yet again to send out a `get` request for data in the URL.

16. Next, you will send a `get` request for the URL obtained above, with the parameter `stream` set to `True` to ensure that the data is not downloaded right away:

    ```
    yield_response = requests.get(crop_yield, stream=True)
    ```

    Run the cell.

17. You will now generate the name of the output file by using Python's `string.split()` method and indexing to get the last value in the resulting list, which is the name of the file as the FAO keeps it. Type in the following:

    ```
    local_file_name = crop_yield.split("/")[-1]
    ```

    Run the cell.

18. In this step, the file will be written to your computer using built-in Python file writing capabilities. Setting the `wb` argument to open allows the file to be written in binary, which will return the data as byte objects, not strings. The data is **chunked**, meaning that the code pulls down only a certain amount of data (a chunk, measured in bytes) at a time; this avoids you having to load the entire response into memory at once.

    Download the ZIP file and name it based on the `local_file_name` you created in the previous step:

    ```
    with open(local_file_name, 'wb') as fd:
        for chunk in yield_response.iter_content(chunk_size=128):
            fd.write(chunk)
    ```

    Run the cell.

19. You now need to unzip the content with `ZipFile` and extract it to a local folder. Type in the following:

    ```
    with ZipFile(local_file_name, 'r') as zipObj:
        zipObj.extractall()
    ```

    Run the cell.

20. Lastly, you will use the unzipped file path in a Pandas function that creates a data frame. Read the downloaded CSV with the encoding `latin1` to preserve the bytes and create a Pandas DataFrame using `read_csv()`:

    ```
    df_yield = pd.read_csv(local_file_name.split(".")[0] +".csv",
    encoding = 'latin1')
    ```

    Run the cell.

This dataset returns a list of crops and the specific yields. However, only rice, potatoes, yams, soybeans, wheat, maize, sorghum, and cassava were used for this analysis. This list was selected as they are some of the most highly produced crops around the world. Lastly, we reduce the dataset to only those crops listed above, as well as some other cleanup tasks.

21. In the next cell, you will reduce the DataFrame so that it only includes certain crops. You will also drop unnecessary columns and finally rename the column containing the yields. Type in the following:

    ```
    df_yield = df_yield.loc[df_yield["Item"].isin(['Rice, paddy',
    'Potatoes', 'Yams', 'Soybeans', 'Wheat', 'Maize', 'Sorghum',
    'Cassava'])]
    df_yield = df_yield.drop(['Area Code', 'Item Code', 'Element Code',
    'Year Code', 'Flag', 'Element', 'Unit'], axis=1)
    df_yield.rename(columns={'Value':'yield(tonnes)'}, inplace=True)
    df_yield.head()
    ```

    Run the cell. The output should look similar to this:

	Area	Item	Year	yield(tonnes)
3526	Afghanistan	Maize	1961	500000.0
3527	Afghanistan	Maize	1962	500000.0
3528	Afghanistan	Maize	1963	500000.0
3529	Afghanistan	Maize	1964	505000.0
3530	Afghanistan	Maize	1965	500000.0

Figure 13.8: Preview of the crop yield data frame

Pesticide and fertilizer use

The pesticide and fertilizer use dataset is obtained in the same way as the crop yield data from the FAO. Pesticide and fertilizer use corresponds to the number 7 in the JSON object and is measured in kilograms per hectare (kg/ha).

22. Take the code in *steps 16-20* above and fill in the information for the pesticide and fertilizer dataset:

```
pest = data['Datasets']['Dataset'][7]['FileLocation']
pest_response = requests.get(pest, stream=True)
local_file_name = pest.split("/")[-1]
with open(local_file_name, 'wb') as fd:
    for chunk in pest_response.iter_content(chunk_size=128):
        fd.write(chunk)
with ZipFile(local_file_name, 'r') as zipObj:
    zipObj.extractall()
df_pest = pd.read_csv(local_file_name.split(".")[0] + ".csv",
encoding = 'latin1')
```

    Run the cell.

This dataset contains around 30,740 records, and for each country and year, it contains the kg/ha of three different fertilizers and pesticides, including nitrogen- (N), phosphorus- (P_2O_5), and potassium- (K_2O) based.

This is a time series dataset from 1961 to the present day. For ease of merging and manipulating the future dataset, you will aggregate all the pesticides and fertilizers based on their year and country, along with performing some other cleanup tasks.

23. In the next cell, you will remove unnecessary columns and use the Pandas `groupby` and `agg` functions to find the total amount of pesticide and fertilizer used each year in each country. Type in the following:

```
df_pest = df_pest.drop(['Area Code', 'Item Code', 'Element Code',
'Year Code', 'Flag', 'Element', 'Unit'], axis=1)
df_pest = df_pest.groupby(['Area', 'Year']).agg({'Value':'sum'})
```

    Run the cell.

24. In the next cell, reset the index following the summation and rename the summed column to 'pestUse(kg/ha)':

```
df_pest.reset_index(inplace=True)
df_pest.rename(columns={'Value':'pestUse(kg/ha)'}, inplace=True)
df_pest.head()
```

Run the cell. The output should look similar to this:

	Area	Year	pestUse(kg/ha)
0	Afghanistan	1961	0.14
1	Afghanistan	1962	0.14
2	Afghanistan	1963	0.14
3	Afghanistan	1964	0.14
4	Afghanistan	1965	0.14

Figure 13.9: Preview of the pesticide and fertilizer use data frame

This results in a dataset that only has one total number of pesticides and fertilizers used every year for each country, resembling the crop yield dataset.

Temperature change

Temperature change is also gathered in the same way as the pesticide and fertilizer use and crop yield data from the FAO. Temperature change is the 15th dataset in the JSON object and is measured as the mean surface temperature change by country, for each month from 1961 to the present year:

25. Take the code in *steps 16-20* above and fill in the information for the temperature change dataset:

```
temp = data['Datasets']['Dataset'][15]['FileLocation']
temp_response = requests.get(temp, stream=True)
local_file_name = temp.split("/")[-1]
with open(local_file_name, 'wb') as fd:
    for chunk in temp_response.iter_content(chunk_size=128):
        fd.write(chunk)
```

```
with ZipFile(local_file_name, 'r') as zipObj:
    zipObj.extractall()
df_temp = pd.read_csv(local_file_name.split(".")[0] +".csv",
encoding = 'latin1')
```

Run the cell.

This temperature data is collected by the FAO from the National Aeronautics and Space Administration Goddard Institute for Space Studies.

The data includes around 537,370 records and includes a record for each month, which we will change to only contain one record for every year for each country.

This will be completed by using the same groupby and agg Pandas functions used for the pesticide and fertilizer use dataset. To receive the average temperature change format, we will divide that column by 12 and finish up with some simple cleanup tasks.

26. In the next cell, you will remove unnecessary columns and use the Pandas groupby and agg functions to find the average temperature change for each year in each country. Dividing that number by 12 will give you the average. Type in the following:

```
df_temp = df_temp.drop(['Area Code', 'Element Code', 'Year Code',
'Flag', 'Element', 'Unit'], axis=1)
df_temp = df_temp.groupby(['Area', 'Year']).agg({'Value':'sum'})
df_temp['Value'] = df_temp['Value']/12
```

Run the cell.

27. In the next cell, you will reset the index following the summation and rename the summed column to 'tempChange(C)':

```
df_temp.reset_index(inplace=True)
df_temp.rename(columns={'Value':'tempChange(C)'}, inplace=True)
df_temp.head()
```

Run the cell. The output should look similar to this:

	Area	Year	tempChange(C)
0	Afghanistan	1961	1.659750
1	Afghanistan	1962	1.332000
2	Afghanistan	1963	2.886083
3	Afghanistan	1964	0.326083
4	Afghanistan	1965	1.539583

Figure 13.10: Preview of the Temperature Change data frame

Now you have successfully obtained all the datasets using the Requests module, ArcGIS Online, and the World Bank API. You have also completed some minor fixes to each individual dataset. Next, we will look at further cleaning up and ultimately combining these datasets to create our final data frame.

Cleaning up and combining the data

Before merging all the data into one dataset, some cleanup needs to take place to ensure the merge is viable. In order to merge datasets, there needs to be a column or multiple columns that match in both data frames. In this case, the merge will occur on the **year** and the **country name** columns.

The year columns have no variations, but the name of the country columns may differ slightly in spelling or if abbreviations are used. The data frame from the World Bank data, `df_wb`, only has an abbreviation to represent countries and contains data for regions, in addition to country names. The actual country names will need to be added and the rows containing region data will need to be removed.

Luckily, the `world_bank_data` API has a readily available dataset containing all the IDs, country names, and information about region data; specifically, the column `'region'`, which specifies whether an entry is a combination of countries.

1. First, you need to obtain a list of all the values that appear in the `country` column for the World Bank data:

   ```
   countries = wb.get_countries()
   ```

   Run the cell.

2. The next step is to remove all the region data using the Pandas `loc` function, which allows you to access a specified group of rows and columns. Mentioned above, `'region'` contains a value that specifies whether a row is for a single country or multiple countries. In this case, you only want single countries, so you don't want any rows that have the value `'Aggregates'` in the `'region'` column.

   Access all the rows that don't have the value `'Aggregates'` in the `'region'` column:

   ```
   df_key = countries.loc[countries.region != 'Aggregates']
   ```

   Run the cell.

3. You want to easily obtain the `'id'` index in the next few steps, so you will reset the index, which moves the `'id'` index into a column. The Pandas `loc` function can be used to reduce the DataFrame to just the ID and the name of the country. The merge can then take place after resetting the index of `df_wb` and then removing redundant columns.

   Type in the following to reset the index, and use the `loc` function again to only return the `'id'` and `'name'` columns:

   ```
   df_key.reset_index(inplace=True)
   df_key = df_key.loc[:,['id', 'name']]
   df_key.head()
   ```

   Run the cell. The output should look similar to this:

	id	name
0	ABW	Aruba
1	AFG	Afghanistan
2	AGO	Angola
3	ALB	Albania
4	AND	Andorra

   Figure 13.11: Preview of the ID and country name data frame

4. You will now merge `df_key` to `df_wb` so that `df_wb` can then be more easily merged on the country name with the data collected from the FAO. First, you need to reset the index on `df_wb` to remove the `'id'` column from the index.

Then, you will use the Pandas merge function to combine the data frames on the abbreviations, 'id' in df_key and 'Country' in df_wb. Once the long-form name is in df_wb, we can drop the redundant columns, 'Country' and 'id'.

Type in the following to reset the index, merge the data frames, and remove redundant columns:

```
df_wb.reset_index(inplace=True)
df_wb = pd.merge(df_wb, df_key, left_on='Country', right_on='id')
df_wb = df_wb.drop(['Country', 'id'], axis=1)
```

Run the cell.

At the moment, there is one data frame containing all the World Bank data, df_wb, and three separate FAO data frames, df_temp, df_pest, and df_yield. Let's merge those three FAO datasets using the Pandas merge function.

5. First, merge df_temp and df_pest and then merge in df_yield:

```
df_fao = pd.merge(df_temp, df_pest, left_on=['Area', 'Year'],
         right_on=['Area', 'Year'], how='outer')
df_fao = pd.merge(df_fao, df_yield, how='right', on=['Area', 'Year'])
df_fao.head()
```

Run the cell. The output should be similar to the following:

	Area	Year	tempChange(C)	pestUse(kg/ha)	Item	yield(tonnes)
0	Afghanistan	1961	1.659750	0.14	Maize	500000.0
1	Afghanistan	1962	1.332000	0.14	Maize	500000.0
2	Afghanistan	1963	2.886083	0.14	Maize	500000.0
3	Afghanistan	1964	0.326083	0.14	Maize	505000.0
4	Afghanistan	1965	1.539583	0.14	Maize	500000.0

Figure 13.12: Preview of the merged FAO datasets

There are now two datasets, df_wb and df_fao. Before merging data frames from different sources, it is important to make sure the columns that the merge will take place on are the same data type; this can be completed using the Pandas astype function.

6. In the next cell, make sure all the columns that the merge will take place on are the same data type, in this case, string:

```
df_wb['name'] = df_wb['name'].astype(str)
df_wb['Year'] = df_wb['Year'].astype(str)
df_fao['Area'] = df_fao['Area'].astype(str)
df_fao['Year'] = df_fao['Year'].astype(str)
```

   Run the cell.

After checking `'name'` and `'year'` from the `df_wb` data frame and `'Area'` and `'Year'` from the `df_fao` data frame, there is only one more step. This is to reduce both data frames down to only those countries that show up in both data frames. This is done to avoid a large number of null values in our data frame. The first step of this process is to collect a list of all the unique countries in each data frame by using the Pandas `unique` function on both data frames.

7. Create two lists of the unique country values in each dataset:

```
fao_list = df_fao['Area'].unique()
wb_list = df_wb['name'].unique()
```

   Run the cell.

8. You will now use a list comprehension to compile a list of countries in both data frames. This loops through the countries in the `fao_list`; if that country is in the `wb_list`, then it will be added to a list called `both_list`. Run the list comprehension to obtain a list of the countries in both data frames:

```
both_list = [x for x in fao_list if x in wb_list]
```

   Run the cell. You will use the resulting list to reduce both the `df_fao` and `df_wb` data frames using the `loc` and `isin` Pandas functions.

9. In the following cell, reduce the data frames to only contain rows that have a country in `both_list`:

```
df_wb = df_wb.loc[df_wb['name'].isin(both_list)]
df_fao = df_fao.loc[df_fao['Area'].isin(both_list)]
```

   Run the cell.

10. Using the Pandas `merge` function, you will now combine `df_wb` and `df_fao`. Once combined, remove all the rows that have no yield associated with them. Type in the following code to run the final merge, remove a redundant column, remove any rows that don't have a yield, and remove unnecessary columns. Return a random sample of the final dataset using the `sample` function:

```
df_master = pd.merge(df_wb, df_fao, left_on=['name', 'Year'],
            right_on=['Area', 'Year'], how='outer')
df_master = df_master.drop(['Area'], axis=1)
df_master = df_master.loc[df_master['yield(tonnes)'].notna()]
df_master.sample(5)
```

Run the cell. The output should be similar to the following:

	Year	population	rainfall(mm/year)	agland(sq/km)	name	tempChange(C)	pestUse(kg/ha)	Item	yield(tonnes)
56244	2000	211513822.0	NaN	471770.0	Indonesia	0.759583	69.27	Maize	27649.0
31028	1997	4534920.0	1113.0	19410.0	Croatia	0.650417	248.96	Potatoes	63189.0
111886	2015	20970000.0	NaN	27400.0	Sri Lanka	1.848667	173.60	Soybeans	4701.0
44131	1970	78169289.0	NaN	190230.0	Germany	1.207583	384.44	Maize	49743.0
6514	1994	17855000.0	NaN	4691430.0	Australia	1.257500	96.06	Wheat	11356.0

Figure 13.13: Preview of all the datasets merged

You now have one data frame consisting of all the datasets gathered earlier. This was achieved by first updating the country names in the World Bank data frame and reducing the countries within each data frame to only contain those that appear in both data frames. Lastly, you merged all the data into one final data frame. This data frame will now be used to fit a random forest model, added to ArcGIS Online, and finally, be displayed in a web app.

Fitting a random forest model

We will use the combined dataset we now have to perform preliminary tests and fit the model. To run these tests and eventually the model, the `sklearn` module will need to be installed using `pip` via the command prompt (or within the Notebook):

1. In the next cell, install the `sklearn` module and import the following (note that the `sklearn` module may already be installed):

```
pip install sklearn
from sklearn.model_selection import train_test_split
from sklearn.ensemble import RandomForestRegressor
from sklearn.metrics import r2_score
```

Run the cell.

Before running any tests, note that the random forest model only accepts numeric variables, meaning all the categorical variables – specifically the `'Item'` field – will need to be changed to numeric. Essentially, each value will be represented by a number. Some cleanup needs to be completed as well, removing null values and dropping columns that are redundant.

2. In the next cell, you will drop redundant columns, turn the `'Item'` column into quantitative values, and lastly drop null values:

   ```
   df_ml = df_master.drop(['name'], axis=1)
   df_ml = pd.get_dummies(df_ml, columns=['Item'], prefix = ['Item'])
   df_ml = df_ml.dropna()
   ```

   Run the cell.

3. Now we want to split the data frame into a **features** section, containing all the columns except for the crop yields, and the **target** section, containing only the crop yield column. Split the data frame by getting a list of all the columns, removing `'yield(tonnes)'`, and then simply indexing based on that list and `'yield(tonnes)'`:

   ```
   col_ind = list(df_ml.columns)
   col_ind.remove('yield(tonnes)')
   X = df_ml[col_ind]
   y = df_ml['yield(tonnes)']
   ```

   Run the cell.

4. Now, we will take the features and target sections and split them each into one training set and one testing set. This will be a 70/30 split, as discussed earlier, which will be expressed in the `test_size` parameter in the `train_test_split` function. Then, we can run `RandomForestRegressor`, which will be fitted based on the training data. Type in the following code to split the features and target sections, run `RandomForestRegressor`, and fit the model based on the training data using `fit`:

   ```
   X_train, X_test, y_train, y_test = train_test_split(X, y,
   test_size=0.3, random_state=42)
   rf = RandomForestRegressor(n_estimators=200, random_state=42,
   n_jobs=-1, verbose=1)
   rf.fit(X_train, y_train)
   ```

   Run the cell.

We now have a model, `rf`, that has been fitted based on our training data. We will use the `predict` function on the model, `rf`, with the training data as our parameter to make predictions on. We will then be able to receive an R squared score, letting us know how well our model can predict yields. We can look at the columns in the features section to rank them by their importance in predicting the yields.

5. In the next cell, you will predict using our model, return an R squared score, and sort and return the most important columns that are used to predict crop yields:

```
y_pred = rf.predict(X_train)
r2_score(y_pred=y_pred,y_true=y_train)
plot_list = df_master.columns
features = X.columns
importances = rf.feature_importances_
feat_imp = pd.Series(importances, features).sort_
values(ascending=False)
feat_imp[:9]
```

Run the cell. The output should be similar to this:

```
population           0.360437
Item_Rice, paddy     0.175942
agland(sq/km)        0.094614
pestUse(kg/ha)       0.075401
tempChange(C)        0.074703
Item_Maize           0.062454
Year                 0.045689
Item_Wheat           0.044650
rainfall(mm/year)    0.025257
```

Figure 13.14: Features ranked by importance in predicting crop yields

In this section, you imported the necessary Python modules to complete the fitting of the random forest model. You then took the final merged data frame and completed the fitting of that model. *Figure 13.14* shows that the most important feature within our data frame for predicting crop yields is **population**. The R squared score lets you know how well your model was able to predict the yields.

In the next section, you will take the data frame you created in the previous section and upload it to ArcGIS Online.

Loading the result into ArcGIS Online

The final merge involves combining the shapefile we collected from ArcGIS Online and the data frame from the FAO and World Bank to give our data frame geometry for visualization. The `df_master` `'name'` column needs to be renamed to match the shapefiles' `'COUNTRY'` column. Then, the two datasets will be merged on that column:

1. In the next cell, you will rename the column containing the country name in `df_master` to match the country name column in the SEDF. Then, you will merge the columns on that column and remove unnecessary columns:

```
df_master.rename(columns={'name':'COUNTRY'}, inplace=True)
sdf_master = sdf.merge(df_master, on='COUNTRY')
sdf_master = sdf_master.drop(['ISO', 'COUNTRYAFF', 'AFF_ISO',
'FID'], axis=1)
```

   Run the cell.

2. The data frame merged above, `sdf_master`, will be exported to a shapefile and then added to ArcGIS Online for use in the web app to be created below. This is completed in a similar way to how the world boundaries shapefile was downloaded and read into a data frame. In the next cell, convert the data frame to a shapefile using the `spatial.to_featureclass` method:

```
shp = sdf_master.spatial.to_featureclass('FoodandAgData.shp')
```

   Run the cell.

3. A shapefile is made up of multiple files. Before we can upload the shapefile created above, those files need to be placed in a ZIP file. We will utilize the `ListFiles` ArcPy function to find all the necessary files related to our shapefile. First, we need to change the current workspace, which can be found in `arcpy.env.workspace`, to where we just wrote the shapefile in *step 2*. We use the `os` module to do this. After changing the workspace, we can utilize the `ZipFile` module again to write the shapefile to a ZIP file. We call the `ListFiles` function with the shapefile name followed by a wildcard, which can represent zero or several characters. For example, if you were searching for states and input `Te*`, it would return `Texas` and `Tennessee`. We also want to check that we are not grabbing any files that have locks on them, so we add an `if` statement to make sure the filename does not contain the word `'lock'`.

The next cell will contain this whole process explained above:

```
arcpy.env.workspace = os.path.dirname(os.path.abspath(shp))
file_list = arcpy.ListFiles('FoodandAgData.*')
with ZipFile('FoodandAgData.zip', 'w') as zipObj2:
    for x in file_list:
        if 'lock' not in x:
            zipObj2.write(x)
```

Run the cell.

4. Once the ZIP file that contains all the data collected has been created, it can be uploaded to ArcGIS Online. This is similar to how the world boundaries dataset was grabbed from ArcGIS Online, except we use the `add()` function and specify the shapefile in the `data` parameter. You can also write in some metadata under the `item_properties` parameter such as the title, type, and tags. Using the `gis` variable created earlier, the content can be added to your account.

   In the next cell, add the content to your ArcGIS Online account, creating some metadata, and then look at the thumbnail of the content you just added:

```
arc_shape = gis.content.add(item_properties={'title': 'CropYields',
'type': 'Shapefile', 'tags':['Food', 'Agriculture']},
data='FoodandAgData.zip')
arc_shape
```

   Run the cell. The output should be similar to this:

Figure 13.15: ArcGIS Online thumbnail of CropYields

5. Once added, the last step in the notebook is to publish the shapefile as a feature service using the `publish()` function and obtain the ID. **This ID will be used to add the shapefile to the web app:**

```
published_service = arc_shape.publish()
published_service.id
```

Run the cell.

Within this section, you did some final cleanup tasks on the data frame and merged it with the world countries' data frame to ensure that our data has a geometry for visualization purposes. You also saved the merged data frame as a shapefile for it to be added to ArcGIS Online and zipped it. Lastly, you obtained the ID of the uploaded shapefile so that it can be added to the web app that you will create in the following section.

Generating an HTML file using ArcGIS API for JavaScript

You now have a shapefile with all of the data collected hosted via ArcGIS Online. This can be used to create a web app using HTML, CSS, and JavaScript, the final stage of this chapter's case study. It's a bit different from what we've been doing so far, but will make for a pretty and interactive web map based on the data you just created.

You can look at the completed HTML file in the code folders for the final result if you encounter any issues.

We will not be writing any Python code in this section, but by using the ArcGIS API for JavaScript and some HTML tags, it will demonstrate how we can extend what we have done so far to create an interface that is very useful to an end user:

1. Create an HTML file in a text editor of your choice.
2. Add in HTML tags, `<html></html>`. Everything will occur between these two tags.
3. Following those tags, add in the head tags, `<head></head>`, which serve as a container for the metadata.
4. Within the head tags, add a title between title tags, add a link to reference a CSS style sheet maintained on the ArcGIS website, and lastly, add in a source link to ArcGIS JavaScript functionality, which is used to bring in all the functionality and tools used below. This is the equivalent of importing `arcpy` in Python:

```
<html>
  <head>
    <title>Agricultural Crop Yields</title>
    <link
```

```
        rel="stylesheet"
        href="https://js.arcgis.com/4.21/esri/themes/light/main.css"
      />
      <script src="https://js.arcgis.com/4.21/"></script>
    </head>
  </html>
```

5. Following the closing head tag, add in a pair of body tags, `<body></body>`, which will contain all the information in the document body. Within the body tags, add in a `div` tag with an `id` labeled as `"viewDiv"`, which will hold the view of the map:

```
<body>
   <div id="viewDiv"></div>
</body>
```

6. Following that `div`, add another with an `id` labeled as `"titleDiv"` and class labeled `"esri-widget"` so that we can reference the information later and place it within its own widget. Within that `div` tag, add another `div` element with an `id` labeled `"titleText"`; between the tags, add a title, such as `Agricultural Crop Yields`. Now, the text between the body tags should appear as follows:

```
<body>
    <div id="viewDiv"></div>
    <div id="titleDiv" class="esri-widget">
        <div id="titleText">Agricultural Crop Yields</div>
      </div>
</body>
```

7. Since we now have some tags in the document's body that also contain IDs, some CSS styling can be added. Between the final head tag and the first body tag, add in a pair of style tags. Within these tags is where the `div` elements created above can be referenced, using a # followed by the name of the ID. The code below shows the process explained above on the div elements, adding simple styling that makes sure our map, or `"viewDiv"`, fills up the whole page and our title, or `"titleDiv"`, is placed 10 pixels off its given position to make sure it is visible:

```
<style>
  #viewDiv {
    padding: 0;
    margin: 0;
```

```
      height: 100%;
      width: 100%;
    }
    #titleDiv {
      padding: 10px;
    }
  </style>
```

8. The only thing left to include is the JavaScript functionality. The final tags that need to be added are a pair of script tags, `<script></script>`, that should go in-between the closing style tag and the first body tag. Specific functions, `MapView`, `Map`, `FeatureLayer`, and `Basemap`, need to be added.

   `MapView` is used to display the map, `Map` stores, and manages the layers to be displayed, `FeatureLayer` is used to create our shapefile created in this chapter, and `Basemap` is used to create the `basemap` object to be displayed:

```
  <script>
    require([
      "esri/views/MapView",
      "esri/Map",
      "esri/layers/FeatureLayer",
      "esri/Basemap",
    ], (MapView, Map, FeatureLayer, Basemap) => {
    });
    </script>
```

   Within those curly brackets right after the arrow is where those functions will be used. There are four steps to adding our ArcGIS Online hosted data to our web map, which will all occur within those curly brackets.

9. First, add in a basemap using the `Basemap` function and specify the following ID, associated with the dark human geography basemap found on ArcGIS Online, `"4f2e99ba65e34bb8af49733d9778fb8e"`:

```
        const basemap = new Basemap({
          portalItem: {
            id: "4f2e99ba65e34bb8af49733d9778fb8e"
          }
        });
```

10. Second, add in our hosted data by using the ID we gathered earlier using the `FeatureLayer` function in *step 5* of the *Loading the result into ArcGIS Online* section. Optionally, you can reduce the data using the `definitionExpression` property:

```
// agricultural crop yield layer queried at start
const layer = new FeatureLayer({
  portalItem: {
    id: "3252811b3f3047298024a8047bcc3b57"
  },
  definitionExpression: "Item = 'Maize' AND Year = '2019'"
});
```

11. Next, we need to add both the basemap and feature layer to our map using the `Map` function:

```
// map
const map = new Map({
  basemap: basemap,
  layers: [layer]
});
```

12. Lastly, using the `MapView` function, display the map and place it in the `div` element with the ID labeled `"viewDiv"`:

```
 // view containing starting extent
 const view = new MapView({
  map: map,
  container: "viewDiv"
});
```

13. In order to place the title, we have to take the `view` element and add in the title div element, choosing where to put the element, which, in this case, is the top right:

```
view.ui.add("titleDiv", "top-right");
```

14. To add more interaction between the user and the web app, a pop-up template can be added so that when the user clicks on a country, information about that country will appear. The `createPopupTemplate()` function can be added to include the information that will be shown in the popup. The example will return the name of the country and the year. This function should be added into the `FeatureLayer` created in *step 10*:

```
        // agricultural crop yield layer queried at start
        const layer = new FeatureLayer({
          portalItem: {
            id: "3252811b3f3047298024a8047bcc3b57"
          },
          definitionExpression: "Item = 'Maize' AND Year = '2019'",
          popupTemplate: createPopupTemplate()
        });

        //popup template
        function createPopupTemplate(){
          return {
            title: "{country}, {Year}"
        };
        };
```

The HTML file should now look like this:

```
<html>
  <head>
    <title>Agricultural Crop Yields</title>
    <link
      rel="stylesheet"
      href="https://js.arcgis.com/4.21/esri/themes/light/main.css"
    />
    <script src="https://js.arcgis.com/4.21/"></script>
  </head>
  <style>
    #viewDiv {
      padding: 0;
      margin: 0;
      height: 100%;
      width: 100%;
    }
    #titleDiv {
      padding: 10px;
    }
  </style>
  <script>
```

```
require([
  "esri/views/MapView",
  "esri/Map",
  "esri/layers/FeatureLayer",
  "esri/Basemap",
], (MapView, Map, FeatureLayer, Basemap) => {
    const basemap = new Basemap({
      portalItem: {
        id: "4f2e99ba65e34bb8af49733d9778fb8e"
      }
    });
    // agricultural crop yield layer queried at start
    const layer = new FeatureLayer({
      portalItem: {
        id: "3252811b3f3047298024a8047bcc3b57"
      },
      definitionExpression: "Item = 'Maize' AND Year = '2019'",
      popupTemplate: createPopupTemplate()
    });

     //popup template
     function createPopupTemplate(){
      return {
        title: "{country}, {Year}",
        content: [{
          type: "fields",
          fieldInfos: [{
            fieldName: "Item",
            label: "Crop",
```

```
                }, {
                  fieldName: "yield_tonn",
                  label: "Yield",
                  format: {
                    places: 2,
                    digitSeparator: true
                  }
                }]
              }]
            };
          };

          const map = new Map({
            basemap: basemap,
            layers: [layer]
          });
           const view = new MapView({
            map: map,
            container: "viewDiv"
          });
          view.ui.add("titleDiv", "top-right");

        });
    </script>
    <body>
        <div id="viewDiv"></div>
        <div id="titleDiv" class="esri-widget">
            <div id="titleText">Agricultural Crop Yields</div>
          </div>
    </body>
  </html>
```

15. Lastly, save the HTML file. To view the web map, just double-click on the filename or right-click on the file and choose which browser to open the file with.

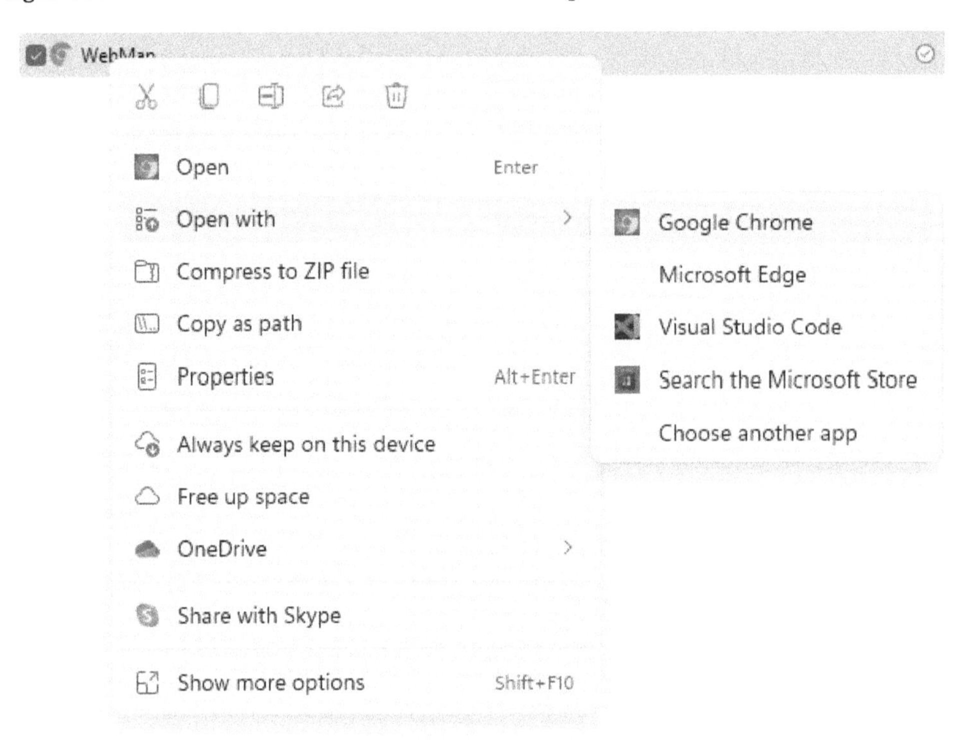

Figure 13.16: Opening the web map

You should see the final result, a working web map that displays the data and is clickable.

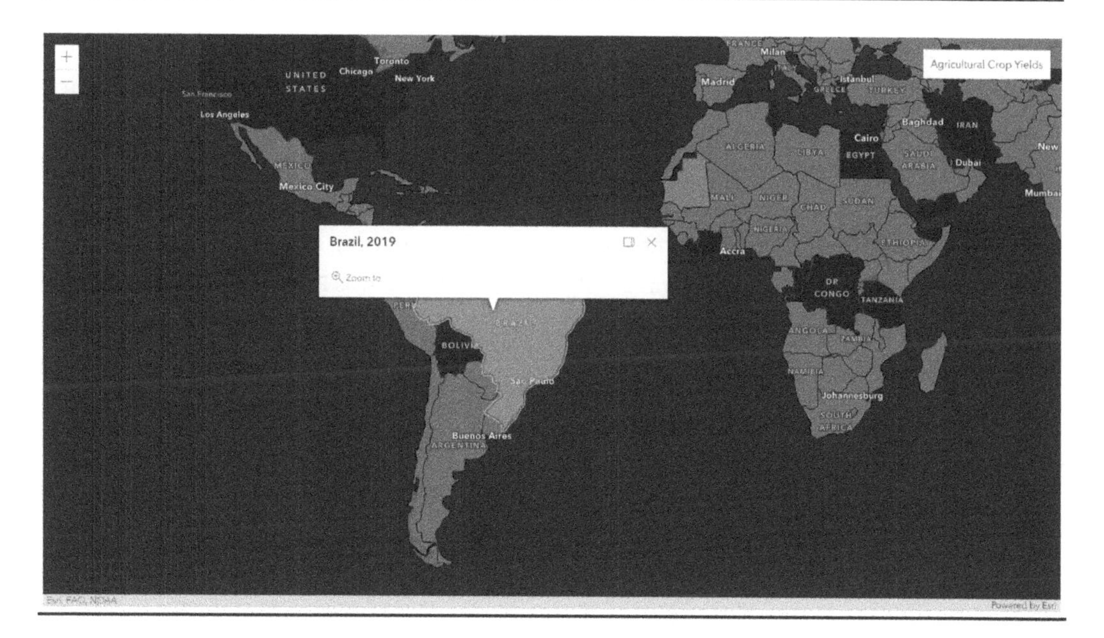

Figure 13.17: Final web mapping application

Summary

In this chapter, you solved a real-world problem by creating an ETL workflow to predict agricultural crop yields using all the concepts and tools you've learned about throughout the book.

Firstly, you learned about the problem, understanding the data and tools needed to complete the tasks at hand. You then downloaded all the data, utilizing the Requests module, ArcGIS Online, and the World Bank API. You cleaned up all those individual datasets to ensure that a merge was possible and then completed the merge.

You took the merged data frame and used it to fit the random forest model, demonstrating its ability to predict yields quickly and efficiently. This Notebook can be used to update all the data to predict crop yields when needed by simply rerunning the whole Notebook again. Lastly, you created a web application to display the shapefile created in the Notebook using HTML, CSS, and JavaScript.

This process could form the framework to provide additional workflows to make sure governments, farmers, policymakers, and many more have the information needed to prevent food insecurity in a timely manner.

*

In this book, you have gone from an introduction to Python, to using ArcPy for complex ArcGIS analyses and cartography, to using advanced Python modules such as Pandas and NumPy for creating custom Python tools to process both vector and raster data. You have even been introduced to a little bit of the ArcGIS API for JavaScript and HTML to create a web map based on your data.

It has been a long journey from the beginning to the end, but you've made it and now you're ready for the next step: using these lessons and tools in your daily work. The best way to learn is to use something over and over in your work; it will give you a great insight into where these tools are best deployed and how you can create new, custom tools that will make your work more fun and interesting.

We applaud you for taking the time to work hard and improve your skills and hope that you feel accomplished and excited for the future: a future full of code.

We wish you the best, and hope you have enjoyed the book!

packt.com

Subscribe to our online digital library for full access to over 7,000 books and videos, as well as industry leading tools to help you plan your personal development and advance your career. For more information, please visit our website.

Why subscribe?

- Spend less time learning and more time coding with practical eBooks and Videos from over 4,000 industry professionals
- Improve your learning with Skill Plans built especially for you
- Get a free eBook or video every month
- Fully searchable for easy access to vital information
- Copy and paste, print, and bookmark content

At www.packt.com, you can also read a collection of free technical articles, sign up for a range of free newsletters, and receive exclusive discounts and offers on Packt books and eBooks.

Other Books You May Enjoy

If you enjoyed this book, you may be interested in these other books by Packt:

Learn Python Programming - Third edition

Fabrizio Romano

Heinrich Kruger

ISBN: 9781801815093

- Get Python up and running on Windows, Mac, and Linux
- Write elegant, reusable, and efficient code in any situation
- Avoid common pitfalls like duplication, complicated design, and over-engineering
- Understand when to use the functional or object-oriented approach to programming
- Build a simple API with FastAPI and program GUI applications with Tkinter
- Get an initial overview of more complex topics such as data persistence and cryptography
- Fetch, clean, and manipulate data, making efficient use of Python's built-in data structures

Learning ArcGIS Pro 2 - Second Edition

Tripp Corbin, GISP

ISBN: 9781839210228

- Navigate the user interface to create maps, perform analysis, and manage data
- Display data based on discrete attribute values or range of values
- Label features on a GIS map based on one or more attributes using Arcade
- Create map books using the map series functionality
- Share ArcGIS Pro maps, projects, and data with other GIS community members
- Explore the most used geoprocessing tools for performing spatial analysis
- Create Tasks based on common workflows to standardize processes
- Automate processes using ModelBuilder and Python scripts

Packt is searching for authors like you

If you're interested in becoming an author for Packt, please visit `authors.packtpub.com` and apply today. We have worked with thousands of developers and tech professionals, just like you, to help them share their insight with the global tech community. You can make a general application, apply for a specific hot topic that we are recruiting an author for, or submit your own idea.

Share your thoughts

Now you've finished *Python for ArcGIS Pro*, we'd love to hear your thoughts! Scan the QR code below to go straight to the Amazon review page for this book and share your feedback or leave a review on the site that you purchased it from.

https://packt.link/r/1803241667

Your review is important to us and the tech community and will help us make sure we're delivering excellent quality content.

Index

A

G

H

I

J

K

L

M

N

Q

R

S

W

Z

Milton Keynes UK
Ingram Content Group UK Ltd.
UKHW051543040324
438721UK00013B/78